ALGAL BIOTECHNOLOGY
New Vistas

ALGAL BIOTECHNOLOGY
New Vistas

— Editor —

Mihir Kumar Das

G.M.College (Autonomous)
Sambalpur – 768 004
Orissa, India

2010

DAYA PUBLISHING HOUSE

Delhi - 110 035

ISBN 81-7035-647-4
ISBN 978-81-7035-647-9

Published by : **Daya Publishing House**
1123/74, Deva Ram Park
Tri Nagar, Delhi - 110 035
Phone: (011) 27383999
Fax: (011) 23260116
e-mail : dayabooks@vsnl.com
website : www.dayabooks.com

Showroom : 4760-61/23, Ansari Road, Darya Ganj,
New Delhi - 110 002
Phone: 23245578, 23244987

Laser Typesetting : **Classic Computer Services**
Delhi - 110 035

Printed at : **Chawla Offset Printers**
Delhi - 110 052

PRINTED IN INDIA

Preface

The term 'algae' is a very difficult one to define. It may appear in text books of Botany, Zoology and Microbiology. In general, algae are organisms that include seaweeds and a number of single celled and multicellular microscopic forms. Algae are ubiquitous; they inhabit oceans, freshwater bodies, rocks, soils and trees. There may be over 50,000 algal species on the earth.

Man's uses of algae have a long history. In China; marine algae were used as food as far as back as 600-800 BC. In recent decades, there has been renewed interest in the utilization of algae as sources of health food and high value chemicals and pharmaceuticals and for aquaculture, agriculture and wastewater treatment. Even so, the biotechnological potential of algae is still far from fully exploited.

The eating patterns of people all over the world have recently undergone marked changes, due to the globalization of markets along with innovation in food technology. Seaweeds are being consumed because they contain considerable quantity of proteins, fatty acids, vitamins and essential micronutrients like Fe, Zn, Mn and Cu needed to maintain a healthy life. They also form a source of antioxidants. There is an increasing interest in the importance of dietary minerals from seaweeds, in the prevention of several diseases like cardiovascular and thrombosis.

Seaweeds have been used as food besides forming a major source for phytochemicals like agar, alginate and carrageenan. Alginate is extracted from brown algae, mainly species of *Ascophyllum, Laminaria* and *Macrocystis*. The value of the world alginate production is more than US $ 180 million and Norway accounts for 25 per cent of this amount. Alginate has traditionally been used in foods, pharmaceuticals and dyes, but now whole new vistas are opening up in the fields of biotechnology, pharmaceuticals and medicine. It is conceivable that in time, Norway's seaweed forests could become as valuable a national resource as her fish farms.

For decades, the algal industry moved forward very slowly. In a world economy with low prices for fossil fuels and foods, alternative food and energy production sources made no economic sense. Interest in algae as a food source spurred considerable research after each of the world wars but production problems and costs ended those efforts. Excitement about algae as an energy source reignited algal research in the 1970s but faded due to production problems and oil prices at $10 a barrel. Recently, the combination of escalating costs for energy and foods combined with climate change has renewed interest in algae as a clean, carbon neutral energy source. Unlike other forms of green energy such as solar, wind, waves, tides and geothermal, algae offer the only practical source for liquid transportation fuels that may displace oil imports. In the U.S. and many other countries, about 97 per cent of oil imports are liquid transportation fuels and increasing costs and availability jeopardize the stability of nations. Disruption of fossil fuel imports would be catastrophic to the economy, food supply, military and transportation for most countries. Consequently, many countries are examining the potential for algae as an energy source.

Research and production experience suggest that algal biomass offer considerable advantages over land-based biofuels such as ethanol. Ethanol production consumes its energy contribution in fossil resources required for growing, harvesting, refining and distribution. Ethanol competes with food crops for cropland, freshwater, fossil fuels and scarce agricultural chemicals thereby increases demand and prices for all agricultural inputs and drives up food and feed costs. Algal production does not compete with food because the biomass can be grown on deserts or wastelands using no freshwater and no or few fossil fuels. Algae produce biomass use abundant resources that are surplus and cheap and will not run out. Algae use sunshine, carbon dioxide, waste or brine water and some nutrients. Algal production can use green energy for supporting biomass growth such as solar or wind.

Algae are harvested to remove the remaining contaminants. A last stage of bioremediation, still in development, will ensure that the water discharge from the process exceeds acceptable quality standards. The water and sludge treatment process offers a clean-up and management service for sewage treatment systems while also generating a low-cost feedstock for conversion to fuel. The result is an algae-based extract that will ultimately be converted to an alternative fuel source. Aquaflow Binomics expects to be able to produce a viable biofuel on a commercial scale.

Algal Biotechnology has a lot of promises for its industrial applications in the field of biological nitrogen fixation, nutraceuticals and other bioactive compounds. Recent advances in cyanobacteria for bioactive compounds have revealed the potency of cyanobacteria to produce a large variety of cytotoxic compounds active against a number of disease causing bacteria, viruses and fungi. The progress in physiological and biochemical study of cyanobacteria disclosed their significant potential for colorants, polysaccharides, pharmaceuticals and pharmacological probes.

The role of algae in the field of environment protection is well established. Further, there is a need for conservation of natural resources and also developing technologies which are environment friendly. Cyanobacterial blooms can produce health and environmental hazards in water and affect potability of water and recreational purposes. How, why, and when? The blooms are produced and how to deal with them are questions whose answers are necessary for the safety of human and animal health in regions where these organisms occur in abundant. The blooms are linked to eutrophication of water *i.e.* the biological response of water to over enrichment by plant nutrients. Some chapters deal with the discussions regarding the growth of cyanobacterial blooms, their nature, distribution and impact on human health.

There is great potential for the use of diatoms in nanotechnology. This potential lies in the pores and channels which give rise to a greatly increased surface area, and the silica structure which lends it to chemical modification. In addition there is a huge variety in the sizes and shapes of diatoms available, providing scope for the selection of a particular species of diatom tailored to a particular requirement. The silica walls of diatoms can be used for the attachment of active biomolecules, such as antibodies, using either primary amine groups or the carbohydrate moiety. These modified structures can, therefore, be used for antibody arrays or for use in techniques such as immuno-precipitation. These small organisms can help biologists and nanotechnologists to work together to further research.

Algal biotechnology has come a long way since the early days of agronomy/ mariculture for food use and the recovery of agars, carrageenans and alginates. Our increasing knowledge of algal physiology and biochemistry, combined with genetic engineering is opening up new vistas. For example there is the genetic engineering of Phaeodactylum by the Carnegie and Martek groups to enable heterotrophic growth. The application of the ability to grow microalgae to very high yields in bioreactors, to a percentage of a theoretical maximum, as opposed to ponds, holds great promise for the future of algal biotechnology. The uses of microalgae have generated a surge of activity in optimising bioreactor technology and design, including some applications of growing algae on foam plants. There is the potential application of some microalgae that produce extra-cellular polysaccharides being used as a means of drag reduction for ship hulls. At the macroalgal side possibilities on the horizon involve their use as a source of antifoulants, the possible use of sulfated polysaccharides in biomedical applications and the possibility of genetically engineering macroalgae to provide deterrence to microbial pathogens in mariculture. In the final analysis the expanded future of applying algal biotechnology for good, new and imaginative use must lie in ensuring that good science is tied to good economics.

The aim of bringing out of this book is to bring together the multi-disciplinary researches going around the world in the area of Algal Biotechnology and to disseminate information on latest technology in mass cultivation and industrial applications of algae.

This book can not be made possible without the help and effort of many. First of all, I am indebted to the authors of the various chapters for their excellent contributions.

Second, I would like to thank the reviewers whose critical comments and constructive suggestions have helped to improve the quality of this book greatly. Third, I would like to thank my wife, Mrs Ratna Choudhury and my daughter, Miss Saptaparna Das for their moral support, without whom it would have been very difficult to complete this book. Finally, I would like to acknowledge the assistance of Daya Publishing House, New Delhi in producing this book.

Mihir Kumar Das

Contents

Preface v

List of Contributors xi

1. **The Algal Industry Survey** 1
Mark Edwards

2. **Prospective in Diatom Nanotechnology** 25
Jannendra Rath, Sikha Mandal and Bijaya Kumar Padhi

3. **Programmed Cell Death: An Integral Event in the Development of Algae and Higher Organisms** 36
Shiv Shanker Pandey, Vivek Ambashtha and Budhi Sagar Tiwari

4. **Spirulina: The Superfood and Medicine** 62
Preeti Das, A.K. Mishra and Mihir K. Das

5. **Application of Seaweeds as Food: A Scenario** 77
P.V. Subba Rao, K. Ganesan and K. Suresh Kumar

6. **UV-B Radiation-Induced Stress and Protection Strategies in Cyanobacteria** 93
R.P. Sinha, M.B. Tyagi, Sushil Kumar and Ashok Kumar

7. **Growth Response of Cyanobacteria from Sandy Soil and Mine Waste Burdened Soil to Different Environmental Variables** 111
Pramila Tripathy and S.P. Adhikary

8. **The Impact of Fungicides on Rice Field Cyanobacteria** 133
Anshuman Das and Mihir Kumar Das

9. **Role of Blue Green Algae in Rice Production** 145
Y.V. Singh

10. **Biotechnological Relevance of Microbes in Agriculture** 159
Rajan Kumar Gupta

11. **Lipids and Fatty Acids from Marine Algae: A Potential Biofuel Resource** 170
S. Chakraborty and T. Bhattacharya

12. **Algal Biodiesel: Procedures and Resources for Laboratory Study** 176
Simrat Kaur, H.K. Gogoi, R.B. Srivastava and M.C. Kalita

13. **Industrial Utilization of Algal Fatty Acids** 194
S.B. Padhi, P.K. Swain, S.K. Behera and G. Behera

14. **Cyanobacterial Toxins** 206
Anjana Pandey and Archana Tiwari

15. **Algae and the Human Affairs in the 21st Century** 223
Surendra Singh, Shipra Das and Anuradha Tiwari

16. **Responses of Rice Field Cyanobacteria to Insecticides** 241
Mihir Kumar Das

17. **Cyanobacterial Toxins and Public Health** 253
Mukesh Kumar

18. **Cyanobacteria for Biofertilizer, Bioremediation and Bioactive Compounds** 260
Kaushal Kishore Choudhary

19. **Production of Nutraceuticals and Antioxidant Enzymes in a Tropical Food Alga *Nostochopsis lobatus*** 277
Usha Pandey

20. **Bioremediation of Heavy Metals by Microalgae** 288
V.D. Pandey

Author Index 297

Subject Index 299

List of Contributors

Adhikary, S.P.
Centre for Biotechnology, Visva-Bharati, Santiniketan – 731 235, West Bengal, India

Ambashtha, Vivek
School of Life Sciences, Jawaharlal Nehru University, New Delhi–110 067, India

Behera, G.
Algal Research Laboratory, Department of Botany, Berhampur University, Berhampur, Orissa, India

Behera, S.K.
Algal Research Laboratory, Department of Botany, Berhampur University, Berhampur, Orissa, India

Bhattacharya, T.
Department of Environmental Science and Technology, Institute of Science and Technology for Advanced Studies and Research, V.V. Nagar – 388 120, Gujarat, India

Chakraborty, S.
Department of Biological and Environmental Science, N.V. Patel College of Pure and Applied Sciences, V.V. Nagar – 388 120, Gujarat, India

Choudhary, Kaushal Kishore
Department of Botany, B.R. Ambedkar Bihar University, Muzaffarpur – 842 001, Bihar, India

Das Mihir Kumar
P.G. Department of Botany, G.M. College (Autonomous), Sambalpur – 768 004, Orissa, India

Das, Anshuman
P.G.Department of Microbiology, Orissa University of Agriculture and Technology, Bubaneswar – 751 003, Orissa, India

Das, Preeti
P.G. Department of Botany, G.M. College (Autonomous), Sambalpur – 768 004, Orissa, India

Das, Shipra
Algal Biotechnology Laboratory, Department of Biological Science, Rani Durgavati University, Jabalpur – 482 001, India

Edwards, Mark
Department of Food Marketing and Sustainability, Morrison School of Management and Agribusiness, Arizona State University, USA

Ganesan, K.
Marine Biotechnology and Ecology Discipline, Central Salt and Marine Chemicals Research Institute (CSIR), Bhavnagar – 364 002, India

Gogoi, H.K.
Bioenergy Division, Defense Research Laboratory (DRDO), Assam

Gupta, Rajan Kumar
Department of Botany, Pt. L.M.S. Government Post Graduate College, Rishikesh – 249 201, Uttarakhand, India

Kalita, M.C.
Department of Biotechnology, Gauhati University, Guwahati, Assam

Kaur, Simrat
Bioenergy Division, Defense Research Laboratory (DRDO), Assam

Kumar, Ashok
School of Biotechnology, Banaras Hindu University, Varanasi – 221 005, India

Kumar, K. Suresh
Marine Biotechnology and Ecology Discipline, Central Salt and Marine Chemicals Research Institute (CSIR), Bhavnagar – 364 002, India

Kumar, Mukesh
Department of Botany, Sahu Jain College, Najibabad – 246 763 U.P., India

Kumar, Sushil
School of Biotechnology, Banaras Hindu University, Varanasi – 221 005, India

Mandal, Sikha
Department of Botany, Visva-Bharati, Santiniketan – 731 235, West Bengal, India

Mishra, A.K.
P.G. Department of Botany, G.M. College (Autonomous), Sambalpur – 768 004, Orissa, India

Padhi, Bijaya Kumar
Centre for Environmental Studies, Visva-Bharati, Santiniketan–731 235, West Bengal, India

Padhi, S.B.
Algal Research Laboratory, Department of Botany, Berhampur University, Berhampur, Orissa, India

Pandey, Anjana
Nanotechnology and Molecular Biology Lab, Department of Biotechnology, University of Allahabad, Allahabad – 211 002, U.P., India

Pandey, Shiv Shanker
School of Life Sciences, Jawaharlal Nehru University, New Delhi–110 067, India

Pandey, Usha
Department of Botany, Faculty of Science and Technology, M.G. Kashi Vidyapith, Varanasi – 221 005, U.P., India

Pandey, V.D.
Department of Botany, Govt. Post-Graduate College Rishikesh – 249201, Dehradun, Uttarakhand, India

Rao, P.V. Subba
Marine Biotechnology and Ecology Discipline, Central Salt and Marine Chemicals Research Institute (CSIR), Bhavnagar – 364 002, India

Rath, Jannendra
Department of Botany, Visva-Bharati, Santiniketan–731 235, West Bengal, India

Singh, Surendra
Algal Biotechnology Laboratory, Department of Biological Science, Rani Durgavati University, Jabalpur – 482 001, India

Singh, Y.V.
Centre for Conservation and Utilization of Blue Green Algae, Indian Agricultural Research Institute, New Delhi – 110 012, India

Sinha, R.P.
Centre of Advanced Study in Botany, Banaras Hindu University, Varanasi – 221 005, India

Srivastava, R.B.
Bioenergy Division, Defense Research Laboratory (DRDO), Assam

Swain, P.K.
Algal Research Laboratory, Department of Botany, Berhampur University, Berhampur, Orissa, India

Tiwari, Anuradha
Algal Biotechnology Laboratory, Department of Biological Science, Rani Durgavati University, Jabalpur – 482 001, India

Tiwari, Archana
Nanotechnology and Molecular Biology Lab, Department of Biotechnology, University of Allahabad, Allahabad – 211 002, U.P., India

Tiwari, Budhi Sagar
School of Life Sciences, Jawaharlal Nehru University, New Delhi–110 067, India

Tripathy, Pramila
P.G. Department of Botany, Utkal University, Bhubaneswar – 751 004, Orissa, India

Tyagi, M.B.
Mahila Maha Vidyalaya, Banaras Hindu University, Varanasi – 221 005, India

Algal Biotechnology: New Vistas (2010) *Pages 1–24*
Editor: **Mihir Kumar Das**
Published by: **DAYA PUBLISHING HOUSE, NEW DELHI**

Chapter 1

The Algal Industry Survey

Mark Edwards*

Department of Food Marketing and Sustainability,
Morrison School of Management and Agribusiness,
Arizona State University, USA

ABSTRACT

The Algal Industry Survey was designed to provide a baseline of information about the emerging industry. Respondents were generally positive about the future of the industry and optimistic about algae's potential to help solve critical social and economic problems. Recommendations to move the industry forward include better access to information, substantial increases in public and private funding for algal R&D, stronger education and training, more information on production issues and better networking and collaboration. Future of algal industry research needs to drill down on production, supply chain and social and economic issues. Improved information on industry needs will support industry participants and provide critical information needed for public policy decisions and support.

Keywords: Algal industry, Biofuels, Survey.

1. Introduction

For decades, the algal industry moved forward very slowly. In a world economy with low prices for fossil fuels and foods, alternative food and energy production sources made no economic sense. Interest in algae as a food source spurred considerable research after each of the world wars but production problems and costs ended those efforts. Excitement about algae as an energy source reignited algal

* E-mail: drmetrics@cox.net

research in the 1970s but faded due to production problems and oil prices at $10 a barrel. Recently, the combination of escalating costs for energy and foods combined with climate change has renewed interest in algae as a clean, carbon neutral energy source. Unlike other forms of green energy such as solar, wind, waves, tides and geothermal, algae offer the only practical source for liquid transportation fuels that may displace oil imports. In the U.S. and many other countries, about 97 per cent of oil imports are liquid transportation fuels and increasing costs and availability jeopardize the stability of nations. Disruption of fossil fuel imports would be catastrophic to the economy, food supply, military and transportation for most countries. Consequently, many countries are examining the potential for algae as an energy source. Research and production experience suggest that algal biomass offer considerable advantages over land based biofuels such as ethanol. Ethanol production consumes its energy contribution in fossil resources required for growing, harvesting, refining and distribution. Ethanol competes with food crops for cropland, freshwater, fossil fuels and scarce agricultural chemicals thereby increases demand and prices for all agricultural inputs and drives up food and feed costs. Algal production does not compete with food because the biomass can be grown on deserts or wastelands using no freshwater and no or few fossil fuels. Algae produce biomass use abundant resources that are surplus and cheap and will not run out. Algae use sunshine, carbon dioxide, waste or brine water and some nutrients. Algal production can use green energy for supporting biomass growth such as solar or wind. Relatively little is known about the algal industry for three reasons: newness, intellectual property protections and extravagant claims. Minimal credible research has examined this new industry and most the major firms are extremely secretive. Intellectual property protections for proprietary strains of algae are carefully protected and proprietary production methods undermine scientific collaboration. Numerous scientists at algal conferences complain that they have signed nondisclosure agreements with their companies and cannot share critical details about what they have learned from their algal production experiences. The consequence of secrecy is that new firms are sentenced to repeat past mistakes.

Companies have gone out of business repeating the same algal production mistakes because prior knowledge was locked up in intellectual property protection. The algal industry lacks credibility because many firms make grandiose claims on their websites and corporate brochures with the intent of attracting capital investors. These firms carefully avoid reporting their actual production figures or production failures in order to maximize the probability of obtaining the next round of funding.

The Algal Industry Survey is intended to provide a baseline of industry information and to assist in creating strategy for research, development and demonstration priorities.

2. Methodology

The Algal Industry Survey examines the critical industry issues including especially algal production. Survey responses came from participants of Algae World 2008 which met in Singapore in November and yielded 137 respondents. Algae World 2008 was among the first few high profile international conferences focused on the algal industry. Algae World had attracted close to 150 participants. The conference emphasized algal biomass production for energy and the speakers emphasized growth, development, extraction and processing issues focused on maximizing algal oils for energy production. The Algae World 2008 Conference in Singapore included many international respondents from all over the world but primarily from Pacific Rim countries and India. About 40 per cent of the respondents identified themselves as scientists, 30 per cent consultants and 20 per cent academics. Roughly 50 per cent of respondents had less than five years experience in the algal industry while about 16 per cent with over 20 years of industry experience. The algal industry is new so it is not surprising that most participants have relatively short experience in the industry. In the U.S., there has been no government funding for algal research for over a decade so most people with algal experience work for international firms.

Survey research includes several limitations. The sample size for the Algal Industry Survey is relatively small which limits its reliability and generalizability. The questions inquire about respondent beliefs so the results are perceptual rather than factual. Future industry surveys need to reflect the insights and beliefs of more experienced participants. Larger sample sizes will enable analysis of possible differences between experienced and less experienced respondents and possible differences between algal producers focused on different products or producers from different countries.

3. Results

The survey began with a question about which biofuels would be produced from algal biomass.

What will be the Top Three Biofuels made from Algae?

Biodiesel, gasoline and jet fuel were the dominant biofuels respondents believe will be made from algal biomass. Several respondents felt that methane or hydrogen would be produced. There was little support for ethanol.

What Feedstock make the most Sense for Algal Oil Production?

	Feedstock	*World*
Several presentations on the use of marine microalgal feedstock which may have skewed this result. Respondents indicated a significant preference for naturally occurring species over genetically modified organisms. GMOs are likely to create ecological and political challenges for producers. (Percentages do not add to 100 because respondents were able to select multiple methods.)	Marine algae	58 per cent
	Carefully selected strains	58 per cent
	Naturally occurring species	42 per cent
	Freshwater algae	28 per cent
	GMO strains	16 per cent

Growing System

	Growing System	*World*
The type of growing system was the most common question most attendees asked. Producers have the most experience with open pond production and open ponds offer the lowest cost production method. Natural algal stands are not favored probably due to their low productivity for target organisms.	Natural stands	13 per cent
	Open or semi-closed ponds	58 per cent
	Semi-closed systems	42 per cent
	Closed systems	28 per cent

Where will Algal Production Occur?

	Location for Algal Production	*World*
Respondents were split on where algae production will occur indicating a slight favoritism to the tropics and mid-latitudes.	Tropics	38 per cent
	Mid-latitudes	35 per cent
	All over the Earth	27 per cent

Critical Variables

	Critical System Parameters	*World*
The critical variables for algal production systems put species selection at the top. Species selection dominates the important production issues. Secondary issues include light penetration, the type of growing system and controlling culture variables. Less emphasis was given to the cost of nutrients, methods of mixing the culture, the type of monitors and the avoidance of algal build up on or in the production containers. Production systems seem to be most critical to the algal industry.	Algae species	4.1
	Light penetration	4.0
	Growing containers	3.9
	Variable controls, *e.g.* temperature and pH	3.6
	Nutrient costs	3.2
	Mixing methods	3.2
	Monitors	2.9
	Fouling	2.9

Critical Issues for the Algal Industry

Production systems represent the most tangible elements of algal biomass growth and development. Speaker emphasized a wide variety of different biomass production models. Numerous people indicate concerns about staffing algal R&D and production because so few people are trained, especially in the U.S. where few universities have algae labs. Developing or selecting the best algal strains is a key issue because many firms use species selection for proprietary differentiation. Cost for algal production inputs was also seen as a critical issue. The set of production issues beginning with component separation, light management, extraction are important but not as critical as production systems, personnel and strain selection.

Most Critical Industry Issues	*World*
Production systems	4.3
Trained personnel	4.2
Algal strain selection	4.2
Input costs	4.0
Develop high lipid strains	3.6
Component separation	3.6
Light management	3.6
Extraction	3.6
Contamination	3.4
Nutrient delivery	3.2
Temperature management	3.1
Intellectual property	3.0
Monitoring systems	3.0
Mixing	2.9
Public image	2.8

Supply-chain

The supply-chain results were consistent with the results from the critical algal industry issues. Conference interviews indicated that many people were concerned about the availability of capable people for staffing the necessary disciplines to build anew industry. It remains unclear whether upstream, input supply, and downstream, processing will be integrated with producers or separated in vertical marketing channels. The level of concern about processing, monitoring, extraction and nutrient supply were about equal.

Supply Chain Issues	*World*
Growing systems	4.3
Trained personnel	4.2
Component separation	4.2
Design/construction	4.0
Processing	3.6
Monitoring systems	3.7
Extraction	3.6
Nutrient supply	3.6
Distribution/logistics	3.4

Water remediation dominated social and economic issues.

Likelihood that Algae will Play Out Significant Role in Social and Economic Issues

Water remediation was the highest rated issue possibly because of the excellent presentation by Professor Avigad Vonshank of Ben Guiron University that included cleaning polluted water. Practically all nations of the world need water remediation and algae provide a low-cost, low energy solution.

Displacing oil imports, fossil fuels and sequestering CO_2 all received high scores. Narrative comments from Algae World indicated a strong focus on algal energy production with significantly less interest in co-products such as animal fodder, pharmaceuticals, fertilizers or moderating world hunger.

Social and Economic Issues	*World*
Cleaning polluted water	4.5
Displacing oil imports	4.1
Displacing fossil fuels	3.8
Sequestering CO_2	3.7
Cleaning polluted air	3.5
Animal fodder	3.4
Health foods	3.3
Medicine/pharmaceuticals	3.0
Moderating world hunger	2.8
Creating organic fertilizer	2.7
Reducing transport costs	2.6
Clean burning cooking fuel	2.5

Respondents were presented with the question: "If your country focused on alga-culture and made it a top priority, how soon could industry replace 100 per cent of imported oil?"

Replace 100 per cent of Imported Oil for your Country

Respondents were optimistic regarding the probability of replacing imported oil with home grown algal oil. Several narrative comments indicated that replacing oil imports was the top priority

Replace 100 per cent of Imported Oil	*World*
10–20 years	47 per cent
21–40 years	34 per cent
Never	19 per cent

The following two questions requested narrative comments.

What Three Key Things Need to Happen to Move the Algal Industry Forward?

The dominant words used to enhance the industry were information, financing, funding sources, education and production. The industry needs government investment because the initial investments and risks are very high. The recommendations are grouped and ordered based on how often respondents recommended the action.

Information

The industry needs information sources to convey the algae story.

- ☆ Why isn't there a website that summarizes key algal sources, links and resources?
- ☆ Is there a profile of businesses and what they're doing in the algal industry? (See Appendix 1)
- ☆ Can we create information sheets, pamphlets and books to inform people about algae?
- ☆ Can we create a PR campaign, a political action committee and lobbyists?

Financing

The industry needs considerably more funding from both public and private sources.

- ☆ Where are existing firms getting their funding?
- ☆ How can the algal industry persuade government to provide more R&D funding?
- ☆ How can investors evaluate algal investment opportunities?

Education and Training

The industry needs to find ways to train people in algal production.

- ☆ Where can algal companies find trained technical and professional personnel?
- ☆ What kind of algal training is available and where?
- ☆ What technical skills and competencies are critical for building an algal industry?
- ☆ Is there a grade school and college curriculums for algae?

Production

The industry needs to describe and define algal production.

- ☆ What are the key production issues and how are they solved?
- ☆ What is the level of investment necessary for algal production?
- ☆ How can production losses be minimized from invasive species or grazers?
- ☆ What are practical methods for algal extraction and biomass processing?
- ☆ Can small community sized production systems be developed?

Collaboration

The industry needs to find ways to enhance networking and shared knowledge.

- ☆ How can information be shared and still maintain a little property protections?

- ☆ What kind of collaborative social network might work for algal professionals?
- ☆ Is there a summary of bloggers and journalists who follow algae?

What Algae Information would Most Benefit You?

Respondents indicated a desire for better information on algae, insights on financing, support for education and training, production information and stronger collaboration.

Additional requests included the ideas below.

- ☆ *Industry summary*. The algal industry may follow the lead of other forms of renewable energy such as solar and wind to create a summary of the industry.
- ☆ *Demonstration units*. The algal industry should build and operate demonstration facilities so that people can see algal production.
- ☆ *Decision support*. Information on production, extraction and processing are too distributed and need to be more accessible.
- ☆ *Products and coproducts*. What is the total product array for algal biomass and what production strategies are used to maximize each product?
- ☆ *Real production numbers*. Actual rather than theoretical production numbers would be a huge breakthrough.
- ☆ *Ideal strains*. What are the ideal strains for various products and what are sources for these strains?
- ☆ *Market trends*. What are market trends in the algal industry?
- ☆ *Independent reviews*. Are there independent reviews of algal production methods?

Appendix 1

Algal Industry Producer Profiles

About 98 per cent of the algal industry focus today is on algal biofuel production but has not produced a hundred barrels of oil. Within three years, the industry will be producing millions of gallons of algal oil and valuable coproducts. Green solar production will expand exponentially. The algal industry has moved from staid to super charged with breakthroughs occurring weekly. Most innovations are coming from the private sector because U.S. government R&D and grants have been zero. Some of the leading players in the industry are getting $10 million in private equity but that is hardly enough to build a lab, let alone a small pilot plant.

The industry needs substantial government investment to achieve Green Independence.

The industry faces two major threats:

1. *Insufficient investment slows R&D* and the world food crisis may degrade into mass migrations and war over insufficient food, clean water and energy.

2. *Commercial firms control R&D* and put a lock on algal production and biotechnology breakthroughs.

If private firms locked up the basic production methods, then sustainable world food solutions would exist but are likely to be beyond the financial means of the people who most need them. Widespread adoption and diffusion require open source, public access to Technologies(www.biofuelsdigest.com)

Three companies–BASF of Germany, Syngenta of Switzerland and Monsanto of St. Louis–have filed applications to control nearly two-thirds of the climate-related gene families submitted to patent offices worldwide. These "climate ready" genes will help crops survive drought, flooding, saltwater incursions, high temperatures and increased ultraviolet radiation–all of which are predicted to undermine food security in coming decades(Weiss, 2008)

Company officials deny the climate-ready seed applications amount to an intellectual property grab. They say GMO seeds will be crucial to solving world hunger but would not be developed without patent protections. Monsanto, for example, makes 60 per cent of its revenue from genetically modified seeds(Marrero, 2008). In 2006, over 78 per cent of U.S. corn used for all purposes came from genetically modified seeds. Big agribusinesses are eager to control access to genetically modified algal strains.

Lack of government R&D investment has pushed the industry towards private investment that focuses predominately on biofuels. Algal solutions for food, medicines, vaccines and fertilizers are getting very little attention. Industry activity is difficult to decipher due to extensive puffery and trade secrets.

Executives talking about their companies at conferences, interviews, press releases and websites tend to make grandiose claims designed to impress investors but too often tend to be more hype and hope than fact. Neither production efficiencies nor costs are typically revealed because they are considered proprietary. The other challenge is that the industry is so new, few participants have either a background in biotechnology or a track record in producing algae.

Algae's potential has seduced many investors, including prominent Silicon Valley venture capitalist firms. De Beers Fuels, a South African company, collapsed in 2007 following several years of false promises. Over 75 per cent of the companies researched for this project that had high hopes in the 1980s and 1990s no longer exist.

The Department of Defence estimates that current production cost of algal oil currently exceeds $20 per gallon. A crude oil barrel contains 42 gallons. Crude oil gets changed into petroleum gas, gasoline, oils, tar and asphalt. Roughly 28 gallons of gasoline is refined from each barrel of crude oil.

Therefore, a $120 barrel of crude yields about $112 worth of gasoline when gasoline costs $4 a gallon. Algal fuel at $20 a gallon would cost $560. Obviously, algal fuel is not economic until the cost of production decreases by a factor of five. In spite of the current cost difference, over 50 companies and 20 universities are working on algae, primarily for the production of algal oil. The organizations profiled here are sorted based on their algae growing strategy, not their investment potential. Growing strategies include open ponds, natural settings and closed algaculture systems.

BiofuelsDigest.com tracks activity in algae and the biofuels industry.

Open Ponds

LiveFuels, based in Menlo Park, California plans to extend the Aquatic Species Program research and use open-pond algae biofactories to commercialize its technology. John Sheehan, who led the Aquatic Species Program, joined Live Fuels in 2007 as VP of Sustainable Development. Instead of attempting to convert algae directly into ethanol or biodiesel, this startup is trying to create green crude that could be fed directly through the nation's current refinery system.

LiveFuels created a national alliance of scientists led by Sandia National Laboratories, a U.S. Department of Energy National Laboratory focused on producing biocrude oil by the year 2010. The alliance is expected to sponsor dozens of labs and hundreds of scientists by the year 2010.

The company's web site displays technical exuberance in predicting that algae can produce up to 20,000 gallons of oil per acre. The company goes on to state that the entire U.S. supply of imported oil could potentially be grown on 20 million acres of marginal land.

OriginOil, Inc. in Los Angeles, California, received its first funding in 2005 and is developing a technology that will transform algae into a true competitor to petroleum. The company claims its patented technology will produce "new oil" from algae, through a costeffective, high-speed manufacturing process. This supply of new oil can be used for many products such as diesel, gasoline, jet fuel, plastics and solvents without the global warming effects of petroleum.

Seambiotic's Algal Ponds

Seambiotic, located in Ashkelon, Israel was founded in 2003 and produces algae for a variety of applications, including health foods, fine chemicals, medical products and biofuels. The firm is working with Inventure Chemical and with the Israeli Electric Company, using IEC's smokestack for a source of CO_2 while it grows algae in eight open algal ponds.

PetroSun based in Scottsdale, Arizona is publicly held and the management team comes from the petroleum industry. PetroSun began their algae-to-biofuel production factory in Rio Honda, Texas, in April 2007. The algae farm is a network of 1,100 acres of saltwater ponds that PetroSun thinks will make 4.4 million gallons of algal oil and 110 million pounds of biomass per year. PetroSun intends to extract algal oil on-site at the farm and transport it to company biodiesel refineries via barge, rail or truck. The company plans to open more farms in Alabama, Arizona, Louisiana, Mexico, Brazil, and Australia in 2008. The company also offers environmentally-friendly energy production: recycled energy. Their recycled-energy technologies capture the energy content of waste exhaust heat from industrial processes and internal combustion engines. PetroSun's markets for the Waste Heat Generator include algae and other biomass facilities, oil refineries and drilling rig power plants. It is estimated that 60 per cent of all fossil fuel burned for these and other industrial uses is wasted in the form of heat, pressure and polluting emissions. Their goal is to capture this wasted energy and convert it to usable electricity. The Company is focusing its initial efforts in Louisiana, Texas, New Mexico and Arizona.

PetroSun BioFuels Refining recently signed a joint venture to develop and operate a 30 million gallon a year algal biodiesel facility in Coolidge, Arizona. Construction is projected to commence in the third quarter of 2008. In 2007, PetroSun announced a letter of intent to supply 54 million gallons of algal oil a year to a new Bio-Alternatives biodiesel plant in south Louisiana. The initial delivery to Bio-Alternatives refinery will be in the third quarter of 2008.

PetroSun created an algae-to-jet fuel team relationship with Science Applications International. The companies are working to transition algal biofuel technology to the commercial sector with government contracts. PetroSun has made twenty acres of ponds available at its Rio Hondo, Texas facility for R&D related to an algae-to-jet fuel refiner.

Neste Oil in Helsinki, Finland is refining imported vegetable oils, palm oil and algae to make 170,000 tons of biodiesel a year in Porvoo, southern Finland. The renewable fuel is suitable for all diesel engines and is the strategic cornerstones for Neste, who say the technology outperforms both existing biodiesel products and crude oil-derived diesel products available.

Neste's renewable fuels goal is to have 70 per cent of its raw materials coming from nonfood feedstocks in ten years. By 2020, they want to have all their raw materials they use from outside the food chain.

Ingrepo, a Netherlands-based biotechnology company specializing in industrial largescale algae production, plans to build algae production facilities in Malaysia. Partnering with Biomac Sdn Bhd, they will provide Malaysians with the opportunity to grow large-scale algal production for biofuels. Biomac CEO Syed Isa Syed Alwi says the algal PowerFarms will be ready for commercialization in the next year. Malaysia was chosen for its good weather conditions to grow algae, infrastructure and government interest in agro-biotechnology. Natural settings the aquaculture industry began when producers enhanced natural settings to produce more oysters, clams and fish in open, semi- and closed-growing environments. Algae producers around the world have similarly been enhancing algal stands in natural settings by reducing predators and improving water mixing. The advantage of finding algae growing and cultivating the growth in natural settings seems obvious: nature provides the growing container and most the nutrients. However, challenges similar to controlled settings occur in natural settings, including especially controlling growth and stability.

Kelco, based in San Diego, harvests natural kelp beds with a specially designed mowing machine. They load the heavy biomass onto barges for transport to the processing facility to produce alginic acid.

Neptune Industries, based in Boca Raton, Florida, creates sustainable, eco-friendly aquaculture with integrated solutions. Dwindling supplies of wild-caught stocks, continued environmental damage, escaping fish and disease from self-polluting net pens have restricted industry growth.

In addition to hydroponically grown vegetables, lettuce, herbs and fish ponds, Neptune's patented Aqua-Sphere system uses fish waste to create additional revenue streams through the growth of algae for biofuels and methane gas.

Blue Marble Energy, based in Seattle, searches for unwanted wild algae growth and has developed methods for cleaning polluted water where excess nutrients lead to algal blooms that plague water systems. BME converts algal biomass to energy by creating, centralizing, and harvesting wild algae blooms.

BME's technology and process of harvesting remediates polluted water through biomass generation. BME technology harnesses nutrients and converts polluted environments into natural biofactories for generation of renewable energy feedstock while cleaning the environment. The company's business model is smart, they get paid to clean water and they produce biomass that can be processed or sold. BME marine technology can be placed in a broad array of geographies:

- Wastewater systems
- Freshwater lakes, rivers, and streams
- Metal remediation for mines and other polluting industries
- Coastal remediation

By addressing wild algae growth versus the traditional mono-culture growth for biomass generation, the company keeps capital costs low and is able to produce a volume output that is multiples above closed- and pond-based systems.

Aquaflow Binomics, based in New Zealand, has a goal to become the first company in the world to economically produce biofuel from wild algae harvested from open air environments. The three-year-old startup sources its algae from algae-infested polluted water systems; cleaning the polluted environment in the process. Aquaflow Binomics harvests algae directly from the settling ponds of standard effluent management systems and other nutrient-rich water. The process can be used in many industries that produce a waste stream including the transport, dairy, meat and paper industries.

The two-step process first optimizes the ponds' productive capacity and then determines the most efficient and economic way of harvesting the pond algae. Algae are provided with full opportunity to exploit the nutrients available in the settling ponds, thereby cleaning up the water. Algae are harvested to remove the remaining contaminants. A last stage of bioremediation, still in development, will ensure that the water discharge from the process exceeds acceptable quality standards.

The water and sludge treatment process offers a clean-up and management service for sewage treatment systems while also generating a low-cost feedstock for conversion to fuel. The result is an algae-based extract that will ultimately be converted to an alternative fuel source. Aquaflow Binomics expects to be able to produce a viable biofuel on a commercial scale.

In 2007, publicly held Aquaflow used its algae-based biodiesel to run a Land Rover driven by New Zealand's Minister of Climate Change. The company has been working with Boeing on algae-to-bio-based jet fuel.

Biofuel Systems, a Spanish company, is developing a system for producing energy from marine algae, with the hope of replacing fossil fuels and reducing pollution. Biofuel Systems predicts the process will produce massive amounts of biopetroleum (their term for biodiesel) from phytoplankton in a limited space and at a very moderate cost. The company says their system produces biodiesel from marine plankton and is very different from existing systems that are producing biodiesel. The company envisions producing biopetroleum using a proprietary energy converter. The system will use phytoplankton as feedstock.

Closed Systems

Closed systems offer far more control over growing parameters than open ponds or natural settings. Stressing algae to create more production of desired components by changing production parameters is practical only in closed systems. Closedsystems also avoid water loss from evaporation.

A2BE Carbon Capture of Boulder, Colorado builds carbon capture and recycle, CCR, systems that take advantage of algae's capacity to profitably recycle industrial CO_2 emissions into fuel and other co-products. Mark Allen, CEO, says their advanced energy conversion system combines algal CO_2 capture technologies with biomass gasification and creates an integrated renewable fuel production system. The CO_2 can be recycled from any source and the biomass feedstock for gasification into syngas may come from wood waste, municipal solid waste or the processed algae waste. The CO_2 produced from the biomass gasification process is recycled to grow algae.

Jim Sears, President and CTO developed the patented system design and notes that the Carbon Capture and Recycle (CCR) biofactories can be scaled from a few acres to large farms that recycle industrial CO_2 emissions into algal biomass that can be further processed into valuable commodities including biofuel, animal feed protein and organic fertilizer. At the core of the technology is the algae growing and harvesting biofactory. Each machine is 450′ long and 50′ wide consisting of twin 20′ wide x 10″ deep x 300′ long, transparent plastic algae water-beds. It holds 150,000 gallons of algae. The biofactories work with any species of algae including cyanobacteria and diatoms.

The harvesting technology is similarly adaptable to fit local needs. A2BE offers a novel bioharvesting technology where brine shrimp feed on the algae and the shrimp are harvested and processed. The CCR machine is climate adaptive due to thermal barriers above or below the culture flow that regulate temperature. This allows deployment nearly anywhere there is sunshine.

The A2BE business model shows how CO_2 recycling is profitable. Their business plan shows each ton of CO_2 may be captured at a cost of about $40 for nutrients and

\$10 for the CCR biofactory and operations. The net revenue of \$200 per ton of CO_2 captured is based on: oil (\$40), protein (\$90), methane (\$25), fertilizer (\$40), oxygen (\$30) and CO_2 credit(\$25).

A2BE has created an even more compelling production attribute than the profitability per ton of CO_2. Their CCR biofactory creates a carbon negative process because each ton of carbon captured and recycled into the various algal co-products displaces and avoids about 1.25 tons of carbon entering the atmosphere. The carbon negative process holds true when the original carbon is fossil sourced and the resulting products are burned as fuel.

A2BE is not only building a company to take on the substantial challenge of carbon capture but they are building a collaborative group of select institutions, corporations, and key researchers to address the spectrum of talents and disciplines needed to rapidly commercialize a solution called algae@work.

GreenFuel Technologies of Cambridge, Massachusetts is led by Bob Metcalfe who has a telecom background. The company reached an agreement in 2008 to build a fuel plant in Europe–worth \$92 M. GreenFuel Technologies evolved from MIT and government grants for research and demonstration projects. The company has a world-class board of directors but has made some serious mistakes in executing strategy. Essentially, the company discovered, similar to many other startups, that growing algae was more expensive than they had planned. In 2007, the company had to change CEOs, lay off a large proportion of their staff and shut down some projects such as the Arizona Public Service greenhouse in Arizona. Recent tests of an algae-based system developed by GreenFuel reported that it could capture about 80 per cent of the CO_2 emitted from a power plant during the day when sunlight is available.

GreenFuel Technologies claims that using its patented technologies for growth on a one acre site the company can produce algal biomass in a year that can be separated to components that include:

- ☆ 7,000 gallons of jet fuel
- ☆ 5,000 gallons of ethanol
- ☆ 1,000 tons of protein for foods
- ☆ 200 pounds of specialized nutrients
- ☆ 20 pounds of pigments

These production parameters lead the industry in hope and hype. GreenFuel Technologies builds algal biofactory systems which use recycled CO_2 to feed the algae. Their process uses the containers to carefully control the algae's intake of sunlight and nutrients. The algae are refined to biofuels. GreenFuel is backed by Polaris Ventures, Draper Fisher Jurvetson and Access Private Equity.

Solazyme Inc., based in San Francisco, is a five year old biotechnology company that harnesses the power of microalgae to produce clean and scalable high performance oils, biofuels, and "green" chemicals. The company focuses on new methods to improve production productivity.

Solazyme ignores the sun and grows algae in the dark in large tanks where they are fed sugar to supercharge their growth. Harrison Dillon, a geneticist and patent lawyer who serves as the company's president and chief technology officer claims it's a thousand times more productive than natural processes. Solazyme says it has already made thousands of gallons of high-grade biodiesel and even light sweet "biocrude" with its processes, which can use anything from chemical waste to wood chips as a source of carbon.

Solazyme, raised $10 million in equity financing and $5 million in debt in 2007, and is experimenting with different feedstocks, algal species and oil extraction methods. The company hopes to reach commercial-scale biodiesel production in two or three years. Refiner Imperium Renewables of Seattle and Chevron have recently signed partnership agreements with the company.

Solazyme is using its technology to make specialty oils for the cosmetics industry in order to meet cash flow commitments. Solazyme demonstrated to the Department of Defence that their algal diesel, Soladiesel has superior cold weather properties to any commercially available biodiesel and is more suitable for cold weather climates where the military has been unable to use biodiesel.

Algenol Biofuels of Fort Meyers, Florida, was founded in 2006 to develop industrialscale algaculture systems to make ethanol from algae on desert land using seawater and CO_2. Algenol uses a patented technology with blue green algae, cyanobacteria that are nitrogen fixing which reduces their fertilizer cost. The firm uses natural and environmental selection combined with molecular biology to produce low cost and environmentally safe biofuels.

Algenol plans to make ethanol with blue green algae that produce oil and then secrete it. They will use 3.5 million biofactories to grow the algae that are three-feet by fifty-feet and shaped like soda bottles. Most algae companies are trying to make biofuels by drying and pressing the biomass to make vegetable oil that can be processed into biodiesel. Algenol will use a process to coax individual algal cells to secrete ethanol. The fuel can be taken directly from the algal tanks while the algae continue to thrive. This process uses significantly less energy than drying and pressing the biomass for oil. Algenol signed an $850 million deal with the Mexican company *BioFields* to grow algae for biofuel. Algenol plans to make 100 million gallons of ethanol annually in Mexico's Sonoran Desert by the end of the 2009. By the end of 2012, Algenol plans to increase production to one billion gallons. The U.S. will produce about 10 billion gallons of corn ethanol in 2008 but will consume 40 million acres of cropland, two trillion gallons of freshwater and 5 billion gallons of fossil fuel.

Algenol operates the world's largest algae library in Baltimore, Maryland to study the organism that can grow in salt or freshwater, and expanding the technique to locations beyond Mexico. The company hopes to build algae-to-ethanol farms on U.S. coasts. Sapphire Energy apphire Energy, based in San Diego was launched in May of 2007 and initiated a new biofuel category called green crude production. CEO Jason Pyle says his team has built a revolutionary molecular platform that converts sunlight and CO_2 into renewable, carbon-neutral alternatives to conventional fossil fuels without the downsides of current biofuel efforts. The end product is not ethanol or biodiesel but biocrude, renewable 91 octane gasoline.

Sapphire's fuel products are chemically identical to molecules in crude oil, making company products entirely compatible with the current energy infrastructure–cars, refineries, and pipelines. Sapphire's scalable production facilities can grow economically because production is modular and transportable. The green crude produces fewer pollutants in the refining process and fewer harmful emissions from vehicle tailpipes. Sapphire will not reveal the type of algae they use but it is most likely a genetically modified cyanobacteria, blue-green algae. The advantage to this form of algae is that the algae secrete the biocrude oil which rises to the top of the tank and can be skimmed. Avoiding harvesting the algae saves time, cost and may be more productive if the plants secrete enough oil.

Inventure Chemical Technology based in Seattle is working on their patent-pending algae-to-jet fuel product and has produced algae-based fuel in 10 gallon tests. The company plans to set up a test plant to produce up to 15 million gallons of biofuel a year. The algae used for biodiesel conversion is sourced from facilities in Israel, Arizona, and Australia. Inventure expects its technology will deliver a viable ROI for companies that use algae technology for sequestering CO_2. Inventure also provides expertise in both process conversion and plant design and construction.

Vertigro Energy, based in San Diego, is a joint venture of eco-technology companies,Valcent Products and Global Green Solutions focused on producing vegetable oil which can be used directly as biodiesel. Valcent's High Density Vertical Growth System maximizes algae growth in a closed loop, vertical system. In addition to biofuel, the algal oil can also be used in foods, feed stocks, pharmaceutical supplies, and beauty products. The company says 90 per cent by weight of the algae is captured carbon dioxide, which is sequestered by this process and contributes significantly to the reduction of greenhouse gases. Valcent has commissioned the first commercial-scale bioreactor pilot project at its test facility in El Paso, Texas. The company believes it can significantly lower costs over oilproducing crops such as palm and soybean.

Solena, based in Washington State, uses its patented plasma technology to gasify algae and other organic substances with high energy outputs. Solena's plants produce clean, reliable electricity, using no fossil fuels and no CO_2 emissions. The company is talking with Kansas power firm *Sunflower* to build a 40-megawatt power plant which will run on gasified algae. The algae would be grown in big plastic containers and fed by sunlight and sodium bicarbonate, which is a byproduct of an adjacent coal plant. Using a plasma gasifier, Solena's technology converts all forms of biomass into a synthetic gas, syngas. The syngas is then conditioned and fed into a gas turbine to produce electricity. Solena's sequestration process recycles CO_2 and in the process produces biomass for a continual renewable source of fuel.

Solix Biofuels, based in Fort Collins, Colorado, was founded in April 2006 and backed by Colorado State University's Engine and Energy Conversion Laboratory. Solix Biofuels intends to use microalgae to create a commercially viable biofuel that will play a vital role in solving climate change and petroleum scarcity without competing with global food supply. The company announced in 2008 that it will build its first large-scale facility at the nearby New Belgian Brewery, where CO_2 produced during the beer-making process will be used to feed the algae. Solix says their success comes from knowing how to select the right algal species, to create an optimal photo biological formula for each species and to build a cost effective biofactory that can precisely deliver the formula to each individual algal cell, no matter the size of the facility or its geographical location.

XL Renewables, based in Phoenix, Arizona, is a 2007 start-up with a patent pending algal production system called Simgae for simple algae. The company changed its name from XL Dairy Group to XL Renewables in 2007 to emphasize its focus on creating renewable energy using dairy waste streams.

XL Renewables uses common agriculture and irrigation components to produce algae at a fraction of the cost of competing systems. The XL Super Trough uses a miniature greenhouse-type process to produce the algae in laser-leveled 18-inch deep, 1,250- foot long troughs. Mechanized equipment installs the specially designed plastic liner sheets with integrated aeration and lighting systems along the six-foot wide troughs.

Depending upon need and customer demand, a plastic sheet can be installed on top of the trough to make it a closed system and increase algae production during cooler temperatures. The XL Super Trough has no moving parts and no connection points except at the end of the troughs. The water used in the process is fortified to enhance production and is pumped through the troughs to a harvest system where the algae are extracted. The water is recycled back through the troughs. Carbon dioxide is injected periodically and after roughly 24 hours the flow leaves the trough with a markedly greater concentration of algae than when it started. Supporting hardware components and processes involved are direct applications from the agriculture industry. Re-use of these practices avoids the need for expensive hardware and costly installation and maintenance. The Super Trough System for algae biomass production is available for $25,000 per acre.

The design is expected to provide an annual algae yield of 300 dry tons per acre. Capital costs are expected to be approximately $45k–$60k, a 2–16 times improvement over competing systems. President Ben Cloud estimates profitable oil production costs of $0.08–$0.12 per pound. These oil costs compare to recent market prices of feedstock oils that range from $0.25–$0.44 per pound. XL Renewables is developing an integrated biorefinery located in Vicksburg, Arizona, 100 miles west of Phoenix. The $260 million project integrates a modern dairy operation with a biofuels plant to produce ethanol, biodiesel, milk, animal feed and compost fertilizer. The integrated biorefinery uses the dairy manure, along with other waste streams to provide 100 per cent of the power, heat and steam needs of the project and significantly lower production costs. The company expects to produce algal fertilizer at about $300 a ton.

XL Renewables plans to sell the XL Super Trough System and Algae Biotape globally for the economical production of algal biomass to be used as an alternative feedstock to corn for biofuels production.

Aurora Biofuels, developed at the University of California at Berkeley, uses genetically modified algae to efficiently create biodiesel. Aurora claims the patented technology, developed by microbial biology professor Tasios Melis, creates biodiesel fuel with yields 125 times higher and have 50 per cent lower costs than current production methods.

Bionavitas, Based in Snoqualmie, Washington, says it has developed patent-pending technology for the high-volume production of algae using biofactories. Their 2007 patent application shows their competitive advantage to be the "lighting system that includes one or more light-emitting substrates configured to light at least some of a plurality of photosynthetic organisms retained in the bioreactor." This sounds like fiberoptic lights embedded in the algaculture system.

Bodega Algae, based in Boston, Massachusetts, is associated with MIT and was founded in 2007. Bodega Algae says it has developed a patent-pending system to grow algae in algaculture systems with light and nutrients that it says is lower cost and more efficient than the current methods. However, Bodega Algae's website is now inactive.

Cellena, based in Hawaii, is a joint venture created by the algae-to-biofuel startup HR Biopetroleum and Shell oil. Shell has majority share of the company, which is in the process of building a demo facility on the Kona coast of Hawaii.

Cellena announced in 2008 a new process for extracting algae oil without using chemicals, drying or an oil press. The company said that its patent-pending technique uses 26 kilowatts of power to produce 12,000 gallons of algal oil per hour with a yield of 50 per cent from the initial algal paste.

The company also constructs and operates algae biofuels plants that use effluent gases from power plants to produce renewable fuels and to mitigate emissions of carbon.

Canada. Backed by oil companies and utilities, Canadian researchers have plans to develop algae farms that convert CO_2 from oil sands projects and coal-fired power plants into biofuels, chemicals and fertilizers. A consortium led by the Alberta Research Council has completed research that suggests the algae would thrive under northern light and temperatures with an appropriate covering for winter months. The $20 million algal project is being funded by major Canadian energy companies, including Petro-Canada, Royal Dutch Shell, EnCana Corp. and Epcor Power, a coal dependent Alberta-based utility.

Their research indicates that for the large industrial emitters, the system could take about 30 per cent of their emissions. Their goal is eliminating 100 million tons of CO_2 emissions a year; about a third of Alberta's current production of greenhouse gas emissions. The researchers include scientists from Alberta, Saskatchewan, Manitoba and Quebec; believe they can boost the productivity of the system so that CO_2 can be removed at a cost of about $25 a ton.

Health Foods and Nutraceuticals

The *Nostoc commune* represents a broad set of patents for a wide variety of algae food, fuel, water, pharmaceutical and health applications.

Nostoc commune: Filed by Fan Lu in North Carolina and others, U.S. Patent 6,667,171 describes a process for producing Nostoc formulations using a plurality of photosynthetic microns including cyanobacteria. U.S. Patent 6,579,741 discloses a method of culturing algae capable of producing large amounts of unsaturated fatty acids and phototrophic pigments and/or polysaccharides (http://www.freepatentsonline.com/70160704.html).

These patents describe methods for cultivating edible *Nostoc commune* formulations and their use for promoting health. In addition, the invention relates to methods for promoting the health of an individual utilizing the Nostoc formulations, dietary supplements, food products and/or pharmacological compositions. This invention also provides a method for cultivating *Nostoc commune* comprising (*a*) isolating and purifying *Nostoc commune*; (*b*) culturing the *Nostoc commune*; and (*c*) conditions suitable for optimal growth of *Nostoc commune*. These Nostoc patents, similar to other broad patents, threaten to lock up a major algal species from public use. It will take years to determine how broadly algal patents will be enforced. Most of the companies in this health food category harvest natural stands of algae or produce Spirulina in ponds.

Earthrise, based in southern California, began producing Spirulina in 1982. Today, Earthrise Nutritionals' farm is the world's largest Spirulina farm. Earthrise products are marketed in 30 countries on six continents.

Hainan DIC Microalgae Co. of China has a joint marketing agreement with Earthrise. The two firms produce over 800 tons of Spirulina each year in open ponds. A third facility in Thailand closed in 2006.

Cyanotech, based Hawaii, produces natural astaxanthin and Hawaiian Spirulina Pacifica–all natural, functional nutrients that the company claims enhance human health and nutrition. The algae is grown at its 90-acre facility in Hawaii using patented and proprietary technology and distributes them to nutritional supplement, nutraceutical, and cosmeceutical makers and marketers in more than 40 countries.

BioEarth Spirulina, based in Italy, and *Green Valley*, based in Germany, get their product, Spirulina Maxima from producers using artificial lakes in Mexico where the product has been grown, harvested and eaten for centuries and more recently from China. The product is sold in tablet form as a health food.

Omega Tech, of Boulder Colorado, markets its patented algal chicken feed rich in omega-3 fatty acids. The chicken feed, called DHA Gold for docosahexaenoic acid, the long-chain fatty acid that it contains Omega Tech president, William Barclay, claims chicken meat from animals raised on DHA Gold contain five to seven times the amount of DHA in normal commercially bred chicken.The feed is made from schizochytrium, a tiny single-cell organism dense with DHA. When harvested and dried, the algae look like wheat flour that has a golden hue. The company grows the algae in stainless steel vats. Monsanto is analyzing the product for vitamins and to fortify food or infant formula. Infant formula, unlike human breast milk, does not contain the omega-3 fatty acids.

Dolphin Sea Vegetable Company in Northern Ireland, established in 1993, harvests red marine algae and markets a variety of products and food supplements directly. They also sell products to kill bacteria and viruses. DSV supplies wild red marine algae to other manufacturers and carries out research into the pharmacological benefits of seaweeds and algae for various medical, pharmaceutical and health foods.

The company is developing PhycoPLEX through clinical trials to examine its effects on immune system function and modulatory action. The clinical trials are conducted at the University of Ulster, supported by a European Union grant.

Other Firms working on Algae

Algae BioFuels (PetroSun)	*Carbon Capture Corp*	*Global Green Solutions*	*Kiwikpower*	*Simplexity*
Algaen	Cell Tech	GreenEnergy	OriginOil	World Oil
Arare	Diversified Energy	GreenShift	PetroAlgae (XL TechGroup)	
Aquaflow	EnAgri	Green Start Products	Plaatts	
Biodiesel	Energy Farms	GS Cleantech	Pelletbase	
Biofuels Digest	Energy Update	Infinifuel	Raytheon	
Biofuel Review	Ethanol India	Inventure	Renewable Energy Magazine	
Bionavitas	Genergetics	Kent Sea Tech	Texas Clean Fuels	

State and University Actions

Some states are taking action to fund entrepreneurial businesses in renewable energy and biofuels.

Scripps Institution of Oceanography, in La Jolla, California, offers undergraduate and graduate degrees in marine biology. Scientists such as Stephen Mayfield in the Department of Cell Biology are studying genetic engineering on algae for biofuel production. Greg Mitchell teaches and researchers biological oceanography.

Woods Hole Oceanographic Institution, on Cape Cod, Massachusetts, offers undergraduate and graduate programs in marine biology. Scientists are working on the impact of climate warming on ice algal production in the Arctic Ocean and others are researching the causes and consequences of red tides. Scientists at the DOE's Argonne National Laboratory in cooperation with the University of Illinois and Northwestern University are working on converting algae to hydrogen gas. They are working with algae that contain an enzyme called hydrogenase, which creates small amounts of hydrogen gas. The objective is to remove the catalyst from the hydrogenase and use it during photosynthesis.

University of Washington, in Seattle and Friday Harbor in the San Juan Islands, offers a marine zoology/botany program for undergraduate and graduate studies. Students may take courses in the San Juan Archipelago doing field studies of natural history, adaptations, evolution, and taxonomy of algae and herbivores.

University of Miami, the Rosenstiel School of Marine and Atmospheric Science offer degree programs in marine biology where students study algae, sea grasses and coral reefs among other topics.

University of New Hampshire, Michael Briggs and the Biodiesel Group in the Department of Physics is working on cost effective algae-based technologies for biodiesel production.

The Emerging Technology Fund in *Texas* will provide $4 million to Texas AgriLife.Research and General Atomics to conduct microalgal research and development.

Minnesota has made similar grants available for renewable energy sources.

North Dakota: DOE has partnered with Chevron to develop higher-oil yield strains of microalgae. The Defense Advanced Research Projects Agency, DARPA, is working on a project with Honeywell, General Electric and the University of North Dakota.

Virginia: Old Dominion University researchers in Virginia have successfully piloted a project to produce biodiesel feedstock by growing algae at municipal sewage treatment plants. The researchers hope that these algal production techniques could lead to reduced emissions of nitrogen, phosphorus and carbon dioxide into the air and surrounding bodies of water. The pilot project is producing up to 70,000 gallons of biodiesel per year.

Arizona State University Polytechnic: LARB, The Laboratory for Algal Research and Biotechnology, works on all phases of algae growth and commercial production for biofuels. The lab produces various species of algae for biofuel feedstocks and tests growing, harvesting and processing variables in the lab and at their field site. LARB directors Professors Qiang Hu and Milton Sommerfeld supervise projects a series of projects such as producing jet fuel from algae and bioremediation of wastewater from human wastes, industrial wastes and dairy waste streams. Engineers built portable algaculture systems that can test the capability of various algal species that may be used to remediate wastewater at a variety of sites. Some algae in lakes and reservoirs are capable of producing toxins that may cause fish kills and affect human health. LARB is involved in several projects where waters are monitored to check for the presence of potential toxic algae. Supported by the Salt River Project and NSF Water Quality Center, the project uses molecular fingerprinting to detect and treat toxic algae that release toxic compounds in water supplies.

ASU BioDesign Institute and the Global Institute of Sustainability: Professors Bruce Rittmann and Wim Vermaas, in life sciences have been studying cyanobacteria for the past 20 years. Currently, they are researching ways to bioengineer cyanobacteria to produce biofuel. This work sponsored by British Petroleum and Science Foundation Arizona and plans to design and create organisms that store more lipids for biodiesel.

Greenindependence.org: The Algae Collaboratory for Sustainable and Affordable Foods and Energy at ASU Polytechnic creates a global social network for green food and fuels. This social marketing collaboratory brings together scientists, academics, practitioners and students who share knowledge and biotechnology tools to bring algae's full promise to the world. The collaboratory operates to create sustainable and affordable foods and fuels for all people on Earth. In addition, projects focus on remediating polluted water and air and creating valuable products such as fertilizers, medicines and vaccines.

The Collaboratory supports sustainability technology development, communication, entrepreneurship and technology transfer for scale-up for commercial algal production. Initiatives are directed toward every stage of algae selection, growth and development, harvest, processing and marketing. The Collaboratory also conducts R3D, research, development, demonstration and diffusion on issues associated with small-scale algaculture systems that may be sited in villages, wasteland, inner cities, roof tops, balconies and back yards.

Patents and other forms of intellectual property are possibly the strongest threat to algal production besides ethanol subsidies. Patents may prevent wide adoption of new production technologies due to the cost of paying patent holders. For example, many of the most productive corn, wheat, rice other food seeds are patented and beyond the means for many farmers. The Collaboratory is dedicated to maximizing open source solutions available to all people on the planet. An excellent potential solution to the intellectual property issues is the development of an IP Pool where industry members share their IP and users pay a reasonable fee based on use and scale. Educational and research opportunities are expanding and institutions and additional programs are posted on the collaboratory site.

Green Algae Strategy: End Oil Imports and Engineer Sustainable Food and Biofuels

Green Algae Strategy shares the fascinating story of extraordinary innovation occurring not in deep space or in deep oceans but simply under our feet. Few people are aware that one of Earth's oldest, tiniest and simplest organisms holds such great potential for desperately needed sustainable solutions for our very hungry, thirsty and needy planet.

Green Algae Strategy engineers hope for a better life for billions of people who lack sufficient and affordable food, freshwater, fresh air, fertilizer and clean burning fuel for cooking and heating fires.

Algal production holds promise for Green Independence:

1. Freedom from oil imports
2. Freedom from world hunger
3. Moderation of climate change

Algae can provide these solutions while consuming no cropland, no freshwater and no fossil fuels. Every pound of algae grown for food and fuel also captures nearly two poundsof CO_2 while releasing pure oxygen into the atmosphere.Freedom from oil imports will occur by growing algae on desert and wasteland about the size of Maine. Freedom from hunger will engage 10 million Green Masterminds globally who have the knowledge and capability for growing nutritious food and high energy biofuel locally with green solar, algaculture. Slowing climate change will occur by diffusing green solar knowledge and technology globally and seeding ocean dead zones with algae in order to capture and sequester massive amounts of carbon dioxide.

4. Summary

Most industry participants believe algal production will focus on three biofuels; green diesel, gasoline and jet fuel, JP-8. There seems to be no industry consensus on a best approach to algal biomass production including growing systems or production locations. Algal producers are experimenting with a diverse set of production models. Production models seem to vary based on the production objectives, type of feedstock and location. International producers tend to use open ponds while U.S. producers are planning to use closed or semi-closed cultivated algal production systems. International producers are using naturally selected algae species while U.S. producers are planning to use a combination of species selection and genetically modified organisms that maximize the production of algal oil.

The critical industry issues are stable production systems, trained personnel, strain selection and costs. Besides the production of liquid transportation fuels, respondents believe algae will play a major role in carbon capture and recycle, water remediation and feed for fish, fowl and animals.

Nearly half of respondents believe that a focused algaculture program could replace oil imports for their country in 20 years. Conversely, roughly 20 per cent of respondents believe that algae will never replace oil imports.

Recommendations to move the industry forward include better access to information, substantial increases in public and private funding for algal R&D, stronger education and training, more information on production issues and better networking and collaboration. Industry participants want to see demonstration projects, decision support systems and independent reviews of algal production systems. Future of algal industry research needs to drill down on production, supply chain and social and economic issues. Improved information on industry needs will support industry participants and provide critical information needed for public policy decisions and support.

Acknowledgments

Several key people assisted in the development of the survey including:

- ✰ Mark Allen, CEO and President, A2BE
- ✰ Milton Sommerfeld, Professor, ASU
- ✰ Jim Sears, CTO and Chief Scientist, A2BE
- ✰ Jeff Mettais, Marketing Officer, A2BE

- ✰ Greg Mitchell, Research Biologist, Scripps Institution of Oceanography
- ✰ Qiang Hu, Assoc. Prof. Arizona State University
- ✰ Al Darzins, Applied Sciences Manager, NREL

References

http: / /americas.irc-online.org/am/5179

http: / /www.BiofuelsDigest.com.

http: / /www.freepatentsonline.com/70160704.html

Weiss, Rick. Firms Seek Patents on 'Climate Ready' Altered Crops, Tuesday, May 13, 2008;

Algal Biotechnology: New Vistas (2010) *Pages 25–35*
Editor: **Mihir Kumar Das**
Published by: **DAYA PUBLISHING HOUSE, NEW DELHI**

Chapter 2

Prospective in Diatom Nanotechnology

Jannendra Rath[1], Sikha Mandal[1] and Bijaya Kumar Padhi[2]*

[1]*Department of Botany, Visva-Bharati, Santiniketan–731 235, West Bengal, India*

[2]*Centre for Environmental Studies, Visva-Bharati, Santiniketan–731 235, West Bengal, India*

ABSTRACT

Diatoms are the major group of hard-shelled algae and one of the most common types of phytoplankton in fresh and marine ecosystems. A characteristic feature of diatom cells is that they are encased within a unique cell wall made of silica. Silicate materials are very important in nature and they are closely related to the evolution of living organisms. Diatom walls show a wide diversity in form, some quite beautiful and ornate, but usually consist of two symmetrical sides with a split between them. There is great potential for the use of diatoms in nanotechnology. This potential lies in the pores and channels which give rise to a greatly increased surface area, and the silica structure which lends it to chemical modification. In addition there is a huge variety in the sizes and shapes of diatoms available, providing scope for the selection of a particular species of diatom tailored to a particular requirement. The silica walls of diatoms can be used for the attachment of active biomolecules, such as antibodies, using either primary amine groups or the carbohydrate moiety. These modified structures can, therefore, be

* E-mail: jrath@visva-bharati.ac.in, jnanendra@gmail.com

used for antibody arrays or for use in techniques such as immunoprecipitation. These small organisms can help biologists and nanotechnologists to work together for further research.

Keywords: Diatom, Nanomaterials, Nanotechnology.

1. Introduction

Diatoms are the major group of eukaryotic algae, and are one of the most common types of phytoplankton. Most diatoms are unicellular, although they can exist as colonies in different shape of filaments or ribbons (*e.g. Fragillaria*), fans (*Meridion*), zigzags (*Tabellaria*), or stellate colonies (*Asterionella*). A characteristic feature of diatom cell is that they are encased within a unique cell wall made of silica (hydrated silicon dioxide) called frustule. These frustules show a wide diversity in form. Some are quite beautiful and ornate, and usually consist of two asymmetrical sides with a split between them, hence the group is named 'Diatom' which literally means "cut into two halfs". Fossil evidence suggests that they originated during, or before, the early Jurassic Period. Diatom communities are a popular tool for monitoring environmental conditions, and are commonly used in studies of water quality.

Diatoms belong to a large group called the heterokonts, including both autotrophs (*e.g.* golden algae, kelp) and heterotrophs (*e.g.* water moulds). Their yellowish-brown chloroplasts are typical of heterokonts, with four membranes and containing pigments such as fucoxanthin. Individuals usually lack flagella, but they are present in gametes and have the usual heterokont structure. Most diatoms are non-motile, although some move via flagellation.

2. The Biology and Ecology of Diatoms

Diatoms are unicellular organisms of the kingdom Protista, characterized by a silica shell of often intricate and beautiful sculpture. Traditionally diatoms are divided into two orders: centric diatoms (Centrales), which are radially symmetric, and pennate diatoms (Pennales), which are bilaterally symmetric. The former are paraphyletic to the latter. A more recent classification divides the diatoms into three classes: centric diatoms (Coscinodiscophyceae), pennate diatoms without a raphe (Fragilariophyceae), and pennate diatoms with a raphe (Bacillariophyceae).

Diatoms are microscopic single-celled plants that live in freshwater and marine ecosystems, from the poles to the tropics. They are found in sea ice and snow; some are even blown into the air. Benthic diatoms live in or on the sediments and on the surface of other plants (epiphytic), animals (epizootic), or rocks (epilithic). Some diatoms are found as unicell; others form long "chains" by linking to the adjacent cell, either by abutting end to end or by joining their protruding spines or setae. They are usually yellowish or brownish, and are found in fresh and saltwater, in moist soil, and on the moist surface of plants. Like all plants, diatoms contain chlorophyll and other pigments that capture the energy of sunlight to convert carbon dioxide and water molecules into carbohydrates via photosynthesis.

Diatom reproduce asexually by cell division. In most species, is a diatom divided to produce two daughter cells, each cell keeps one of the two valves and develops a smaller valve within it. As a result, after each division cycle the average size of diatom cells gets smaller. Once such cells reach a certain minimum size rather than simply divide vegetatively, they reverse this decline by forming an auxospore. This expands in size give rise to a much larger cell, which then returns to size-diminishing divisions. Auxospore production is always linked to meiosis and sexual reproduction. There are more than 200 genera of living diatoms and it is estimated that there are approximately 1, 00,000 species (Round *et al.*, 1990). For the most part they lack flagella. Most diatoms are autotrophic though some heterotrophic or symbiotic species can be found in particular habitats.

The living matter of each diatom is enclosed within a shell of silica. These shells are marked by minute pores or depressions that allow the living material access to its environment. When present in abundance in water they impart golden- brown color to that, therefore also known as Golden Brown Algae. Diatoms are highly polymorphic on the basis of their size and shapes. Their survival also requires nutrients such as nitrogen, phosphorus, silicon, and trace metals. Diatoms are a major source of the world's oxygen, producing amounts equivalent to all the tropical rain forests (Field *et al.*, 1998). They are also responsible for 40 to 45 per cent of the total production of organic carbon compounds in the oceans (Mann, 1999), or 20 to 25 per cent for the entire Earth. As such, they are a significant source of food for other organisms within the marine food web and play an important role in the biogeochemical cycle of elements (Sarthou *et al.*, 2005). Because of their large cell size and rapid sinking rate, they also play an important role as "biological pump" that removes carbon dioxide from the atmosphere and places it at depth, thus helping to moderate the increasing levels of carbon dioxide associated with climate change.

2.1 Valve Morphogenesis

Diatoms cells are contained within a unique silicate (silicic acid) cell wall comprised of two separate valves (or shells). This cell wall is also called frustule and its two parts typically overlap one another like the two halves of a Petri dish. They cling together by means of a gelatinous substance to form zigzag rows of ribbon shaped colonies. The biogenic silica that compose the cell wall is synthesized intracellularly by the polymerization of silicic acid monomers. This material is then extruded to the cell exterior and added to the wall. Characterization of diatom biogenic silica has been performed at the nanoscale level. Solid-state ^{29}Si nuclear magnetic resonance spectroscopy techniques confirmed the amorphous nature of the frustule and were used to estimate the coordination state of the silicone (Gendron-Badou *et al.*, 2003; Bertermann *et al.*, 2003). Gas adsorption techniques were used to determine the shell surface area (Vrieling *et al.*, 1999).

The valve surfaces of diatoms have large pores (foramen) on the outer surface. These pores lead to a channel through the silica shell with a sieve plate at the base. In vivo, this system allows the passage of nutrients and gases to the protoplasm whilst providing a protective outer shell. The intricate 3D microstructure of diatoms results in a high surface area formed by myriad pores and channels (Figure 2.1). The

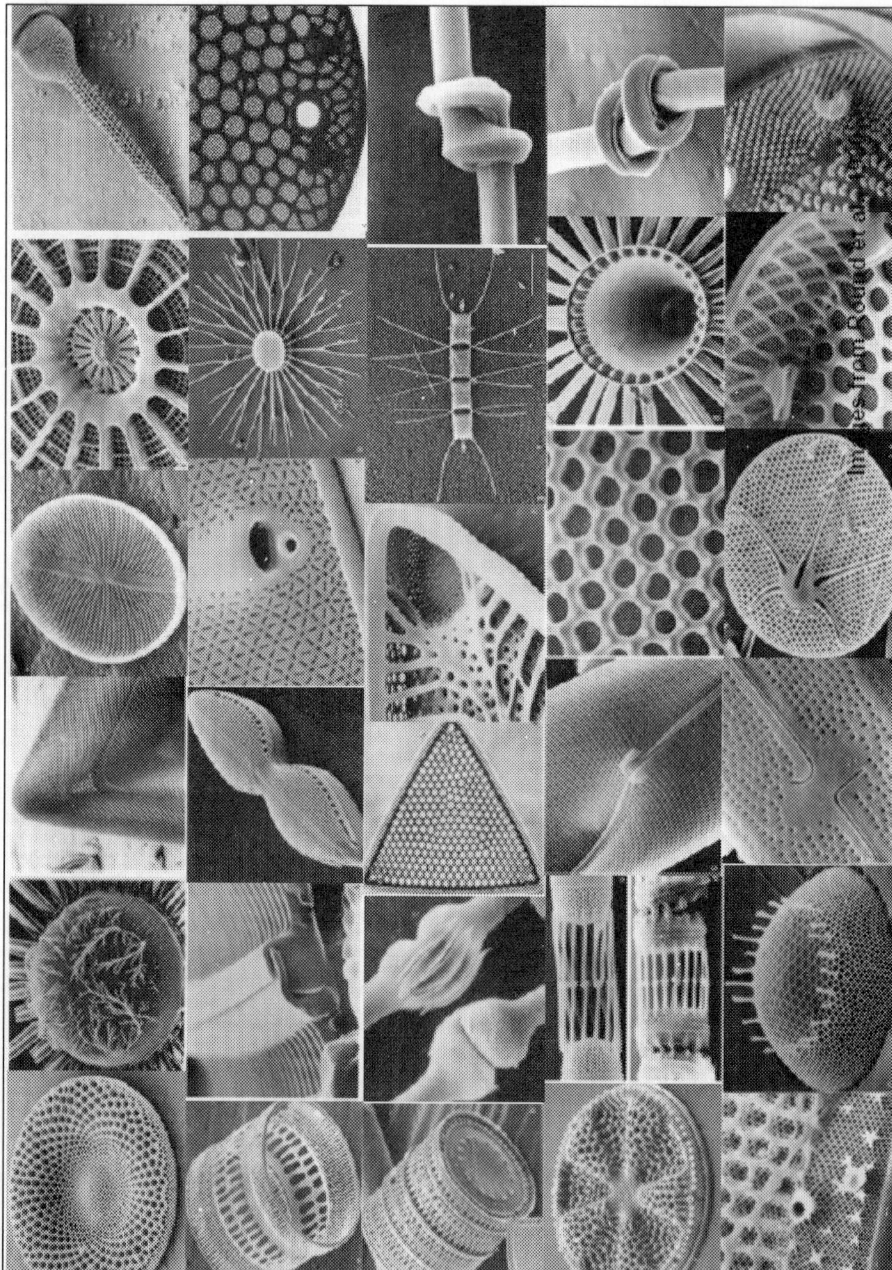

Figure 2.1: Morphogenesis of Diatom showing Myriad Pores and Channels

combination of the silica chemistry of the frustule coupled with the high surface area makes it particularly suitable for applications in nanotechnology (Townley *et al.*, 2008; Gordon and Parkinson, 2005; Sterrenburg, *et al.*, 2005; Madou, *et al.*, 2002; Hamm,*et al.*, 2002). Diatoms produce diverse three-dimensional structures, due to their exponential rate of growth, may be used in the manufacture of components for nanotechnology as an alternative to present linear lithographic techniques (Gordon and Parkinson, 2005). The frustule is composed of silica and poorly characterized organic material (principally glycoproteins and carbohydrates). Diatoms are interesting for nanotechnologists because of the nano scale architecture of their silica shell, which exceeds the capabilities of present-day human engineering. There is a need for a systematic based catalog, improved diatom culture techniques, better understanding of the mechanisms of diatom morphogenesis and motility, for selection of industrially useful nano materials.

3. Application in Nanotechnology

A major goal for nanotechnology is the development of demanding systems which are cheaper, smaller, cleaner, and faster. An ideal solution is to exploit the cheap nanostructures which are provided by Nature. Unlike most manufacturing processes, diatoms achieve a high degree of complexity and hierarchical structure under mild physiological conditions. Diatoms also show remarkable mechanical strength which is useful for industrial applications. In addition they are easy to culture, have a fast reproductive rate, a dominant vegetative state of reproduction, low culture costs and present a virtually unlimited renewable source which provides an opportunity of diatoms in nanotechnology application. In this chapter we are highlighting some of the potential uses of diatoms and diatom-specific processes for nanotechnology.

Historically diatoms were used by Alfred Nobel for the adsorption of nitroglycerin to create dynamite (Pulz and Gross, 2004). The deposition of silica by diatoms may also prove to be of utility to nanotechnology. Diatom cells repeatedly and reliably manufacture valves of different shapes and sizes and potentially allowing diatoms to manufacture micro or nano-scale structures which may be use in a range of "widgets" including optical systems; semiconductor nanolithography; and even using diatom valves as vehicles for drug delivery. Using an appropriate artificial selection procedure, diatoms that produce valves of particular shapes and sizes could be evolved in the laboratory and then used in chemostat cultures to produce mass as nanoscale components. An advantage of diatoms is that they allow direct fabrication of three-dimensional structures instead of the layer by layer manipulation common in present day nanotechnology which is based primarily on the lithographic methods used to make microelectronics (Madou, 2002). Furthermore, diatoms reproduce and can outperform any factory production method via exponential growth of their numbers (Sandhage *et al.*, 2002).

3.1 Diatom Nanobiotechnology in Medical Application

Biologists and particularly diatomists have long studied the properties of diatoms. Just now engineers are discovering how to exploit features unique to these organisms. Their uniform nano pore structure, micro channels, chemical inertness, and silica microcrystal structure suggest many nano scale applications such as embedding diatom frustules in a metal-film membrane; magnetizing frustules for pinpoint drug delivery; and producing silica nano powders from frustules. There is great potential for the use of diatoms in nanotechnology. Researchers in the UK (Townley *et al.*, 2008) have demonstrated that the silica walls of diatoms can be used for the attachment of active biomolecules, such as antibodies, using either primary amine groups or the carbohydrate moiety. These modified structures can, therefore, be used for antibody arrays or for use in techniques such as immunoprecipitation. Since attachment of bioactive molecules to a solid support is the key to lab-on-a-chip of technology for covalently tethering antibodies to the diatom silica matrix and assaying their biological activity. It is possible to tether the antibody to the diatom surface via the antibody amino groups, conjugation via these groups gives the potential for multiple attachments with resulting conformational distortions. Therefore investigation on an alternative tethering approach–to attach via the carbohydrate rather than protein components of antibody molecules is required. The next step for the scientists is to incorporate these modified nanostructures into a workable design which could provide a cheap and disposable drug delivery vehicles to treat diseases such as targeting a single cancer cell.

3.2 Bioactive Compounds from Diatom

Lincoln *et al.* (1990) reviewed the biologically active compounds from diatoms are: antibiotics, enzyme inhibitors (to treat diseases caused by excessive enzyme activity, *e.g.* carbohydrate disorders), pharmacology active compounds and toxins. The diatoms with these potential applications belonged to the following genera: *Amphiprora, Asterionella, Bacillaria, Bacteriastrum, Chaetoceros, Coscinodiscus, Cyclotella, Fragilaria, Gyrosigma, Hemidiscus, Licmophora, Lithodesmium, Navicula, Nitzschia, Odontella, Phaeodactylum, Rhizosolenia, Skeletonema* and *Thalassiosira* (Lebeau and Robert 2003a). Antibacterial and antifungicidal effects have been noted in all the algal classes and notably in diatoms (Lincoln *et al.*, 1990; Pesando 1990; Viso *et al.*, 1987). Antibacterial compounds are generally present among a complex of fatty acids, as shown by Viso *et al.* (1987) and Imada *et al.* (1992). Biologically active compounds extracted from diatom cells have been proposed for a range of nano technological applications.

3.3 Engineering Application

Major goal for bionanotechnology is the development of microscale total analysis systems (µTAS) or lab-on-a-chip. An ideal solution is to use the cheap nanostructures which are provided by Nature. Unlike most manufacturing processes, diatoms achieve a high degree of complexity and hierarchical structure under mild physiological conditions. Diatoms also show remarkable mechanical strength which is useful for industrial applications. In addition they are easy to culture, have a fast reproductive rate, a dominant vegetative state of reproduction, low culture costs and present a virtually unlimited renewable source. What is also interesting from a fabrication point of view is that these silica structures are produced in diatoms without the need of industrial chemical processes, using only light and minimal nutrients to generate an exceptionally cheap and renewable material. The semiconductor industry has been able to double the density of transistors on computer chips in every few years. They've been doing it using photolithographic techniques for the past 30 years. But they are actually hitting a wall now because they're getting down to the resolution of visible light.

Using a chemical process that converts the shell's original silica (silicon dioxide, SiO_2) into the semiconductor material silicon, researchers have created a new class of gas sensors based on the unique and intricate three-dimensional (3-D) shells produced by diatoms. The converted shells which retain the 3-D shape and nanoscale detail of the originals, could also be useful as battery electrodes, chemical purifiers–and in other applications requiring complex shapes that nature can produce better than humans. Recently a diatom-derived silicon sensor has been produced by Prof Kenneth H. Sandhage, at the Georgia Institute of Technology for detection of nitric oxide. This diatom-derived silicon sensor possessed a combination of speed, sensitivity, and low voltage operation that exceeded conventional sensors. Therefore the unique diatom-derived shape, high surface area and nanoporous, nanocrystalline silicon material all contributed towards such attractive gas sensing characteristics.

Diatoms are fabulous for making very precise shapes, and making the same shape over and over again by a reproduction process that, under the proper growth conditions, yields microshells at a geometrically-increasing rate. Diatoms can produce three-dimensional structures that are not easy to produce using conventional silicon-based processes. The potential here is for making enormous numbers of complicated 3-D shapes and tailoring the shapes genetically, followed by chemical modification convert the shells into functional materials such as silicon. Silicon can form compounds that have a high lithium content. Because diatom-derived silicon structures have a high surface area and are thin walled and highly porous, the rate at which lithium ions get into and out of such silicon structures can be high. Therefore for a given battery size, it could store more power, or recharge the battery faster by using such structures as electrodes.

Diatoms create their shells by employing special proteins and subcellular organs to first assemble silica nanoparticles which are composed of just a few hundred atoms. The proteins and subcellular organs then orchestrate the assembly of those nanoparticles into shells. Therefore we have two levels of structure, these nanoparticles and more interesting thing to that we can take these particles as a little brick, and they are assembled into ornate microstructures. If we can incorporate elements such as silicon, germanium, titanium, and gallium into the diatoms' silica shells at the nano scale, these elements follow the laws of quantum mechanics instead of Newtonian physics, giving them unique and commercially desirable properties. At the nano scale, for example, the metal germanium glows blue when energy is applied to it. This has a host of applications in electronic and medical imaging technologies. The process to incorporate germanium nano particles in silica is possible but difficult with existing technology. The conventional process involves vaporizing a germanium crystal in a vacuum with a high-energy laser beam and coaxing the vaporized atoms to glom onto a silica surface, that has to be done at a high temperature

[and] at a high vacuum and [with] all the equipment associated with the control of that. However in diatom we need to add just enough dissolved metal to the vat to absorb it without dying.

Another advantage of using diatoms is that when the algae divide, they make a perfect copy of themselves, meaning we can make a gazillion of these and they are all the same shapes. In addition to the ability of the diatoms to absorb these metals and create nano structured materials, each diatom species makes shells with unique designs. And there are tens of thousands of diatom species. Therefore tens of thousands of micro-templates can be produced. Some have holes, some ribs, some oval, some square and all the micro-fabrication has been done by the organisms. We just put additional material on it. Therefore, in the future, the researchers hope they can use these diatoms to make intricate designs at the microscale that are currently not possible with existing technology. Therefore many products including flexible computer screens, cheap and efficient solar cells, filtration devices, and drug delivery vehicles could be a reality using diatom nanotechnology. Silica cell wall of the diatom *Coscinodiscus granii* can be regarded as a photonic crystal slab waveguide. Furthermore, they present models to show that light may be coupled into the waveguide and give photonic resonances in the visible spectral range. In testing, the researchers showed that the silicon they produced was photoluminescent–*i.e.* it glows when illuminated by certain wavelengths of light. That shows the fabrication process produced a nanoporous, nanocrystalline structure- and may have interesting photonic applications in addition to the electronic ones.

3.4 Silica Biomineralization

Another interest is the understanding the artistry of the amorphous silica shell and the underlying biomineralization process. The cell wall of diatoms, (the frustule), consists of two thecae overlapping like a Petri dish, each half being made up of the valve and several girdle bands spanning the circumference of the cell. The frustule is composed of silica and poorly characterized organic material (principally glycoproteins and carbohydrates). Diatoms are interesting for nanotechnologists because of the nanoscale architecture of their silica shell, which exceeds the capabilities of present-day human engineering. Indeed, in diatoms 'bio-glass' formation does not require high temperatures, high pressure or the use of caustic chemicals, and it is accomplished under mild physiological conditions.

3.5 Biomimetic Approaches

Materials scientists and chemists are now exploiting diatoms to develop new biomimetic approaches and to create silicon-based microdevices with specific features. Experiments using diatom bio-inspired molecules as synthetic biomimetic templates have been performed. For example, using the pR5 peptide, which comprises the amino acid sequence of one repeat unit of silaffin-1, it was shown that the lysine modifications are not necessary to induce silica formation below pH 7, but have a strong influence on precipitate morphology (Naik *et al.*, 2003). By incorporating the pR5 peptide into a polymer hologram created by two-photon-induced photo-polymerization, nanopatterning of silica spheres could be achieved. In fact, the silica spheres formed a two-dimensional array with the periodicity of the hologram (Brott *et al.*, 2001). The pR5 peptide was also found to be effective for the immobilization of enzymes inside silica mesoporous material (Luckarift *et al.*, 2004). Phage-display libraries provided another fruitful approach to select biomimetic peptides of interest. A peptide library was tested against silica particles, allowing the identification of molecules that have significant binding activity to the mineral surface. The selected peptides also proved to have silica precipitation activities (Naik *et al.*, 2002).

4. Genetic Manipulation of Diatoms and Tailored made Nanomaterial

Research has shown that uniquely modified proteins called silaffins and extremely long polyamine chains play a role in the structural design of the novel silica nanostructures. Genetic manipulation of diatoms will increase the understanding of their cellular biochemistry and potentially enable the use of these organisms for the production of commercially valuable compounds and nanomaterials. But inserting a gene through the strong silica cell wall is difficult. The wall must be penetrated, but not broken, and the foreign gene must be accepted into the diatom's genome. To insert the genes, such as those that encode different silaffins, through the diatom cell wall, microparticle bombardment might be a proven technology. DNA-coated tungsten particles can be shot on the diatoms under high helium pressure, thus enabling them to penetrate the strong diatom cell wall. The diatom incorporates the introduced DNA into its genome. Selection of the transfected cells is achievable using the antibiotic nourseothricin. When new genes are introduced they can be expressed constantly or be turned on and off when necessary. Knowing the genome sequence and having established a method for genetic modification of this organism means we can, in principle, analyze the function of every gene and the protein that it encodes. This will eventually enable us to identify the key cellular biomolecules involved in creating the strong, intricately patterned diatom cell walls and novel nanostructures.

Diatoms build their hard cell walls by laying down submicron-sized lines of silica, a compound related to the key material of the semiconductor industry–silicon. If we can genetically control that process, we would have a whole new way of performing the nanofabrication used to make computer chips. It has been reported that a set of 75 genes specifically involved in silica bioprocessing in the diatom *Thalassiosira pseudonana.* Therefore manipulating the genes responsible for silica production and potentially harness them to produce lines on computer chips. This could vastly increase chip speed, because diatoms are capable of producing lines much smaller than current technology allows.

5. Conclusion

To a nanotechnologist, diatoms seem almost heaven-sent. Not only do they turn out precision-made, three-dimensional structures quickly, cheaply and in enormous quantities, they do so at room temperature and pressure, and without a toxic chemical in sight. Further progress in this field will require identifying the molecular entities involved in pattern formation, which is actively being pursued by biochemical, proteomic, genomic, and genetic manipulation approaches. We have yet to fully understand the processes that build these structures and to learn how to create conditions to reproduce them. Research along these lines is paramount and of the utmost importance to find out how diatoms assemble these nanometer-patterned, intricate micro-architectures to find better methods for creating nanomaterials in the laboratory. Understanding and subsequent exploitation of the mechanisms by which diatoms are able to fabricate such structures at ambient temperatures and pressures may have significant applications in silicon chemistry, materials science and nanotechnology. Possible future applications of atomic force microscopy and spectroscopy techniques in biology, and especially in diatom research are manifold, interesting and challenging: *e.g.* watching cell division in real time at the nanoscale, finding out how diatoms produce their amorphous silica frustules, revealing what tricks they apply to generate strong adhesives or how their lubricants minimize friction in moving parts. Biological materials are simultaneously smart, dynamic, complex, self-healing, and multifunctional, characteristics difficult to achieve in purely synthetic systems. Biomimetics, the use of biological principles in materials synthesis and assembly, may be a path for realizing nanotechnology, such as molecular and nanoscale electronics. It is wonderful to dive deeper and deeper into the nanoworld, to watch biological processes at unprecedented resolution and to consider how their secrets might be applied in technological innovations. However, one should bear in mind that this focused approach teaches us only about biomolecular interactions. Like all other organisms, diatoms are more than just an assemblage of simply interacting biomolecules. With their emerging complex properties, they can teach us about life itself. After all to say it with, the English philosopher and mathematician "physics has to be explained in terms of a generalized theory of the organism."

Acknowledgement

We thank to the head of the department of Botany for his valuable suggestions to improve the manuscript.

References

Bertermann R, Kroger N, and Tacke R (2003) Solid-state 29Si MAS NMR studies of diatoms: structural characterization of biosilica deposits. Anal Bioanal Chem 375: 630-634.

Bradbury J (2004) Nature's Nanotechnologists: Unveiling the Secrets of Diatoms. PLoS Biology 2: 306.

Brott LL, Naik RR, Pikas DJ, Kirkpatrick SM, Tomlin DW, Whitlock PW, Clarson SJ and Stone MO (2001) Ultrafast holographic nanopatterning of biocatalytically formed silica. Nature 413: 291-293.

Gendron-Badou A, Coradin T, Maquet J, Frohlich F and Livage J (2003) Spectroscopic characterization of biogenic silica. J Non-Cryst Solids 316: 331-337.

Gordon R, Sterrenburg FAS and Sandhage KH (2005) Preface to the Diatom Nanotechnology. J Nanosci Nanotech 5: 1–4.

Hamm CE, Merkel R, Springer O, Jurkojc P, Maier C, Prechtel K, and Smetacek V (2003) Architecture and material properties of diatom shells provide effective mechanical protection. Nature 421: 841-843.

Lebeau T and Robert JM (2003a) Diatom cultivation and biotechnologically relevant products. Part I: Cultivation at various scales. Appl Microbiol Biotechnol 60: 612–623.

Lebeau T and Robert JM (2003b) Diatom cultivation and biotechnologically relevant products. Part II: Current and putative products. Appl Microbiol Biotechnol 60: 624–632.

Lincoln RA, Strupinski K, Walker JM (1990) Biologically active compounds from diatoms. Diatom Res 5: 337–349.

Lopez PJ, Descles J, Allen AE and Bowler C (2005) Prospects in diatom research. Curr Opinion in Biotech 16: 180–186.

Luckarift HR, Spain JC, Naik RR and Stone MO (2004) Enzyme immobilization in a biomimetic silica support. Nat Biotechnol 22: 211-213.

Naik RR, Brott LL, Clarson SJ and Stone MO (2002) Silica-precipitating peptides isolated from a combinatorial phage display peptides library. J Nanosci Nanotech 2: 95-100.

Naik RR, Whitlock PW, Rodriguez F, Brott LL, Glawe DD, Clarson SJ, and Stone MO (2003) Controlled formation of biosilica structures in vitro. Chem Commun 238-239.

Patrick R (1977) Ecology of freshwater diatoms and diatom communities. In D. Werner (Ed.), The Biology of Diatoms 285–317.

Pulz O, Gross W. (2004) Valuable products from biotechnology of microalgae. Appl Microbiol Biotechnol 65: 635–648.

Round EF, Crawford RM and Mann DG (1990) The Diatom Biology and Morphology of Genera. Cambridge University Press, Cambridge, England: 7-20.

Round FE, Crawford RM and Mann DG (1990) The Diatoms, biology and morphology of the genera. London, Cambridge University Press pp. 460.

Round FE, Crawford RM, Mann DG (1990) The Diatoms. Cambridge, Cambridge University Press pp 356.

Townley HE, Parker AR and White-Cooper H (2008) Exploitation of Diatom Frustules for Nanotechnology: Tethering Active Biomolecules. Adv Funct Mater 18: 369–374.

Vrieling EG, Beelen TPM, van Santen RA and Gieskes WW (1999) Diatom silicon biomineralization as an inspiration source of new approaches to silica production. J Biotechnol 70: 39-51.

Wee KM, Rogers TN, Altan BS, Hackney SA and Hamm C (2005) Engineering and medical applications of diatoms. J Nanosci Nanotechnol 5: 88-91.

Algal Biotechnology: New Vistas (2010) *Pages 36–61*
Editor: **Mihir Kumar Das**
Published by: **DAYA PUBLISHING HOUSE, NEW DELHI**

Chapter 3

Programmed Cell Death: An Integral Event in the Development of Algae and Higher Organisms

Shiv Shanker Pandey, Vivek Ambashtha and Budhi Sagar Tiwari*

School of Life Sciences, Jawaharlal Nehru University, New Delhi–110 067, India

ABSTRACT

Programmed cell death (PCD) has been recognized as a fundamental cellular process conserved in plants, animals and several unicellular eukaryotes. It is used for defense and developmental mechanisms. While cell death pathways in animals have been well characterized, relatively little is known about the molecular mechanism of such a strategy in plants and phytoplanktons. Although, very few regulatory proteins or protein domains have been identified as conserved across all eukaryotic PCD forms, still plants and animals share many hallmarks of PCD, both at cellular and molecular levels. Morphological and biochemical features like chromatin condensation, nuclear DNA fragmentation, and participation of caspase like proteases in plant PCD appear to be similar across the eukaryotic kingdom and in conformity with the process in metazoans as well. Different ecological

* E-mail: budhi@rediffmail.com

factors induce PCD in algae. In this chapter we tried to give all known examples of phytoplanktons PCD. In case of *Dunaliella tertiolecta*, the unicellular chlorophyte light deprivation induces PCD. Heat stress induces PCD in *Chlorella saccharophila* cells. *Chlamydomonas reinhardtii* cells exhibit cell death process akin to that of apoptosis when exposed to ultraviolet (UV)-C irradiation (1–100 J/m^2). Depletion of dissolved CO_2 stimulates formation of reactive oxygen species (ROS) and induces cell death in both natural and cultured *Peridinium* populations. Current evolutionary explanations for phytoplankton PCD are also summarized. A major challenge for apoptosis research in a variety of protists remains the identification of the molecular cell death machinery.

Keywords: *PCD, Algae, DNA fragmentation, ROS, Metacaspases, Autophagy.*

1. Introduction

Programmed cell death (PCD) is a genetically regulated sequence of (potentially interruptible) events that lead to the controlled and organized destruction of the cell (Lockshin and Zakeri, 2004). It is crucial for defence responses to restrict the spread of pathogens and for proper development of the multicellular body plan (Lam, 2004). Level of stress on a cell is a main determining factor for selecting different death machinery by the cell. If a population of cell is exposed to low level stresses then the majority of cells survive, if subjected to moderate levels of stress the majority execute PCD, while high levels of stress induce widespread necrosis.

"By the time I was born, more of me had died than survived. It was no wonder I cannot remember; during that time I went through brain after brain for nine months, finally contriving the one model that could be human, equipped for language." This quote from Lewis Thomas (1992) speaks to the importance of programmed cell death (PCD) in human development.

The concept of "programmed cell-death" was used by Lockshin and Williams in 1964 in relation to insect tissue development; around eight year before "apoptosis" was coined in relation to fall of leaves in plants. Since then, PCD has become the more general of these two terms. PCD has been the subject of increasing attention and research efforts. This trend has been highlighted with the award of the 2002 Nobel Prize in Physiology or Medicine to Sydney Brenner (United Kingdom), H. Robert Horvitz (US) and John E. Sulston (UK).

The type of cell death in animal system can be classified as apoptosis (Type I), autophagic cell death (Type II), programmed necrosis (Type III) (Lockshin and Zakeri, 2004; Bras *et al.*, 2005).

Apoptosis is characterized by cell shrinkage, membrane blebbing, chromatin condensation, DNA fragmentation, fragmentation of the nucleus, vesicle formation, externalization of phosphatidylserine from the inner to the outer leaflet of the plasma membrane, oxidative burst, loss of the mitochondrial membrane potential ($\Delta\Psi$), the release of proteins from mitochondria (*e.g.* cytochrome c, endonuclease G, apoptosis-inducing factor), activity of proteases (*e.g.* caspases).

Autophagy, or Type II cell death, appears distinct from apoptosis or necrosis (Schweichel and Merker, 1973). The term autophagy is derived from the Latin meaning 'self-eating' and describes the process whereby cytoplasmic materials are degraded through the lysosomal machinery within the cell. It is characterized mainly by the formation of autophagic vacuoles and also dilation of the mitochondria and endoplasmic reticulum and slight enlargement of the Golgi (Bras *et al.*, 2005). However, cells with enlarged vacuoles and undergoing autophagy do not necessarily die as the process is involved in the routine turnover of cell constituents. It is important, therefore, to use the term 'autophagy/type II cell death' only when there is a fatal destruction of the cell.

In contrast to apoptosis, necrosis has been described as an uncontrolled form of cell death which often follows overwhelming cellular stress where the cell is unable to activate its apoptotic pathways. During necrosis, swelling rather than shrinkage is the defining feature of the morphological change. This swelling is due to the cell loosing the ability to osmoregulate, resulting in water and ions flooding into the cell (Lennon *et al.*, 1991).

2. Programmed Cell-Death in Animals

Programmed cell death (PCD) play a very important role in both development and homeostasis in adult tissues for the removal of either superfluous, infected, transformed or damaged cells by activation of an intrinsic suicide program. Cells undergoing apoptosis usually exhibit a characteristic morphology, including fragmentation of the cell into membrane-bound apoptotic bodies, nuclear and cytoplasmic condensation and endolytic cleavage of the DNA into small oligonucleosomal fragments (Steller, 1995). The cells or fragments are then phagocytosed by macrophages. Signals that can trigger apoptosis can include lineage information, damage due to ionizing radiation or viral infection or extracellular signals. Extrinsic signals may either suppress or promote apoptosis, and the same signals may promote survival in one cell type and invoke the suicide program in others (Steller, 1995). Invocation of the suicide program involves the synthesis of specific messenger RNA molecules and their translation. PCD can sometimes be suppressed by inhibiting transcription or translation (Steller, 1995), which provides evidence that cell death is mediated by intrinsic cellular mechanism.

Cell death is currently the subject of considerable research activity. This interest stems, in part, from the potential for understanding oncogenesis and the possibility of exploiting the cell death program for therapeutic purposes. For example, inhibition of cell death might contribute to oncogenesis by promoting cell survival instead of death. Likewise, triggering cell death might provide the means for eliminating unwanted cells (*e.g.*, tumor cells). This might be accomplished by harnessing tumor necrosis factor (TNF), which triggers apoptosis in some target cells.

Recognition of PCD as a developmental mechanism dates back to the 1930's. Developmental processes that involve PCD include:

Elimination of Transitory Organs and Tissues

Examples include phylogenetic vestiges (pronephros and mesonephros in higher vertebrates), anuran tails and gills and larval organs of holometabolous insects.

Tissue Remodeling

Vertebrate limb bud development (Saunders, 1982; Saunders, 1966) is an example. If PCD fails in the information of the digits, digits remain joined by soft tissue. Compare, for example, the situation in the chick and duck hind limbs. If chick limb mesoderm is combined with duck ectoderm, PCD fails and the digits remain joined (Saunders, 1966). This observation implicates the ectoderm in providing the signal to trigger PCD. Another example is formation of heart loops during vertebrate development. Depletion of cells in spinal ganglia occurs during development of the chick embryo. There is precise chronological and spatial control over this process. Interestingly, injections of nerve growth factor reduce the frequency of cell death in the spinal ganglia. This observation provides a link between growth control and PCD.

Much of our current knowledge about the molecular genetics of PCD comes from work on *Caenorhabditis elegans*. The adult hermaphrodite *C. elegans* forms 1090 somatic cells, of which 131 die by apoptosis. There are four stages in apoptosis in worms (Steller, 1995):

- ☆ Decision whether a cell should die or assume another fate;
- ☆ Death;
- ☆ Engulfment of the dead cell by phagocytes;
- ☆ Degradation of the engulfed corpse.

A number of genes have been identified that regulate these processes in worms (Steller, 1995). Mutations affecting the final three stages affect all somatic cells, whereas genes affecting the death verdict affect very few cells. Execution itself is mediated by *ced-3*, *ced-4* and *ced-9* (*ced* = cell death defective). *ced-3* and *ced-4* act autonomously in the doomed cells to kill them, whereas *ced-9* acts as a brake on the suicide program. If either *ced-3* or *ced-4* is inactivated by mutation, cells survive that would normally die. Constitutive expression of *ced-9* also inhibits cell death. On the other hand, if *ced-9* is inactivated by mutation, there is supernumerary cell death, resulting in early embryonic death of the worm.This observation suggests that *ced-9* prevents cell death in some cells in which *ced-3* and *ced-4* have the potential to function; *i.e.*, the cell death program is ready to run but is prevented from doing so. Mutations in either *ced-3* or *ced-4* suppress such ectopic cell deaths; such double mutants lack all PCD and have phenotypes equivalent to *ced-3* or *ced-4* single mutants.This result implies that they act downstream of *ced-9*. Cloning of the apoptotic genes of *C. elegans* and their characterization have led to considerable understanding of the molecular events of apoptosis and to the identification of mammalian homologues of the apoptotic effectors.

How the death sentence is delivered, and–if delivered–how cells can avoid it to gain a reprieve? Genetic studies on *Drosophila* have led to the identification of a gene called *reaper*, which plays a central role in initiation of cell death (White *et al.*, 1994).

Reaper transcripts are found in doomed cells, and their presence precedes morphological manifestations of apoptosis by 1 to 2 hours (White *et al.*, 1994). The *reaper* gene apparently encodes a small peptide with no known homologies to known proteins. Deletions of *reaper* suppress apoptosis in response to apoptotic stimuli. Thus, *reaper* mutant embryos contain many extra cells (White *et al.*, 1994). *Reaper* embryos have an intact apoptotic pathway that fails to be initiated. *Reaper* is thought to integrate the information from diverse signaling pathways and put a cell on the apoptotic pathway. At this point, the presence of *reaper* homologues in other organisms is unknown.

Why do some cells die and others survive? There is evidence from work on higher organisms that extrinsic signals may protect cells from apoptosis by suppressing the suicide program (Raff, 1992). For example, the survival of developing neurons may depend upon neurotrophic factors secreted by their targets; a failure to receive sufficient stimulation results in death. What is the advantage to the organism of using extrinsic signaling to sustain cell survival? One possibility is that it could provide a simple system to eliminate cells that end up in the wrong place; without a signal to sustain them, rogue cells would be eliminated (Raff, 1992). Consider primordial germ cells, for example. In mammals, they originate in the hindgut and must migrate to the genital ridges, where they form the gametes. Those that fail to reach the genital ridges are eliminated, presumably because they are deprived of the signal (Steel factor) that is required for their survival in the genital ridges (De Felici and Pesce, 1994). A disadvantage to the organism of the mechanism that necessitates signalling to prevent apoptosis is that its failure by mutation can lead to the survival of unwanted cells, which paradoxically–can lead to death of the organism itself. On the other hand, an opportunity is presented by such a mechanism to allow investigators to devise means for targeting unwanted cells for destruction. Prostate cancer is an example. The survival of prostate cells is dependent upon androgens; androgen depletion leads to a reduction in cell number by apoptosis. Recently, the dependence of prostate cells on androgens to avoid cell death has been exploited therapeutically by the use of androgen ablation to invoke apoptosis in prostate cancer cells and prolong survival in men with prostate cancer. Significantly, resistance to androgen depletion correlates with overexpression of *bcl-2*, the human *ced-9* homologue that acts as a brake on prostate cancer cell apoptosis (Raffo *et al.*, 1995). Thus, escape from androgen sensitivity by overexpression of *bcl-2* in a subset of prostate cancer cells leads to proliferation of these cells and, ultimately, death of the patient. If *bcl-2* could be down-regulated, apoptosis of these cells could be invoked and the cancer controlled. Ironically, there may be an inverse correlation between cell death and survival of the organism. Exploitation of this relationship holds much promise for therapeutic control of diseases such as cancer.

3. Programmed Cell-Death in Plant Tissue

PCD plays a number of important roles in plant developmental pathways, including xylogenesis-formation of woody tissues in trees and perennials, leaf abscission, self-incompatibility, and defense-responses to a wide variety of pathogens and environmental stresses. Plants have at least two types of programmed cell death. Apoptosis is one form of PCD, some features of which may be evolutionarily conserved in both plants and animals.

In "APL regulates vascular tissue identity in Arabidopsis", Bonke and colleagues state that one of the two long-distance transport systems in vascular plants, xylem, consists of several cell-types "the differentiation of which involves deposition of elaborate cell-wall thickenings and programmed cell-death." The authors emphasize that the products of plant PCD play an important structural role (Martin Bonke *et al.*, 2003).

Basic morphological and biochemical features of PCD have been conserved in both plant and animal kingdoms. Cysteine proteases are involved in a key step in animal PCD pathways, and have recently been demonstrated to also be important in plant PCD (Solomon *et al.*, 1999). Soybean cells cultured in hydrogen-peroxide, induce PCD and resulting accumulation of high levels of specific proteases (*e.g.* caspases). Intercellular signalling may regulate specific protease inhibitor genes (*e.g.* cystatin), which in turn, regulate protease activity. Salicylic acid inhibits cystatin expression and may do so under high hydrogen-peroxide concentrations–effectively promoting PCD. Conversely, jasmonate is found to inhibit PCD. Oxidative stress from hydrogen peroxide may result through mechanical damage of cells by the feeding of herbivores. Solomon *et al.* (1999) have postulated that downregulation of PCD by jasmonate may occur in cases where there is no survival advantage for PCD (*e.g.* when there is attack by large herbivores).

The morphological changes in tobacco caused by PCD have been described by Mittler *et al.* (1997). Tobacco plants infected with tobacco mosaic virus (TMV) were used to study the hypersensitive response (HR) activated by PCD. When tobacco plants are grown at 30°C, PCD is inhibited which allows TMV infection to spread. TMV-infected plants were shifted from 30°C to 25°C to induce PCD. Uninfected plants also temperature shifted, and TMV-infected plants grown at 30°C, were used as controls. Also, freeze-thaw induced necrosis was used as a control for cell death not caused by PCD (*i.e.* cell death not requiring the activation of some protein). Mittler *et al.* (1997) observed stem and leaf sections using light microscopy, and they found that PCD occurs simultaneously in most leaf cells, but asynchronously in stem cells. Transmission electron microscopy (TEM) was used to visualise the morphology of cells in stem sections of TMV-infected plants and the control plants. In contrast to animal cells, no apoptotic bodies were seen–the authors suggest this may be due to the cell wall which would prevent engulfment by neighbouring cells. However, vacuolization of the cytoplasm and condensation of chromatin was observed, which are similar to events in apoptosis of animal systems. Also, chloroplasts from TMV-infected cells were found to contain starch granules not present in uninfected cells, necrotic cells, or in cells infected by a bacterial pathogen (*Pseudomonas syringae*).

The degradation of DNA also appears to be a common feature of PCD in plants and animals. In some animal systems, large DNA fragments of 300kb and 50kb are produced during PCD, and 180bp segments indicating internucleosomal fragmentation may also be generated. Mittler *et al.* (1997) detected 50kb nuclear DNA fragments in both PCD (at 48 hours) and freeze-thaw (in one hour) treatments using field inversion gel electrophoresis (FIGE) and DNA hybridization (with probes specific for nuclear, chloroplast, or mitochondrial DNA). Thus, nuclear DNA degradation may be a response to mechanical injury and can be mediated by PCD in the hypersensitive response. Fragments of 180 kb were not detected, suggesting there is no internucleosomal cleavage in plants during PCD. The chloroplast genome normally exists in multimeric forms, but levels of the monomeric form were found by Mittler *et al.* (1997) to increase early on during TMV-induced PCD.

The identities of most of the key elements involved in plant PCD remain unknown. Iterative homology searches suggest that caspase-related proteases (metacaspases, first identified in yeasts) are encoded in the Arabidopsis genome (Uren *et al.*, 2000). Recent evidence from inhibitor studies and biochemical approaches suggests that caspase-like proteases may also be involved in cell death control in higher plants (Korthout *et al.*, 2000) and may constitute a common pathway for cell death in both metazoans and plant cells. However, the absence of homologs of metazoan cell death genes in Arabidopsis (*see* Arabidopsis Genome Initiative, 2000) suggests that cell death in higher plants has diverged from that found in the animal kingdom. This controversy can only be addressed by examining cell death in extant organisms that are representative of divisions that emerged before the divergence of metazoans and higher plants.

Janneke Balk and Christopher J. Leaver, of the Department of Plant Sciences, University of Oxford, carried out research on mutations in the mitochondrial genome of sun-flower cells. Results of this research suggest that mitochondria play the same key role in vascular plant PCD as in other eukaryotic cells.

3.1 PCD in Pollen Prevents Inbreeding

During pollination, plants enforce self-incompatibility (SI) as an important means to prevent self-fertilization. Research on the corn poppy (*Papaver rhoeas*) has revealed that proteins in the pistil on which the pollen lands, interact with pollen and trigger PCD in incompatible (ie. *self*) pollen. The researchers, Steven G. Thomas and Veronica E. Franklin-Tong, also found that the response involves rapid inhibition of pollen-tube growth, followed by PCD.

Researchers have noted a number of similarities between apoptosis in animals and PCD in plants (Greenberg, 1996), and comparing the functions of key components involved in these processes between kingdoms may serve fruitful for understanding the regulation of PCD in both. For example, cytochrome *c* release from mitochondria is a key event in apoptosis in animal cells and has also been shown to be an early event in PCD in plants (Balk *et al.*, 1999; Balk and Leaver, 2001).

4. Mitochondria as a Central Regulator of Cell Death in Animals and Plants System

The mitochondria have a central regulatory role in integrating stress and/or PCD signals in both plant and animal PCD (Green and Reed, 1998; Jones, 2000; Tiwari *et al.*, 2002. Bras *et al.*, 2005; Diamond and McCabe, 2007). The release of apoptogenic proteins from the mitochondrial inner membrane space (IMS) triggers apoptosis in mammalian cells (Wang, 2001; Adrain and Martin, 2001) and release of these apoptogenic proteins occurs via a Bax/Bcl-2 controlled pore (Shimizu *et al.*, 1999) or the opening of the permeability transition (PT) pore (Green and Reed, 1998). In mammalian cells the Bax/Bcl-2 pore can selectively release IMS proteins that activate caspases. However, there is no evidence that the Bax/Bcl-2 family normally operates in plant cells. Although there is evidence for a mitochondrial release of PCD activating molecules following a moderate level 55°C heat shock (Balk *et al.*, 2003). The other way that mitochondria may release PCD activating molecules in animal cells is through the formation of the PT pore (Green and Reed, 1998). So, while it appears that plants lack the classical caspase family required for true apoptotic death, there is strong evidence that the PT pore operates in plant cells (Arpagaus *et al.*, 2002; Maxwell *et al.*, 2002; Yu *et al.*, 2002), and three important components of the PT pore in animals (voltage-dependent anion channel, adenine nucleotide transporter, and cyclophilin D) are present in plants (Yao *et al.*, 2004). It has been proposed that, in animal cells, a moderate loss of membrane potential would lead to a release of mitochondrial IMS proteins such as cytochrome c, leading to activation of caspase-like molecules and resulting in apoptosis. However, a more severe loss of membrane potential would lead to catastrophic damage to the cell and result in necrosis (Hirsch *et al.*, 1997). The activation of the PT pore in plant cells would explain the abrupt switch from PCD to necrosis seen in the heat-shocked cells when the level of stress is increased, as the thresholds for living/PCD/necrotic states would be regulated by opening of a mitochondrial pore. Another hypothesis by Lemasters *et al.* (1998) suggests that cells respond to mitochondrial permeability transition (MPT) in a graded manner, with autophagy resulting when MPT occurs only in small number of cells leading to lysosomal degradation of the affected mitochondria and removal of the signals required for stimulation of autophagy. Apoptosis occurs when larger numbers of mitochondria are permeabilized, probably through increased release of apoptogenic factors such as cytochrome c and apotosis-inducing factor (AIF). When virtually all mitochondria are permeabilized and ATP becomes depleted, necrotic death is the outcome. In plants, with apoptotic machinery not available, alternative modes of cell death are switched on. Permeabilization of mitochondria during a moderate stress in plant cells, resulting in the release of apoptogenic factors (but no canonical caspases), could explain why the release of cytochrome c has been widely observed during PCD in plant systems (Balk *et al.*, 1999, 2003; Stein and Hansen, 1999; Sun *et al.*, 1999; Balk and Leaver, 2001; Krause and Durner, 2004; Thomas and Franklin-Tong, 2004; Yao *et al.*, 2004; Vacca *et al.*, 2006), but there is little evidence supporting cytochrome c as a direct activator of plant PCD (Yu *et al.*, 2002; Balk *et al.*, 2003; Vacca *et al.*, 2006). Interestingly, caspase-independent cell death (CICD) in mammalian systems resulting in alternative death morphotypes is the focus of much attention (Chipuk and Green, 2005; Kroemer and Martin, 2005) as it has become clear that pharmacological inhibitors of caspases do not always prevent cell death. In plants, unravelling the pathways of PCD may reveal an evolutionary conserved pathway in eukaryotic cells and aid in the understanding of CICD in animals.

5. Programmed Cell-Death in Algae

The origin of cell death programs in eukaryotes remains obscure (Aravind *et al.*, 2001) but is probably ancient; apoptosis-like phenomena have been reported in unicellular organisms, including Chlorophytes (Berges and Falkowski, 1998), dinoflagellates (Vardi *et al.*, 1999), yeast (Frohlich and Madeo, 2000), and bacteria (Lewis, 2000). However, the pathways through which cell death proceeds have not yet been identified in these organisms.

PCD as a differentiating mechanism is an attractive idea for explaining colony development in cyanobacteria: *Trichodesmium* may use PCD as a way of generating small groups of dispersal cells (hormogonia) that are released from parent colonies after death and during unfavourable conditions (Berman-Frank *et al.*, 2004). Dispersal is also thought to be the adaptive basis for PCD in *Calothrix*, in which colony fragmentation is achieved through the PCD of specific cells (Adamec, 2005). Colony differentiation involving mortality has also been noted in *Anabaena variabilis*. *A. variabilis* produces heterocysts that are viable for only 2–3 days, independent of media and growth conditions, and that show a loss of cell contents during the loss of viability (Reddy *et al.*, 1987). Heterocysts of *A. variabilis*, therefore, appear to undergo a process of terminal differentiation similar to that shown by higher plants. This case, similar to that of the colonial green alga *Volvox*, suggests that some cyanobacteria can differentiate through mortality, leading to the corollary that control of single cell mortality may be an emergent property of cell groups, and under genetic control. Like *Trichodesmium*, another species of *Anabaena* (*A. flos-aquae*) displays 'markers' of PCD such as DNA fragmentation, increased protease activities and plasma membrane permeability (Ning *et al.*, 2002). Furthermore, *A. flos-aquae* shows a population death rate that is a function of nutrient controlled growth rate, and is under circadian, genetic control and therefore programmed in nature (Lee and Rhee, 1999; see also Brussard *et al.*, 1997). Data on the 'background' mortality that may be present in growing phytoplankton populations is patchy; it is a common assumption that, if a (usually laboratory) population is growing, then all the cells are in good health. This assumption is confounded by the considerable metabolic heterogeneity of both natural and laboratory microalgal populations, as revealed through physiological assays (Veldhuis *et al.*, 2001; Brussaard *et al.*, 2001; Agusti 2004). Added to this, kinetic analysis shows that populations of symmetrically dividing cells can simultaneously contain dividing, non-dividing and dying cells whilst showing exponential growth, provided that the death rate is constant (Sheldrake, 1974). Thus, considerable metabolic heterogeneity is possible in phytoplankton populations, although little is known about it, and what does exist is likely to have been overlooked up to now.

A competing idea on the cause of cyanobacterial cell death is based on the dependence of bacterial mortality on growth phase (Aldsworth *et al.*, 1999; Mason *et al.*, 1986). This concept, perhaps more familiar to phycologists, suggests that, when growth is arrested by some form of sublethal stress (*e.g.* nutrient limitation in culture), growth becomes uncoupled from metabolism, leading to cell death through an oxidative burst (Aldsworth *et al.*, 1999). This idea has been used to explain cell death in cultures of phytoplanktonic *Synechococcus* (Sakamoto *et al.*, 1998) and *Synechocystis* (Suginaka *et al.*, 1999); metabolic imbalance (unbalanced growth) causes cell death, via unavoidable generation of oxidative stress. That the uncoupling of growth and metabolism causes unavoidable oxidative stress is an attractively simple explanation for cell death, which does not need to encompass an evolutionary explanation of PCD–cells are simply killed due to 'design constraints'. However, uncoupling of growth and metabolism cannot explain the mechanism of cyanobacterial PCD during differentiation. Differentiation clearly occurs in cyanobacteria and, since metazoan PCD is regulated through cell signalling, the role of signalling in microalgal PCD needs greater attention, and will benefit from studies into the signalling molecules produced by microalgal cells in culture, and also in biofilms. Signalling within bacterial biofilms can lead to PCD via quorum-sensing (Oleskin *et al.*, 2000), or through the induction of prophages and autotoxins (Webb *et al.*, 2003). Quorum sensing (density-dependent gene expression) in bacteria also provides a plausible mechanism for ontogenetic differentiation.

Studies of bacterial populations suggest hitherto unrealized complexity in the way microbial populations are regulated. Whether such complexity can be shown amongst microalgal populations is an open and exciting question, which could considerably complicate existing notions of the factors that regulate food-web structure. For example, intercellular signaling (which is allelopathic in nature) between cyanobacteria and dinoflagellates may be sufficiently sophisticated to harness cell death pathways (Vardi *et al.*, 2002); and the exciting discovery of prokaryote–eukaryote signalling (Joint *et al.*, 2002) shows that there is much to be learned about signalling within microbial communities. PCD in chlorophytes: 'functional' and 'non-functional' PCD Work on chlorophytes has revealed two instances of PCD, with two evolutionary interpretations. Older literature suggests that chlorophyte cell death can occur when a type of multicellularity (*i.e.* differentiation) is an objective of coordinated cell division. In the colonial alga Volvox, somatic cells die after a set number of divisions in a process of 'programmed senescence' (Pommerville and Kochert, 1981; Kirk, 1998).'Senescent' *Volvox* cells shrink, lose chlorophyll, accumulate lipid and then lyse. However, up to the point of lysis (during 'senescence'), the nucleus is unchanged and new protein expression occurs (Hagen and Kochert, 1980; Pommerville and Kochert, 1981). On a morphological basis, these changes appear similar to what happens in 'old' (*i.e.* nutrient-limited) algal cultures (Freudenthal, 1962; Prezelin, 1982; Fogg and Thake, 1987). The similarities between cell death during *Volvox* differentiation, which involves nuclear persistence and protein expression, and the cell death that is routinely observed in old cultures, might suggest a common pathway, involving self-mediation via PCD. In this sense, algal senescence could be functionally equivalent to multicellular plant senescence (which is controlled via PCD) with protease expression and cellular dissolution occurring in order to fulfil a functional role. One possible functional role of such a pathway in unicellular algae (and yeast) is to recycle the organic remains of cells for use by conspecifics (Gomez *et al.*, 1974; Frohlich and Madeo, 2000). This idea is controversial and requires a convincing demonstration of conspecific nutritional (or other) benefit resulting from the cell death of some proportion of the cells (Franklin and Berges, 2004). That cell death could provide nutritional benefit for remaining conspecifics is an idea that is easier to accept for yeast, because yeast cell division can be clearly asymmetric (*e.g.*

in budding yeast; Laun *et al.*, 2001). Asymmetric division is also possible in dinoflagellates (Silva and Faust, 1995), although whether the production and death of different types of cells gives nutritional benefit to remaining cells is unknown, and the utility of such a strategy is obscure. It is also difficult to see how nutritional benefit could be restricted to conspecifics in the planktonic environment. Another problem is that asymmetric division in dinoflagellates occurs during stressful conditions (Reguera and Onzalez-Gil, 2001), when sexual cycles may also be induced, leading to difficulties in interpreting the causes of changed cell morphology. 'Small cells' are thought to be capable of acting as gametes in some dinoflagellates (Silva and Faust, 1995), but the evidence for this is not always conclusive (Reguera and Gonzalez-Gil, 2001). Returning to chlorophytes, a competing viewpoint on the origin of unicellular PCD has been suggested by detailed work on the chlorophyte D. J. Franklin *et al.*, *Dunaliella tertiolecta*. D. tertiolecta, when placed in darkness, undergoes mass cell death, with the apoptotic morphology of dead cells suggesting the existence of a PCD-like process. Insertion of viral lysogenic genes into the *D. tertiolecta* genome at some point in evolutionary history may explain the existence of this pathway (Berges and Falkowski, 1998; Segovia *et al.*, 2003).

Under environmentally stressful conditions, when silencing or repression systems fail (Baulcombe, 2002), these lysogenic genes may be activated in a maladaptive fashion. In this scheme the PCD pathway in unicells is an 'accident of history' and offers no contemporary functional advantage. In the case of *D. tertiolecta*, the 'metabolic imbalance' hypothesis on the cause of cyanobacterial cell death (see above) could also be invoked; thus giving a good example of the difficulty of ascribing single causes to observations of microalgal mortality. The lysogenic gene hypothesis (Berges and Falkowski, 1998; Segovia *et al.*, 2003) is unique in neatly encompassing the latter possibility in an evolutionarily coherent fashion. Whether this hypothesis can be applied to other instances of microalgal cell death remains to be seen. Population crashes and cell death in dinoflagellates Attempts to 'balance' dinoflagellate populations by quantifying cell production and cell loss (as grazing and sedimentation) have not succeeded (Pollingher, 1988; Sanderson and Frost, 1996); unexplained population 'crashes' are a feature of many dinoflagellate bloom populations (Heiskanen, 1993; Heiskanen and Kononen, 1994; Lindahl, 1983; Usup and Azanza, 1988). In the freshwater dinoflagellate Peridinium gatunense, PCD is thought to be induced by CO_2 limitation because, during such stress, cell death can be blocked by cysteine protease (caspase) inhibitors, and dying cells show positive DNA fragmentation (Vardi *et al.*, 1999). The symbiotic dinoflagellate *Symbiodinium* sp. is also thought to undergo PCD (displaying apoptotic morphology) due to heat stress (Dunn *et al.*, 2002). However, in the dinoflagellate *Amphidinium carterae*, cell death in darkness, and during culture senescence, results in similar morphological changes without definitive apoptotic characteristics (Franklin and Berges, 2004). Mortality in *A. carterae* shows greatest morphological similarity with paraptosis (Table 3.1)–an alternative outcome of PCD (Sperandio *et al.*, 2000), which was originally observed in mammalian nerve tissue. Interesting work on cell death in the coccolithophore *Emiliania huxleyi* also reveals a paraptotic morphology resulting from cell death (Bidle *et al.*, 2005). Further molecular and cytological characterization of cells in nature and in culture is needed to identify PCD and to test its ecological importance in a wide range of species. Given the often rapid disappearance of dinoflagellate blooms, PCD in response to environmental stress or energy limitation could potentially explain the enigmatic population crashes seen in nature. Noji *et al.* (1986) considered that dinoflagellate (*Ceratium* spp.) bloom declines in the Baltic are caused by the exhaustion of division potential, coupled with an inability to enter a sexual cycle due to the unfavourable environmental conditions that exist at the end of blooms. Recent work highlighting the importance of environmental stress in the sexual induction of *Volvox* (Nedelcu *et al.*, 2004) suggests that sex and death result from differing magnitudes of environmental stress, which produce a common intracellular consequence (oxidative stress)–'mild' oxidative stress triggers the induction of sex, whereas 'severe'oxidative stress kills the cell. In metazoan cells, oxidative stress is considered to be a key 'trigger' of PCD (Jabs *et al.*, 1996) and the role of oxidative stress is attracting increasing attention in investigations into microalgal cell death. In dinoflagellates, cell death can also be caused by direct interspecific contact (Costas *et al.*, 1993; Uchida *et al.*, 1995), and the importance of this interaction in nature is poorly known. Allelopathic interactions within and between taxa may be

Table 3.1: Simple Definitions of Terms Used to Describe the Causes of Cell Death. The terms used should reflect the evidence demonstrated; terms like 'PCD' should only be used for cells where specific molecular and morphological markers have been demonstrated.

Terms	*Defintion*	*Comment*
Natural cell death	Death that results from an exhaustion of division potential (age), energy limitation (starvation), or intolerance physical conditions (*e.g.* desiccation)	To avoid exhaustion of division potential, most phytoplankton probably undergo rejuvention through sex. Amongst phytoplanktonic eukaryotes, sexual cycles are incompletely characterized.
Programmed cell death (PCD)	Cell death (colourfully called 'cell suicide') resulting from gene expression within the moribund cell. PCD is mediated by a set of enzymes called capases and results in a set series of biochemical events as the cell dies, which produce a distinctive dead cell morphology (*e.g.* apoptosis)	PCD is essential for the successful construction of complex body plans (differentiation) and for homeostasis. PCD is thought to be activated by intercellular signalling, or by environmental stress (cf. natural cell death), causing capase activation and cell death. PCD can be a cause (though not the *ultimate* cause) of cell death once triggered.
Apoptosis	A set of morphological character seen in dying/dead metazoan cells during/ after the activation of PCD. These include DNA (and nuclear organelle) fragmentation, compartmentalization of cell contents into apoptotic bodies, cell shrinkage, membrane 'blebbing' loss of membrane integrity, and phosphatidylaserine inversion.	A poptotic cells are thought to be produced during the death of phytoplankton and yeast cells and therefore PCD is thought to occur in these taxa. Apoptosis cannot be a cause of cell dealth; apoptosis is a *consequence* of PCD.

Contd...

Table 3.1–Contd...

Terms	*Defintion*	*Comment*
Lysis	Rupture of the cell membrane, dissolution of cell contents.	Recognizable as cell loss/ disappearance. Quantification of phytoplankton cell lysis has been attempted with the dissolved enterase method. Lysis can be caused by pathogens and parasites. Viruses may subvert cellular machinery to produce 'PCD-like' processes perior to lysis.
Necorisis	Cell death that does not involve gene expression. In metazona cells, necrosis involves loss of membrane integrity, cell swelling and then lysis.	Necrosis results when cell are killed in the absence of any metabolic self-involvement
Paraptosis	A form of PCD, triggered by a molecular signal, which results in extensive vacualization and mitochondrial swelling in the absence of caspase activation.	Paraptosis produce an alternative morphological outcome to apoptosis.
Senescence	The process of growing old; difficult to recognize in unicellular organisms. In phycology, 'senescence' has been used to refer to the appearance of cells after exponential growth, *i.e.* when many cells are probably dead due to natural cell death.	In microalgal work, the use of 'senescence' is problematic the appearance of cells that have died is used to explain the process that caused cell death. However, an outcome of a process cannot simultaneously by the cause of the process; the cause of death in these instances is not senescence. 'Senescence' in multicellular plants involves PCD.

Source: Franklin *et al.*, 2006.

sophisticated enough to harness PCD mechanisms, but first we need to be able to recognize the extent and nature of PCD in the divergent taxa of the phytoplankton.

In case of algae PCD has been studied in *Dunaliella tertiolecta, Chlamydomonas reinhardtii, Chlorella saccharophila* and *Peridinium gatunense. D. tertiolecta* was the first unicellular plant revealing apoptotic markers during cell death (Berges *et al.*, 1998).

5.1 Programmed Cell Death in *Peridinium gatunense*

During the bloom of the photosynthetic freshwater dinoflagellate *Peridinium gatunense* the pH of the patches rise more than 9 which cause limitation of CO_2 concentration facilitating diversion of photosynthetic electron from CO_2 fixation to oxygen, resulting in the production of ROS to a level sufficient to trigger programmed cell death (PCD). Production of ROS depends on concentration of CO_2, even under a light intensity lower than that required to reach maximal photosynthetic rate (Berman *et al.*, 1995). PCD in *Peridinium* is mediated by H_2O_2 and specific protease(s). Catalase inhibited cell death in culture, implicating hydrogen peroxide (H_2O_2) as the active

ROS. Cysteine protease inhibitor E-64 causes blocking of Cell death stimulating cyst formation. This shows that Cysteine protease(s) are involved in processes that determine whether a cell is destined to die or to form a cyst. Moreover, *P. gatunense* also exhibited other hallmarks of apoptosis, such as protoplast shrinkage and nuclear DNA fragmentation.

Berges and Falkowski (Berges *et al.*, 1998) reported that nutrient constraints or darkness induced a cell-death process in batch cultures of *Thalassiosira* and *Dunaliella*, respectively, which also coincided with induction of a protease (Berges *et al.*, 1998). The cell-death process could be blocked by a cysteine protease inhibitor, but not by other protease inhibitors.

This study shows that CO_2 limitation resulted in the generation of ROS to a level that induced programmed cell death, which resembles apoptosis in animal and plant cells. Our results also indicate that Cysteine protease(s) are involved in processes that determine whether a cell is destined to die or to form a cyst. PCD allows only the best adapted individuals to establish cysts, while eliminating less healthy members of the community; PCD confers a selective advantage to a population during subsequent seasons (Vardi *et al.*, 1999).

5.2 Programmed Cell Death in *Chlorella saccharophila*

Chlorella saccharophila is a unicellular green alga with a wide spread distribution and an ability to occupy extreme environments. It belongs to a group, the Chlorophytes, which is considered to be one of the primary lineages of photosynthetic eukaryotes, together with red algae, land plants and glaucophytes (Delwiche. 1999). It is likely that they have a common origin and that they obtained their plastids through a single primary endosymbiosis with a cyanobacterium (Rodriguez-Ezpeleta *et al.*, 2005).This allows studies which can also provide insight into the evolution of different cellular pathways.

Heat stress induces PCD in *Chlorella saccharophila* cells. This PCD pathway occurs with characteristics features such as chromatin condensation, DNA fragmentation, cell shrinkage and detachment of the plasma membrane from the cell wall. Caspase 3-like activity is also observed in HS-treated *C. saccharophila* cells and this is inhibited by mammalian caspase 3 inhibitor (Zuppini *et al.*, 2007). Photosynthetic pigment contents also reduce which is associated with alteration of chloroplast morphology and the ribulose-1, 5- bisphosphate carboxylase/oxygenase large subunit and the light-harvesting complex of PSII also rapidly disappear (Zuppini *et al.*, 2007).

HS triggers in *C. saccharophila* a caspase-like-dependent PCD pathway exhibiting many key markers of metazoan PCD. Study of heat shock induced PCD in *Chlorella saccharophila* cells shows that unicellular Chlorophytes, higher plants and metazoans share conserved crucial elements of the caspase-mediated cell death cascade (Zuppini *et al.*, 2007).

5.3 Programmed Cell Death in *Dunaliella tertiolecta*

The obligate photoautotroph chlorophyte *Dunaliella tertiolecta* can be cultured in artificial seawater containing inorganic nutrients. When placed in darkness, algae cultures lost their photosynthetic capacity and colour and died after approximately six days. Indeed, *D. tertiolecta* was the first unicellular plant revealing apoptotic markers during cell death.

When the unicellular chlorophyte alga *Dunaliella tertiolecta* is placed in darkness it undergoes a form of cell death reminiscent of apoptosis in metazoans. Many morphological criteria of apoptotic cell death are met, including an increase in chromatin margination, degradation of the nucleus, and DNA fragmentation. Biochemical assays of the activities of cell death-associated proteases, caspases, measured using highly specific fluorogenic substrates, increased with time in darkness and paralleled the morphological changes. The caspase-like activities is inhibited by caspase-specific inhibitors. Antibodies raised against mammalian caspases cross-reacted with specific proteins in the alga. The pattern of expression of these immunologically reactive proteins was correlated with the onset of cell death. The occurrence of key components of apoptosis, and particularly a caspase-mediated cell death cascade in a relatively ancient linage of eukaryotic photoautotrophs, argues against current theories that cell death evolved in multicellular organisms. It is hypothesized that key elements of cell death pathways were transferred to the nuclear genome of early eukaryotes through ancient viral infections in the Precambrian Ocean before the evolution of multicellular organisms and were subsequently appropriated in both metazoan and higher plant lineages (Segovia *et al.*, 2003).

5.4 Programmed Cell Death in *Chlamydomonas reinhardtii*

Chlamydomonas reinhardtii (Ehrenberg) cells exhibit cell death process akin to that of apoptosis when exposed to ultraviolet (UV)-C irradiation (1–100 J/m^2) (Moharikar *et al.*, 2006). This PCD pathway shows typical hallmarks of apoptosis including cell shrinkage, associated nuclear morphological changes, flipping of phosphatidylserine, and DNA fragmentation. Interestingly, fluorescence imaging of DNA changes in UV-C exposed cells, following PicoGreen staining, revealed that extra-nuclear DNA disintegrates before that of nuclear changes, where the latter extensively diffuses out of the nuclear compartment, spreading into the whole cell and reaching the periphery of dying cells. Antibodies against a mammalian caspase-3 share epitopes with a protein of 28 kDa; whose pattern of expression correlates with the onset of cell death. Moreover, growth experiments indicate that spent medium recovered from UV-C exposed cells exhibit a protective effect against cell killing of fresh cultures of *C. reinhardtii* cells by UV irradiation. The protective effect of UV-spent medium is not a general growth promotional response on normal cells, but rather, is specific to UV-exposed cells. A model is proposed that *C. reinhardtii* cells exposed to UV elicit apoptotic-like changes, which in turn lead to an adaptive response in neighbouring cells against fresh rounds of UV exposure, thereby promoting survival of the cell population (Moharikar *et al.*, 2006).

5.5 Programmed Cell Death in *Micrasterias denticulata*

Micrasterias denticulate is a unicellular green alga, served as a cell biological model system for many years and details on its growth properties; physiology and ultrastructure throughout the cell cycle are well known. So this 'immortal' alga

Micrasterias is an ideal object for study of effect of different environmental condition on PCD. Treatments with low concentrations of H_2O_2 causes induction PCD in these organisms, resulting in severe ultrastructural changes to organelles, as observed in TEM. These include deformation and part disintegration of mitochondria, abnormal dilatation of cisternal rims of dictyosomes, occurrence of multivesicular bodies, an increase in the number of ER compartments, and slight condensation of chromatin. Additionally, a statistically significant increase in caspase-3-like activity is detected, which is abrogated by a caspase-3 inhibitor. Photosynthetic activity measured by fast chlorophyll fluorescence decreased as a consequence of H_2O_2 exposure, whereas pigment composition, except for a reduction in carotenoids, is the same as in untreated controls. TUNEL positive staining and ladder-like degradation of DNA, both frequently regarded as a hallmark of PCD in higher plants, could only be detected in dead *Micrasterias* cells(Darehshouri *et al.*, 2008).

6. Algal Toxins in Programmed Cell Death

Toxins produced by marine and freshwater algae represent a significant environmental hazard. Okadaic acid, synthesized by dinoflagellates and concentrated in mussels, is a major causative agent of diarrhetic shellfish poisoning (Bialojan *et al.*, 1998, and Cohen *et al.*, 1990) and microcystins, made by blue-green algae which sometimes infest drinking water supplies, are strong hepatotoxins that have caused sickness and death in both humans and animals (Jochimsen *et al.*, 1999, Hawkins *et al.*, 1985, Runnegar *et al.*1986, and Carmichael *et al.*, 1994). In isolated rat hepatocytes, algal toxins have been shown to induce a disruption of various cytoskeleton elements (Eriksson *et al.*, 1989, Falconer *et al.*, 1992, Ohta *et al.*, 1992 and Blankson *et al.*, 1995) as well as to elicit apoptotic cell death (Boe *et al.*, 1991).The algal toxins are protein phosphatase inhibitors which cause over phosphorylation of cellular proteins (Bialojan *et al.*, 1988 and Haystead *et al.*, 1989).

The cyanobacterial hepatotoxins, microcystin and nodularin, are produced by a wide range of cyanobacteria. Microcystin production has been reported in the four cyanobacterial orders: Oscillatoriales, Chroococcales, Stigonematales, and Nostocales. The production of nodularin is a distinct characteristic of the genus *Nodularia* of Nostocales.

Microcystin-LR (lysine-arginine) is the most investigated cyanobacterial peptide toxin because it is frequently present in cyanobacterial blooms in rivers and lakes. Nodularin occurs in brackish waters. The concentration of toxin in water depends on the cell content of microcystins or nodularins and the concentration of cells in the water. The peptide toxins are contained primarily within the cyanobacterial cells and are rarely released before the cells die or are destroyed (for instance, following water treatment or in the stomach, after ingestion). Due to widespread eutrophication in many settings, these toxins may occur more frequently and at unnaturally high concentration. In natural water bodies and in water storage reservoirs, the concentrations of toxin vary widely from undetectable to several milligrams per litre in cyanobacterial scums.

The specific toxic substances that have been identified to date from different genera of cyanobacteria, together with their primary target organs in humans, are given in Table 3.2.

Table 3.2: Cyanobacterial Toxins and General Features (Metting and Pyne, 1986; Chorus and Bartram, 1999)

Cyanobacterium	*Toxin(s)*	*Structure*	*Primary Target Organ in Mammals*
Microcystis	Microcystin	Cyclic peptide	Liver
aeruginosa	*Microcystis*-type-c	Peptide	Liver
	Microcystin-like-toxins	Peptides	Liver
	Microcystin-like	Peptide	Liver
Aphanizomenon	Aphantoxins	Alkaloids	
flosaquae	Neosaxitoxin	Alkaloids	Nerve axons
	Saxitoxin	Alkaloids	Nerve axons
Anabaena	Anatoxin-a	Alkaloids	Nerve synapse
flosaquae	Anatoxin-b	?	Nerve synapse
	Anatoxin-c	?	Nerve synapse
	Anatoxin-c	?	Nerve synapse
Schizothrix calcicola	Aplysiatoxins	Alkyl phenols	Skin
Lyngbya gracilis	Debromoaplysiatoxin	Alkyl phenols	Skin, gastrointestinal tract
L. majuscule	Debromoaplysiatoxin	Alkyl phenols	Skin, gastrointestinal tract
	Lygbyatoxin	?	Skin, gastrointestinal tract
Oscillatoria nigroviridis	Aplysiatoxin	Alkyl phenols	Skin
Calothrix crustacean	Aplysiatoxin	Alkyl phenols	Skin
Nostoc muscorum	Aplysiatoxin	Alkyl phenols	Skin
S. muscorum	Aplysiatoxin	Alkyl phenols	Skin
Nodularia	Nodularin	Peptides	Liver
Cylindrospermopsis	Cylindrospermopsins	Alkaloids	Liver
All	Lipopolysaccharides (LPS)	Alkaloids	Potential irritant: affect any exposed tissue

Microcystin and nodularin inhibit protein phosphatase and induce apoptotic cell death. Their inhibitory effect on protein phosphatases 1 and 2A can be utilized as tools to understand the phosphorylation-dependent regulatory mechanism underlying the early stage of apoptosis. The incubation of freshly isolated hepatocytes with these toxins results in a rapid hyperphosphorylation of cellular proteins before any morphological signs of apoptosis appears. Proteins subjected to phosphorylation in this early phase of apoptosis may play key roles in this cellular process and become valuable targets for drug development. The ultra-rapid apoptosis-induction by microcystin and nodularin provides a unique amount of synchronized apoptotic cells with "large" amounts of mainly serine/threonine phosphorylated proteins (Solstad *et al.*, 2006).

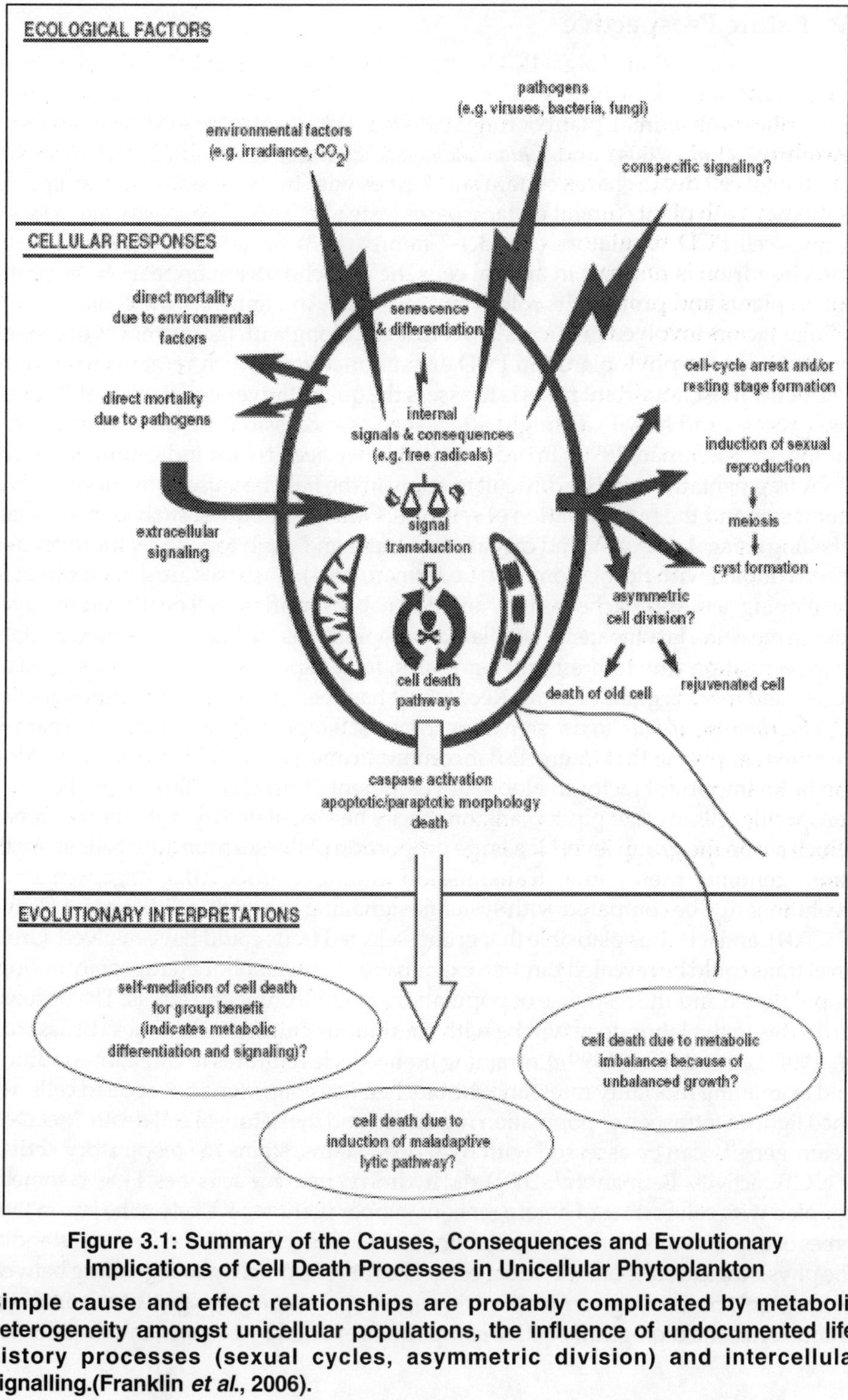

Figure 3.1: Summary of the Causes, Consequences and Evolutionary Implications of Cell Death Processes in Unicellular Phytoplankton

Simple cause and effect relationships are probably complicated by metabolic heterogeneity amongst unicellular populations, the influence of undocumented life-history processes (sexual cycles, asymmetric division) and intercellular signalling.(Franklin *et al.*, 2006).

7. Future Prospective

The investigation of algal PCD is highly novel. It is, at present, difficult to make generalizations about how PCD operates in algae, and whether it has greater similarities with animal, plant or fungal models. Whole genome work in *Thalassiosira* (Armbrust *et al.*, 2004) and *Chlamydomonas* (Shrager *et al.*, 2003) indicates that microalgal cell death shares certain similarities with higher plants, such as enzyme homology with plant/fungal metacaspases instead of animal caspases, and a lack of animal cell PCD regulators (*e.g.* Bcl-2 family, p53). In addition, the role of the mitochondrion is unclear: in animal cells the mitochondrion appears to be central but, in plants and protists, its role is far from clear. In Figure 3.1 the ecological and cellular factors involved in microalgal cell death, along with the current evolutionary explanations for phytoplankton PCD are summarized. Much remains to be done. One of the most important tasks is to assess the quantitative significance of PCD as a loss process, and how PCD might affect species successions and the conversion of carbon (c.f. Kirchman, 1999). In order to do this, we need robust indicators. Some, like DNA fragmentation, will be difficult to apply in the field because of the need for high biomasses and the manipulation of samples. Other techniques, such as cell-specific labelling using Annexin-V and caspase substrates and inhibitors, show more promise when coupled with flow cytometry (Lederman, 2004). Caspases are a good target for developing new approaches as they are reasonably specific to cell death and relatively easy to measure. Freshwater phytoplankton populations are likely to be more tractable to investigation due to their smaller scales, fewer species, high biomasses, easier access and more regular blooms. Recently, it has been shown that the dinoflagellate *Alexandrium ostenfeldii* can use signalling to encyst temporarily and thus avoid parasite infection, implying that intercellular communication evoked by biotic interactions can be an important factor in bloom development (Toth *et al.*, 2004). Can the fate of conspecific cells in other phytoplankton species be modulated by water borne signals, which act on the group level? If a large proportion of the surrounding cells share the same genome, then signal transmission amongst unicellular organisms can, evolutionarily, be compared with systemic signalling in a multicellular plant (Toth *et al.*, 2004), and it is thus plausible that group-selected traits could have evolved. Group level traits could be revealed through examination of metabolic heterogeneity in single populations, and the response of populations to environmental stress. The best way to do this in the laboratory will be with continuous culture techniques (Brussard *et al.*, 1997; Lee and Rhee, 1999).Entraining monospecies cultures to constant conditions and examining mortality rates, and the biochemical characteristics of dead cells, will shed light on intraspecies population regulation and the nature of cell death. Metabolic heterogeneity can be assessed with live/dead stains, stains for respiratory activity (*e.g.* CTC activity; Berman *et al.*, 2001) and hydrolytic enzyme activities. Flow cytometry, coupled with cell sorting of heterogeneous subpopulations, is likely to be key in these types of experiments, and counterstaining techniques will be useful in understanding the physiological state of individual cells. The role of extracellular signalling between phytoplankton needs examination through experiments designed to isolate and characterize the biologically active compounds exuded by phytoplankton cells, and their relationship with the induction of cell death programmes. A good example of the complexity in trophic interactions which can be seen with such an approach is found in Casotti *et al.* (2005). Understanding of surface receptors and signalling pathways will increase as genomic data is generated, and these data should prove useful in predicting the role of signalling. This field has great potential for generating major new insights and, given the revolution in understanding in bacteriology gained from observations of quorum sensing and cell death processes, microbial ecology in its entirety could be set to become much more complex.

References

Adamec F, Kaftan D &Nedbal L (2005) Stress-induced filament fragmentation of *Calothrix elenkinii* (Cyanobacteria) is facilitated by death of high-fluorescence cells. J Phycol 41: 835–839.

Adrain C and Martin SJ (2001) The mitochondrial apoptosome: a killer unleashed by the cytochrome seas. Trends in Biochemical Sciences 26: 390–397.

Agusti S (2004) Viability and niche segregation of *Prochlorococcus* and *Synechococcus* cells across the Central Atlantic Ocean. Aquat Microb Ecol 36: 53–59.

Aldsworth TG,Sharman RL and Dodd CER (1999) Bacterial suicide through stress. Cell Mol Life Sci 56: 378–383.

Aravind L, Dixit VM and Koonin EV (2001) Apoptotic molecular machinery: vastly increased complexity in vertebrates revealed by genome comparisons. Science 291: 1279–1284.

Armbrust JC., Berges JA, Bowler C, Green BR, Martinez D and Putnam NH (2004) The genome of the diatom *Thalassiosira pseudonana*: ecology, evolution and metabolism. Science, 306: 79–80.

Arpagaus S, Rawyler A, Braendle R (2002) Occurrence and characteristics of the mitochondrial permeability transition in plants. Journal of Biological Chemistry 277: 1780–1787.

Balk J, Chew SK, Leaver CJ, McCabe PF (2003) The intermembrane space of plant mitochondria contains a DNase activity that may be involved in programmed cell death. The Plant Journal 34: 573–583.

Balk J and Leaver CJ (2001) The PET1-CMS mitochondrial mutation in sunflower is associated with premature programmed cell death and cytochrome *c* release. Plant Cell **13:** 1803–1818.

Balk J, Leaver CJ and McCabe PF (1999) Translocation of cytochrome c from the mitochondria to the cytosol occurs during heat-induced programmed cell death in cucumber plants. FEBS Lett 463**:** 151–154.

Berges JA and Falkowski PG (1998) Physiological stress and cell death in marine phytoplankton: Induction of proteases in response to nitrogen or light limitation. Limnol Oceanogr 43: 129–135.

Berman-Frank I, Bidle KD, Haramayy L and Falkowski PG (2004) The demise of the cyanobacterium, *Trichodesmium* spp., via an autocatalyzed cell death pathway.Limnol Oceanogr 49: 997–1005.

Berman T, Kaplan B, Chava S, Viner Y, Sherr BF and Sherr EB (2001) Metabolically active bacteria in Lake Kinneret. Aquat Microb Ecol 23: 213–224.

Berman-Frank I, Kaplan A, Zohary T and Dubinsky Z (1995) Carbonic anhydrase activity in the bloom-forming dinoflagellate *Peridinium gatunense*. J Phycol 31: 906-913.

Bialojan C and Takai A (1988) Inhibitory effect of a marine-sponge toxin, okadaic acid, on protein phosphatases- specificity and kinetics. Biochem J. 256: 283-290.

Blankson H, Holen I and Seglen PO (1995)) Disruption of the cytokeratin cytoskeleton and inhibition of hepatocytic autophagy by okadaic acid. Exp. Cell Res 218: 522-530.

Bøe R, Gjertsen BT, Vintermyr OK, Houge G, Lanotte M and Døskeland SO (1991) The protein phosphatase inhibitor okadaic acid induces morphological changes typical of apoptosis in mammalian cells. Exp. Cell Res 195: 237-246.

Bras M, Queenan B, Susin SA. (2005). Programmed cell death via mitochondria: different modes of dying. Biochemistry (Moscow) 70: 231–239.

Brussard CPD., Noordeloos AAM and Riegman R (1997) Autolysis kinetics of the marine diatom *Ditylum brightwellii* (Bacillariophyceae) under nitrogen and phosphorus limitation and starvation. J Phycol 33: 980–997.

Brussaard CPD, Marie D, Thyrhaug R and Bratbak G (2001) Flow cytometric analysis of phytoplankton viability following viral infection. Aquat Microbial Ecol 26: 157–166.

Carmichael WW (1994)) The toxins of cyanobacteria. Sci. Am 270: 78-86.

Casotti R, Mazza S, Brunet C, Vantrepotte V,Ianora, A and Miralto A (2005) Growth inhibition and toxicity of the diatom aldehyde 2-trans, 4-trans-decadienal on *Thalassiosira weissfloggi* (Bacillariophyceae). J Phycol 41: 7–20.

Chipuk JE, Green DR (2005) Do inducers of apoptosis trigger caspase-independent cell death? Nature Reviews Molecular Cell Biology 6: 268–275.

Chorus I, Bartram J (1999) Toxic cyanobacteria in water: a guide to their public health consequences, monitoring and management. E&FN Spon Lond pp 41-90.

Cohen P, Holmes CFB and Tsukitani Y (1990)) Okadaic acid: A new probe for the study of cellular regulation. Trends Biochem Sci 15: 98-102.

Darehshouri A, Affenzeller M, Lütz-Meindl U (2008) Cell death upon HO induction in the unicellular green alga *Micrasterias*. Plant Biol (Stuttg) 10(6): 732-745.

De Felici M and Pesce M(1994) Growth factors in mouse primordial germ cell migration and proliferation. Prog Growth Factor Res 5: 135-143.

Delwiche CF (1999) Tracing the thread of plastid diversity through the tapestry of life. Amer. Nat. 154: S164–S177.

Diamond M, McCabe PF (2007) The mitochondrion and plant programmed cell death. In: Annual Plant Reviews; Plant mitochondria–Ed: Logan DC Oxford: Blackwell Publishing. pp 308–329.

Eriksson JE, Paatero GIL, Meriluoto JAO, Codd GA, Kass GEN, Nicotera P and Orrenius S (1989) Rapid microfilament reorganization induced in isolated rat hepatocytes by microcystin-LR, a cyclic peptide toxin. Exp. Cell Res 185: 86-100.

Falconer IR and Yeung DSK (1992)) Cytoskeletal changes in hepatocytes induced by *Microcystis* toxins and their relation to hyperphosphorylation of cell proteins. Chem–Biol Interact 81: 181-196.

Franklin DJ, Corina PD, Brussaard CPD and Berges JA (2006) What is the role and nature of programmed cell death in phytoplankton ecology? Eur J Phycol 41: 1–14.

Freudenthal H (1962) *Symbiodinium* gen. nov. and *Symbiodinium microadriaticum* sp. nov, a zooxanthella: Taxonomy, life cycle and morphology. J Protozool 9: 45–52.

Frohlich KU and Madeo F (2000) Apoptosis in yeast–a monocellular organism exhibits altruistic behaviour. FEBS Letters 473: 6–9.

Green DR and Reed JC. (1998) Mitochondria and apoptosis. Science 281: 1309–1312.

Greenberg JT (1996) Programmed cell death: A way of life for plants. Proc Natl Acad Sci USA **93:** 12094–12097.

Hagen G and Kochert G (1980) Protein synthesis in a new system for the study of senescence. Exp Cell Res 127: 451–457.

Hawkins PR, Runnegar MTC, Jackson ARB and Falconer IR (1985)) Severe hepatoxicity caused by the tropical cyanobacterium (blue-green alga) *Cylindrospermopsis raciborskii* (Woloszynska) Seenaya and Subba Raju isolated from a domestic water supply reservoir. Appl Environ Microbiol 50: 1292-1295.

Haystead TAJ, Sim ATR, Carling D, Honnor RC, Tsukitani Y, Cohen P and Hardie DG (1989)) Effects of the tumor promoter okadaic acid on intracellular protein phosphorylation and metabolism. Nature 337: 78-81.

Hengartner MO., R.E. Ellis and H.R. Horvitz. (1992) *Caenorhabditis elegans* gene *ced-9* protects cells from programmed cell death. Nature 356: 494-499.

Hirsch T, Marchetti P, Susin SA, Dallaporta B, Zamzami N, Marzo I, Geuskens M, Kroemer G (1997) The apoptosis-necrosis paradox. Apoptogenic proteases activated after mitochondrial permeability transition determine the mode of cell death. Oncogene 15: 1573–1581.

Jochimsen EM, Carmichael WW, An JS, Cardo DM, Cookson ST, Holmes CE, Antunes MB, de Melo Filho DA, Lyra TM, Barreto VS, Azevedo SM and Jarvis WR (1999)) Liver failure and death after exposure to microcystins at a hemodialysis center in Brazil. New Engl J Med 338: 873-878.

Joint, I, Tait K, Callow ME, Callow JA, Milton D, Williams P and Camara M (2002) Cell-to cell communication across the prokaryote-eukaryote boundary. Science 298: 1207.

Jones A (2000) Does the plant mitochondrion integrate cellular stress and regulate programmed cell death? Trends in Plant Sciences 5: 225–230.

Kirchman DL (1999) Phytoplankton death in the sea. Nature 398: 293–294.

Kirk DL (1998) Volvox Molecular-genetic Origins of Multicellularity and Cellular Differentiation. Cambridge University Press, Cambridge.

Korthout HAAJ, Berecki G, Bruin W, Van Duijn B, Wang M (2000) FEBS Lett 475: 139-144

Krause M, Durner J. (2004) Harpin inactivates mitochondria in *Arabidopsis* suspension cells. Molecular Plant–Microbe Interactions 17: 131–139.

Kroemer G, Martin SJ (2005) Caspase-independent cell death. Nature Medicine 11: 725–730.

Lam E (2004) Controlled cell death, plant survival and development. Nature Reviews Molecular Cell Biology 5: 305–315.

Lederman L (2004) Apoptosis: one way to die. BioTechniques.37: 520–521.

Lee D and Rhee G (1999) Circadian rhythm in growth and death of *Anabaena flos-aquae* (Cyanobacteria). J Phycol 35: 694–699.

Lemasters JJ, Niemine A-L and Qian T (1998) The mitochondrial permeability transition in cell death: a common mechanism in necrosis, apoptosis and autophagy. Biochimica et Biophysica Acta 1366: 177–196.

Lennon SV, Martin SJ, Cotter TG (1991) Dose-dependent induction of apoptosis in human tumour cell lines by widely diverging stimuli. Cell Proliferation 24: 203–214.

Lewis K (2000). Programmed cell death in bacteria. Microbiol Mol Biol Ann Rev 64: 503–514.

Lockshin RA, Zakeri Z(2004) Apoptosis, autophagy, and more. The International Journal of Biochemistry and Cell Biology 36: 2405–2419.

Lockshin RA and Williams CM (1964) Programmed cell death. II. Endocrine potentiation of the breakdown of intersegmental muscles of silkmoths. J Insect Physiol *10*: 643-649.

Maxwell DP, Nickels R and McIntosh L(2002) Evidence of mitochondrial involvement in the transition of signals required for the induction of genes associated with pathogen attack and senescence. The Plant Journal 29: 269–279.

Mazal Solomon *et al*(1999) The Involvement of Cysteine Proteases and Protease Inhibitor Genes in the Regulation of Programmed Cell Death in Plants. The Plant Cell 11: 431-444.

Metting B, Pyne JW (1986). Biologically active compounds from microalgae. Enz Microb Technol 8: 386-394.

Mittler R, Simon L and Kam E (1997) Pathogen-induced programmed cell death in tobacco. Journal of Cell Science 110: 1333-1344.

Moharikar S, D'Souza JS, Kulkarni AB, Basuthkar J and Rao BJ (2006) Apoptotic-like cell death pathway is induced in unicellular chlorophyte *Chlamydomonas reinhardtii* (Chlorophyceae) cells following UV irradiation: Detection and functional analyses. Journal of Phycology 42 (2): 423–433.

Ning SB, Guo HL, Wang L and Song YC (2002) Salt stress induces programmed cell death in prokaryotic organism *Anabaena*. J Appl Microbiol 93: 15–28.

Ohta T, Nishiwaki R, Yatsunami J, Komori A, Suganuma M and Fujiki H (1992) Hyperphosphorylation of cytokeratins 8 and 18 by microcystin-LR, a new liver tumor promoter, in primary cultured rat hepatocytes. Carcinogenesis 13: 2443-2447.

Oleskin AV, Botvinko IV and Tsavkelova EA (2000) Colonial organisation and intracellular communication in microorganisms. Mikrobiologiya 69: 249–265.

Pommerville JC and Kochert GD (1981) Changes in somatic cell structure during senescence of *Volvox carteri*. Eur J Cell Biol 24: 236–243.

Raff MC (1992) Social controls on cell survival and cell death. Nature 356: 397-400.

Raffo AJ, Perlman H, Chen MW, Day ML, Streitman JS and Buttyan R (1995) Overexpression of *bcl-2* protects prostate cancer cells from apoptosis *in vitro* and confers resistance to androgen depletion *in vivo*. Cancer Res 55: 4438-4445.

Reddy PM, Yumnam DD and Imchen TY (1987) Regulation of differentiation of multiple heterocysts in a blue-green alga, *Anabaena variabilis*. Arch Hydrobiol, 73: 559–574.

Rodriguez-Ezpeleta N, Brinkmann H, Burey SC, Roure B,Burger G, Loffelhardt W, Bohnert HJ, Philippe H and Lang BF (2005) Monophyly of primary photosynthetic eukaryotes: green plants, red algae and glaucophytes. Curr Biol 15: 1325–1330.

Runnegar MTC, Falconer IR, Buckley T and Jackson ARB (1986)) Lethal potency and tissue distribution of ^{125}I-labelled toxic peptides from the blue-green alga *Microcystis aeruginosa*. Toxicon 24: 506-509.

Sakamoto T, Delgaizo VB and Bryant DA (1998) Growth on urea can trigger death and peroxidation of the cyanobacterium *Synechococcus* sp. strain PCC 7002. Appl Environ Microbiol 64: 2361–2366.

Saunders JW (1966) Death in Embryonic Systems. Science 154: 604-612.

Saunders JW (1982) Developmental Biology. Macmillan. New York.

Schweichel J-U, Merker HJ (1973)'The morphology of various types of cell death in prenatal tissues. Teratology 7: 253–266.

Segovia M, Haramaty L, Berges JA, Paul G and Falkowski PG(2003) Cell Death in the Unicellular Chlorophyte *Dunaliella tertiolecta*. A Hypothesis on the Evolution of Apoptosis in Higher Plants and Metazoans. Plant Physiol 132: 99-105

Sheldrake R (1974) The ageing, growth and death of cells.Nature 250: 381–385.

Shimizu S, Narita M and Tsujimoto Y (1999) Bcl-2 family proteins regulate the release of apoptogenic cytochrome *c* by the mitochondrial channel VDAC. Nature 399: 483–487.

Shrager J, Chang CW, Hauser E, Harris R, Tamse R,Davies J, Zhang Z.and Grossman AR (2003) *Chlamydomonas reinhardtii* genome project: sequence analysis, assembly and annotation. Plant Physiol 131: 401–408.

Solomon M, Belenghi B, Delledonne M, Menachem E and Levine M (1999) The involvement of cysteine proteases and protease inhibitor genes in the regulation of programmed cell death in plants. The Plant Cell. 11: 431-443.

Solstad T, Fladmark KE (2006) Algal toxins as guidance to identify phosphoproteins with key roles in apoptotic cell death. Curr Pharm Biotechnol 7(3): 209-15.

Stein JC, Hansen G (1999) Mannose induces an endonuclease responsible for DNA laddering in plant cells. Plant Physiology 121: 71–79.

Steller H (1995). Mechanisms and genes of cellular suicide. Science 267: 1445-1449.

Suginaka K, Yamamoto K, Ashida H, Sawa Y and Shibata H (1999) Effect of intracellular glutathione on heat-induced cell death in the cyanobacterium, *Synechocystis* PCC 6803. Biosci Biotechnol Biochem 63: 1112–1115.

Sun YL, Zhao Y, Hong X, Zhai ZH (1999) Cytochrome *c* release and caspase activation during menadione-induced apoptosis in plants. FEBS Letters 462: 317–321.

Thomas L (1992) *The Fragile Species.* Macmillan, New York.

Thomas SG, Franklin-Tong VE (2004) Self-incompatibility triggers programmed cell death in *Papaver* pollen. Nature 429: 305–309.

Tiwari BS, Belenghi B and Levine A (2002) Oxidative stress increased respiration and generation of reactive oxygen species, resulting in ATP depletion, opening of mitochondrial permeability transition, and programmed cell death. *Plant Physiol* **128**: 1271–1281.

Toth, GB, Nore NF, Selander E and Pavia H (2004) Marine dinoflagellates show induced life-history shifts to escape parasite infection in response to water-borne signals. Proc R Soc Lond 271: 733–738.

Uren AG, O'Rourke K, Aravind L, Pisabarro MT, Seshagiri S, Koonin EV and Dixit VM (2000) Identification of paracaspases and metacaspases: two ancient families of caspase-like proteins, one of which plays a key role in MALT lymphoma. Mol Cell 6: 961–967.

Vacca RA, Valenti D, Bobba A, Merafina RS, Passarella S, Marra E(2006) Cytochrome *c* is released in a reactive oxygen species-dependent manner and is degraded via caspase-like proteases in tobacco Bright-Yellow 2 cells en route to heat shock-induced cell death. Plant Physiology 141: 208–219.

Vardi A, Schatz D, Beeri K, Motro U, Sukenik A, Levine A and Kaplan A (2002) Dinoflagellate cyanobacterium communication may determine the composition of phytoplankton assemblage in a mesotrophic lake. Curr Biol, 12: 1767–1772.

Vardi A,Berman-Frank I, Rozenberg T,Hadas O,Kaplan A and Levine A (1999) Programmed cell death of the dinoflagellate *Peridinium gatunense* is mediated by CO_2 limitation and oxidative stress. Current Biology 9: 1061–1064.

Veldhuis MJW, Kraay GW and Timmermans KR (2001) Cell death in phytoplankton: correlation between changes in membrane permeability, photosynthetic activity, pigmentation and growth. Eur J Phycol 36: 167–177.

Wang X (2001) The expanding role of mitochondria in apoptosis. Genes and Development 15: 2922–2933.

Webb JS, Thompson LS, James S, Charolton T, Tolker-Nielsen T, Koch B,Givskov M and Kjelleberg S (2003) Cell death in *Pseudomonas aeruginosa* biofilm development.J Bacteriol, 185: 4585–4592.

White K, Grether ME, Abrams JM, Young L Farrell K and Steller H (1994) Genetic control of programmed cell death in *Drosophila*. Science 264: 677-683.

Yao N, Eisfelder BJ, Marvin J, Greenberg JT (2004) The mitochondrion: an organelle commonly involved in programmed cell death in *Arabidopsis thaliana*. The Plant Journal 40: 1000–1007.

Yu XH, Perdue TD, Heimer YM, Jones AM(2002) Mitochondrial involvement in tracheary element programmed cell death. Cell Death and Differentiation 9: 189–198.

Zuppini A,Andreoli C and Baldan B (2007) Heat Stress: an Inducer of Programmed Cell Death in *Chlorella saccharophila*.Plant Cell Physiol 48(7): 1000–1009.

Algal Biotechnology: New Vistas (2010) *Pages 62–76*
Editor: **Mihir Kumar Das**
Published by: **DAYA PUBLISHING HOUSE, NEW DELHI**

Chapter 4

Spirulina: The Superfood and Medicine

Preeti Das*, A.K. Mishra and Mihir K. Das

P.G. Department of Botany, G.M. College (Autonomous), Sambalpur – 768 004, Orissa, India

ABSTRACT

Spirulina are the earliest forms of life, nearly 3.5 billion years old and are the source of all life in the oceans. It provided an evolutionary bridge between bacteria and green plants. Spirulina occurs naturally in tropical and subtropical lakes with high pH and high concentration of carbonate and bicarbonate. Spirulina is not a medicine, but when used as a good source of supplementary food, one can avoid nutrient deficiencies causing illness. Spirulina is totally digestible and is the richest food source available to us. It is also absorbed very quickly after being consumed. Because of its richness, this tiny plant is capable of providing food for humans with almost no wastage and satisfies hunger very quickly when eaten in only small amounts. The topic of Spirulina is currently quite hot for its therapeutic applications. Medical research has already shown that Spirulina can provide benefits to our body. This health food has wide range of nutrients, vitamins and proteins in its composition. Hundreds of scientific studies reveal that Spirulina and its unique phytonutrients and extracts, boost the immune system and improve health in many ways.

Keywords: Spirulina, Food, Nutrients, Medicine.

* E-mail: preetidas81@gmail.com

1. Introduction

'Let your food be your medicine and your medicine your food'-

(Hippocrates, 460-370 BC)

Spirulina is a single-celled, filamentous, spiral-shaped blue green microalgae grown in tropical salt lakes. This microalgae is discovered by Dr. Clement of France in 1962. Dr. Clement found that the Ganimou people living around Lake Chades in Africa had stronger bodies than other civilized people at that time, despite poor living conditions and few resources. Upon further studies, he observed that the Ganimou eats a particular blue green algae found floating on the lake's surface. This alga is spirulina. Such simple one-celled organism already exists when life first appeared on earth. Biologically speaking, they are the earliest forms of life, nearly 3.5 billion years old and are the source of all life in the oceans. It provided an evolutionary bridge between bacteria and green plants. This water plant has renewed itself for billions of years and has nourished many cultures throughout history, in Africa, in the Middle East and in the Americas.

After many years of research, Dr. Clement got the attention of the worldwide medical field on his discovery. Medical researchers in Japan and Mexico were interested with the many benefits that spirulina can bring to the human body. It is being used as part of other treatments due to its nutritional benefits. However, researchers have observed that Spirulina also appears to exhibit properties which promote self-healing. In 1976, Prof. Keiun Kodo (Huang Ching Yun) and Dr. Watanabe had the world first pure clean spirulina cultivation pool constructed. This is used to mass-produce unpolluted top-grade spirulina. Pro. Keiun Kodo has since accumulated 30 years of experience in spirulina cultivation techniques. He was bestowed the rank of Commandorian Cross of the Sovereign Greek order of Saint Dennis of Zante in 2001 for his enormous contributions to society. He was also awarded the Social Culture Merit Award and International Culture Honour Award in the same year for his invaluable research in spirulina.

An article written by Christopher Hills, Ph.D, D.Sc., titled "*Spirulina as nutrition for the brain*" says that "*the interesting part...........is the remarkable effects of this vegetable plankton on our state of consciousness particularly while fasting on nothing else...........most people feel increased energy and clarity of mind after taking the Spirulina as a vitamin supplement.*"

According to Professor Keiun Kodo on this topic, "*Spirulina's rich and balanced nutrients are perfect for the development of brain cells. It is a well-known source of nutrient for children and adults for the development and maintenance of brain cells.*"

2. Biology

Spirulina is the common name for human and animal food supplements similar to *Chlorella* and *Aphanizomenon flos-aquae* (Blue Green Algae).Spirulina comes from two different species of cyanobacteria: *Arthrospira platensis*, and *Arthrospira maxima*. These and other *Arthrospira* species were once classified in the genus *Spirulina*. There is now agreement that they are a distinct genera and that the food species belong to about 35 species of *Arthrospira*; nonetheless, the older term Spirulina remains the popular name.

Scientific Classification

Domain: Bacteria

Phylum: Cyanobacteriaassis = Chroobacteria

Order: Oscillatoriales

Family: Phormidiaceae

Genus: *Arthrospira*

Spirulina are free-floating cyanobacteria characterized by cylindrical multicellular trichome in an open left hand helix (Figure 4.1). Spirulina occurs naturally in tropical and subtropical lakes with high pH and high concentration of carbonate and bicarbonate. *A. maxima* occur in Central America, where as *A. platensis* is confined to Africa, Asia and South America.

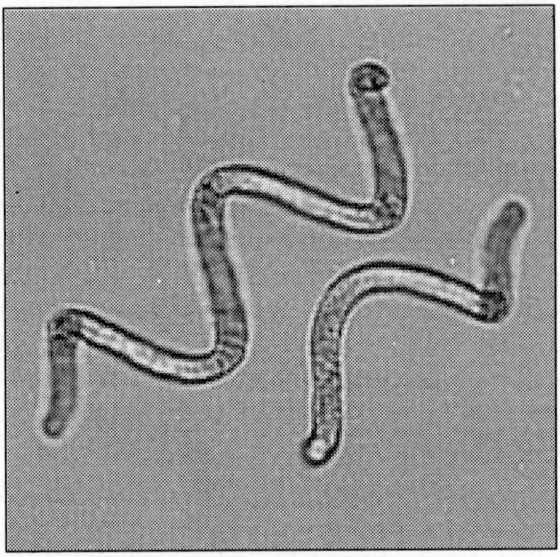

Figure 4.1: Structure of Spirulina

3. Production of Spirulina

The spirulina production involves three major steps, *viz.*, cultivation, harvesting and processing.

3.1 Cultivation

There are four major conditions for growing Spirulina.

1. Tropical weather
2. Strong sunshine
3. Pure water resource
4. Pollution free environment

It is not possible to grow Commercial Spirulina culture in a cold or temperate area. Spirulina needs consistent high temperature which helps it's growth. Spirulina willn't grow anywhere that has constant low temperature (under 25°C).Under 20°C Spirulina will stop reproducing and die in a short time.The optimum temperature should be 30°C to 35°C in mineral rich water(from unpolluted natural sources such as lakes and rivers)

Spirulina absorbs sunshine and then creates a reaction in it's cells. When this reaction starts, Spirulina will produce the nutrients in the cell and convert carbon dioxide into oxygen. Strong sunshine helps Spirulina to produce more nutrients.

Selected strains are used for cultivation of alga in specially constructed ponds. Spirulina grows in alkaline saline water. Because Spirulina easily absorbs nutrients from water.Clean freshwater and nutrients must be supplied regularly to feed the Spirulina. The constant agitation of water is one of the important parameters in cultivation of spirulina. Agitation of algal culture is necessary to keep nutrients evenly dispersed and mix evenly. Additional nutrients can be added to the waters, and this includes carbon dioxide, nitrogen, potassium and iron. These can improve the quality of the Spirulina cultivated. One reason why Spirulina is being farmed in artificial pools is that the natural lake where it can be found have an uncontrollable growing environment.Such salt lake have many other microorganisms such as other types of algae growing in it. These can contaminate the Spirulina and affect its quality.

If the water contains pollutants or heavy metals, these will be highly concentrated in the Spirulina cell. If this happens, then this kind of Spirulina is no longer suitable for human consumption. Hence, no preservatives, herbicides, pesticides or artificial products are added to the pool to remove weed algae or increase production.This is to ensure that the Spirulina is organic, safe to consume and the most of its nutritional content is retained.

3.2 Harvesting

The algal biomass is carefully harvested using specially made filters to recover biomass.

3.3 Processing

The washed biomass is dried and is pulverized to get desired particle size and packed appropriately for commercial use. In the context of rising input costs and low returns in agriculture, spirulina (*Arthrospira platensis*), seems to promise farmers good income than other regular crops with very little investment.

The viability of the industrial production of three strains of *Spirulina platensis* was tested in Malaga, South Spain. In a pre-industrial trial using raceway ponds from laboratory-scale to 450mt^2, all three strains displayed a satisfactory growth. In a 10 month industrial trial in 450mt sq ponds, production was equivalent to 30–32 metric tons of dry powder per hectare per annum. In conclusion, intensive industrial production of Spirulina is viable in certain Mediterranean climates, a region previously thought to be outside its geographic limits (Jimenez *et al.*, 2003).

4. Nutrients and Other Chemicals

Nutritional Values of Elken Spirulina

Elken Spirulina is a type of spirulina called *Arthrospira platensis*, one of the larger dimensions variant that can be consumed by humans. Our spirulina has been consumed by more than 10 million people throughout the world, and is the first to mass-produce unpolluted and top-grade spirulina.

This health food has

- ☆ 1.4 times more nutritional values than cow's milk
- ☆ 2 times more protein than soya bean, 6 times more than egg and 3.5 times more than beef
- ☆ The richest source of iron and folic acid
- ☆ 68 times higher than spinach
- ☆ 29 times higher than raw beef
- ☆ 5 times higher than soya bean
- ☆ Wide range of nutrients, vitamins and proteins in its composition that it is used for space food

Elken Spirulina have rich concentration of GLA, (gamma-linolenic acid) and can be used to replenish inadequate nutrients in a mother's milk. The GLA is in fact 3 times higher than that of Evening Primrose Oil.

Functions of GLA includes:

- ☆ Prevent sensitive dermatitis in children
- ☆ Moderate blood pressure (to normalise blood pressure)
- ☆ Prevent asthma
- ☆ Prevent tumour growth
- ☆ Prevent clotting in our blood vessels
- ☆ Slows down the process of ageing (properties of an anti-oxidant)

The algae are highly digestible food. It contains:

Protein

Spirulina platensis, a blue green microalga, has been used since ancient times as a source of food because of its high protein (about 70 per cent) and nutritional value (Dillon, Phuc, and Dubacq, 1995; Khan *et al.*, 2005; Simpore *et al.*, 2006). Spirulina contains an unusually high amount of protein, between 55 per cent and 77 per cent by dry weight, containing all essential amino acids,though with reduced amounts of methionine, cysteine, and lysine when compared to proteins of meat, eggs, and milk. Also superior to typical plant protein, such as that from legumes. Spirulina is a nutritious protein food supplement and is also used in the manufacture of several medicines, and cosmetics. Its cultivation on a commercial scale is slowly catching up with many farmers in India, particularly in Tamil Nadu.

Essential Fatty Acids

Spirulina is rich in gamma-linolenic acid (GLA), and also provides alpha-linolenic acid (ALA), linoleic acid (LA), stearidonic acid (SDA), eicosapentaenoic acid (EPA),docosahexaenoic acid (DHA) and arachidonic acid (AA).Furthermore, GLA in vivo changes easily to arachidonic acid, and then to prostaglandin E2. The end product can be developed as a blood pressure regulating agent in our body.

Vitamins

Spirulina contains vitamin B1 (thiamine), B2 (riboflavin), B3 (nicotinamide), B6 (pyridoxine), B9 (folic acid), B12 (cobalamin) vitamin C, vitamin D and vitamin E.

B12

The bioavailability of vitamin B12 in Spirulina is in dispute. Several biological assays have been used to test for the presence of vitamin B12. The most popular is the US Pharmacopeia method using the *Lactobacillus leichmannii* assay. Studies using this method have shown Spirulina to be a minimal source of bioavailable vitamin B12. However, this assay does not differentiate between true B12 (cobalamin) and similar compounds (corrinoids) that cannot be used in human metabolism. Cyanotech, a grower of spirulina, claims to have done a more recent assay, which has shown Spirulina to be a significant source of cobalamin. However, the assay is not published for scientific review and so the existence of this assay is in doubt. The American Dietetic Association and Dietitians of Canada in their position paper on vegetarian diets state that spirulina can not be counted on as a reliable source of active vitamin B12.

Tests done on Australian grown spirulina by the Australian Government Analytical Laboratory (AGAL) show Vitamin B12 (cobalamin) levels of 659.1ug/ per100g. A one gram tablet could provide more than three times the recommended daily intake of B12.

Minerals

Spirulina is a rich source of potassium, and also contains calcium, chromium, copper, iron, magnesium, manganese, phosphorus, selenium, sodium and zinc.

Photosynthetic Pigments

Spirulina contains many pigments including chlorophyll-a, xanthophylls, beta-carotene, echinenone, myxoxanthophyll, zeaxanthin, canthaxanthin, diatoxanthin, 3'-hydroxyechinenone, beta-cryptoxanthin, oscillaxanthin, plus the phycobillproteins c-phycocyanin and allo phycocyanin.

5. Spirulina: A Food or a Medicine?

As we all know, some of our illnesses are caused by having insufficient nutrients in our body. These illnesses are just the symptoms to show us that we may be lacking in some nutrients. If we replenish these nutrients in time, the symptoms usually disappear. If not, we can lower the function of our immune system causing further problems.In most cases people will go to consult their doctor and may be prescribed some medicine. Spirulina is not a medicine, but when used as a good source of supplementary food, one can avoid nutrient deficiencies causing illness.The topic of

Spirulina is currently quite hot for it's therapeutic applications. Medical research has already shown that Spirulina can provide benefits to our body. Spirulina can help you to have reasonable levels of nutrients in the body, which will in turn give less of the chance to get sick.

Affirms Charanjit Singh, an Indian Physicist who has been eating spirulina, off and on, for the past 17 years: "My health is remarkably good when I am taking spirulina, I do not even catch a cold. I take it in the form of flakes, sprinkled over any edible. My kids love it too."

Singh's private theory is that 60 to 70 per cent of spirulina comprises DNA, which helps the immune system. When Singh returned to India from the USA seven years ago, he "hunted around for it, and finally tracked it down to Transtech, a company in Madras which was manufacturing and distributing spirulina under the brand name Progen", Transtech's Natarajan Sundaram had started producing spirulina as a village development project in 1989. His output is still only about 20 kg a month.

Spirulina has slowly been gaining in visibility in India and is marketed under the brand names Natoxid, Multinals, and Progen, and is available in powder, flakes or tablet form. A bottle of 60 capsules costs about Rs 95.

There are half-a-dozen players already in the spirulina market-a subsidiary of the Thapar group with a capacity of 120 tonnes is perhaps the biggest. A recent entrant is Lucky Laboratories, an associate of Dabur India, which is currently test-marketing Sunova Spirulina in Karnataka and Goa.

6. Spirulina: Vegetable Protein vs Animal Protein

Spirulina contains more than 60 per cent vegetable protein, which is much higher than fish, pork, or beef (which contains about 15 ~20 per cent). Animal protein is a much bigger molecule than vegetable protein, and is much harder for our system to digest. Most modern people overindulge in animal protein, by eating fish, beef, pork etc. When too much animal protein is eaten, it is deposited in our body as fat. Too much fat will cause high cholesterol levels and may impact our heart and blood vessels. Vegetable protein is water soluble, and is much smaller than animal protein. If you eat too much vegetable protein, it is simply discharged by your system as waste and not stored as fat.

7. Health Benefits

An international detective hunt has been underway for 20 years, Researchers in Japan, China, India, Europe and the USA are discovering how and why the microalgae Spirulina is so effective for human and animal health. Hundreds of Scientific studies reveal how Spirulina, and its unique phytonutrients and extracts, boost the immune system and improve health in many ways.

There are several new peer reviewed scientific studies about the ability to inhibit viral replication, strengthen both the cellular and humoral arms of the immune system and cause regression and inhibition of cancers. While these studies are preliminary and more research is needed, the results so far are exciting.

7.1 Potent Anti-viral Activity

Spirulina platensis was also reported to present antimicrobial activity (Demule, Decaire and Decano, 1996; Ozdemir, Karabay, Dalay, and Pazarbasi, 2004) as well as to inhibit replication of several viruses, such as Herpes simplex and HIV-1 (Ayehunie,Belay, Baba, and Ruprecht, 1998; Herna´ndez-Corona, Nieves,Meckes, Chamorro, and Barron, 2002). In April 1996, scientists from the Laboratory of Viral Pathogenesis, Dana-Farber Cancer Institute and Harvard Medical School announced on–going research, saying "Water extract of *Spirulina platensis* inhibits HIV-1(the AIDS virus), replication in human derived T-cell lines and in human peripheral blood mononuclear cells. A concentration of 5-10μg/ml was found to reduce viral production ".

HIV-1 is the AIDS virus. Small amounts of Spirulina extract reduced viral replication while higher concentrations totally stopped its reproduction. Importantly, with a therapeutic index of >100, Spirulina extract was non-toxic to the human cells at concentrations stopping viral replication.

Another group of medical scientists in Japan has published new studies regarding a purified water extract unique to Spirulina named *Calcium-Spirulan*. It inhibits replication of HIV-1, Herpes Simplex, Human Cytomegalovirus, Influenza A virus, Mumps virus and Measles virus in-vitro yet is very safe for human cells. It protects human and monkey cells from viral infection in cell culture. According to peer reviewed scientific journal reports this extract, "holds great promise for treatment of HIV-1, HSV-1, and HCM infections, which is particularly advantageous for AIDS patients who are prone to these life-threatening infections."

Calcium-Spirulan is a polymerized sugar molecule unique to Spirulina containing both Sulfur and Calcium. How does it work? When attacking a cell, a virus first attaches itself to the cell membrane. However, because of Spirulina extract, the virus cannot penetrate the cell membrane to infect the cell. The virus is stuck, unable to replicate. It is eventually eliminated by the body's natural defences.

7.2 Anti-Cancer Effects

Several studies show that Spirulina or its extracts can prevent or inhibit cancers in humans and animals (Hirahashi *et al.*, 2002; Subhashini *et al.*, 2004). Some common forms of cancer are thought to be a result of damaged cell DNA causing uncontrolled cell growth. Cellular biologists have defined a system of special enzymes called Endonuclease which repair damaged DNA to keep cells alive and healthy. When these enzymes are deactivated by radiation or toxins, errors in DNA go un-repaired and cancer may develop.*In vitro* studies suggest the unique polysaccharide of Spirulina enhance cell nucleus enzyme activity and DNA repair synthesis. This may be why several scientific studies, observing human tobacco users and experimental cancers in animals, report high levels of suppression of several important types of cancer. The subjects were fed either whole Spirulina or treated with its water extracts.

7.3 Anti-Oxidant Activity

Beta carotene of Spirulina supports the body's antioxidant defenses. It helps to stay young. Moreover, Spirulina contains a whole spectrum of natural mixed carotene and xanthophyll phytopigments which, together with phycocyanin, seem to be related to its antioxidant activity (Bhat and Madyastha, 2000; Miranda, Cintra, Barros, and Manchini, 1998; Pineiro Estrada, Bermejo Besco´s, and Villardel Fresno, 2001). *Spirulina platensis* contains phycobilisomes as light-harvesting protein pigment complexes. Phycobilisomes are mainly composed of polypeptides named phycobiliproteins. The two more important phycobiliproteins which occur in this microalga are phycocyanin and allophycocyanin; both of them have the same chromophore group. Moreover, Spirulina contains a whole spectrum of natural mixed carotene and xanthophyll phytopigments which, together with phycocyanin, seem to be related to its antioxidant activity (Bhat and Madyastha, 2000; Miranda *et al.*, 1998). Previous reports have shown that the protein extract of *S. platensis* is a potent radical scavenger (hydroxyl and peroxyl radicals), has chelating ability and inhibits microsomal lipid peroxidation (Bermejo *et al.*, 2008; Pinero Estrada *et al.*, 2001).

7.4 Strengthen Immune System

Medical scientists find it not only stimulates the immune system, it actually enhances the body's ability to generate new blood cells. Important parts of the immune system, Bone Marrow Stem Cells, Macrophages, T-cells and Natural Killer cells, exhibit enhanced activity. Spleen and Thymus glands show enhanced function. Scientists also observe Spirulina causing macrophages to increase in number, become "activated" and, more effective at killing germs.

Feeding studies show even small amounts build up both humoral and cellular arms of the immune system. Spirulina accelerates production of the humoral system (antibodies and cytokines), allowing it to better protect against invading germs. The cellular immune system includes T-cells, Macrophages, B-cells and the anticancer Natural killer cells. These cells circulate in the blood and are especially rich in body organs like the liver, spleen, thymus etc. Spirulina up-regulates these key cells and organs, improving their ability to function in spite of stress from environmental toxins and infectious agents.

7.5 Phycocyanin Builds Blood

Spirulina has a dark blue green colour, because it is rich in a brilliant blue polypeptide called phycocyanin. Studies show it affects the stem cells found in bone marrow. Stem cells are "Grandmother" to both the white blood cells that make up the cellular immune system and red blood cells that oxygenate the body. Chinese scientists document phycocyanin stimulating hematopoisis,(the creation of blood),emulating the affect of the hormone erythroprotein (EPO). EPO is produced by healthy kidneys and regulates bone marrow stem cell production of red blood cells.

Chinese scientists claim phycocyanin also regulates production of white blood cells, even when bone marrow stem cells are damaged by toxic chemicals or radiation.

Based on this effect, Spirulina is approved in Russia as a "medicine food" for treating radiation sickness. The Children of Chernobyl suffer radiation poisoning from eating food grown on radioactive soil. Their bone marrow is damaged, rendering them immunodeficient. Radiation damaged bone marrow cannot produce normal red or white blood cells. The children are anemic and suffer from terrible allergic reactions. Children fed just five grams in tablets each day made dramatic recoveries within six weeks.

7.6 Other Potential Health Benefits

Spirulina is a diet supplement and

- ☆ World's richest natural source of protein, iron, vitamins and minerals. Helps weight loss.
- ☆ World's richest natural source of Pro-Vitamin 'A', helps to protect your eyesight.
- ☆ World's richest natural source of GLA, reduces cholesterol and benefits the heart, helps to develop healthy babies. Studies show nutritional deficiencies can block GLA production in the body.
- ☆ Only vegetable source of vitamin B12, calms your nerves and fights stress.
- ☆ Helps in rejuvenation, increase of energy level, reducing body acidity, cardiovascular support and detoxification. The soft cell walls of spirulina allow it to be digested easily and assimilated, making it ideal for detoxifying.

8. Spirulina Dosage

Spirulina products come in the form of spirulina capsules, tablets, powder and flakes. About 10 (to 20) 500mg spirulina tablets a day for adults and 5 (to10) for children under 12 are usually recommended (Figure 4.2). The dose for spirulina capsules and tablets can vary from 250 mgs to 5 gms a day as well. About a spoonful of spirulina extract or spirulina powder (about 3-5 gms) is enough to last a day.

8.1 Spirulina Should be Taken By

1. Children who don't like or get enough vegetables and or have an imbalanced food intake.
2. Teenagers during their rapid growing period need a sufficient injection of nutrients.
3. Pregnant mums who need extra nutrients.
4. Seniors who have difficulty in having reasonable average 3 meals per day.
5. Sport lovers or athletics who need extra nutrients to keep their energy levels up.
6. Modern busy people who don't have the time to eat good meals.
7. Patients or people who need high volumes of nutrients to assist recovery(with consulting the doctor)
8. Vegetarians who require extra nutrient sources.

Figure 4.2: Spirulina Tablets

8.2 Spirulina Shouldn't be taken too much by:

1. People with hyperparathyroidism.
2. People who have serious allergies to seafood or seaweed.
3. Patients current experiencing high fever.

9. Side Effects of Spirulina

Spirulina is a totally natural product and will not normally cause any problems to the body. Even if too much is taken, there will be no harm to the body, but doing this is a waste. However some people may experience some of the following symptoms after taking Spirulina.

1. Slight fever due to the body's need to burn the extra protein from Spirulina.
2. Slight dizziness, lethargic or constant hunger pangs when a person suffering from hypoglycemia or anemia consumes Spirulina, there may be a drop in the blood sugar level to burn the excessive fat.
3. Nausea, diarrhoea constipation due to poor digestive system, hepatitis, gastric hyperacidity.
4. Pimples breakout, itchiness, red spots appearing due to detoxification (cleansing of your colon) or poor liver function or allergies in the body.
5. Increase in weight, after oxidation of body fats, over weight people may have increased bodily fluid (edema), whereas underweight people may have increased muscle mass. These are symptoms of weak kidneys.

10. The State of Spirulina Research

The serious scientific research on the nutritional benefits of spirulina began in the 1960s, when the French firm Sosa Texcoco was investigating it prior to opening their spirulina harvesting plant at Lake Texcoco in Mexico. During the 1970s, many scientists were hopeful that spirulina's exceptionally high protein content would make it a candidate to feed people in the protein-starved third world. But spirulina research since the 1970s has focused more on its disease-fighting- and-prevention potential than its starvation-prevention potential.

While there have been only a handful of spirulina studies conducted on humans, its test tube and animal study results are certainly interesting and might one day merit a look from pharmaceutical companies. Many medical and educational institutions are looking at spirulina, and following its progress on their websites:

- ☆ Oregon State University reports on a 2005 study headed by The University of South Florida's Paula Bickford, Ph.D. Dr. Bickford and her colleagues at James A. Haley Veteran's Hospital and the National Institute on Drug Abuse found that, following induced strokes, rats which had been given diets supplemented with spirulina showed brains lesions 75 per cent smaller than those of control group rats. The rats fed spirulina also recovered mobility to a greater extent.
- ☆ The Health Information Center at Cedars-Sinai Hospital in Los Angeles references a second study by Dr. Bickford's team in which the brains of aged rats fed spirulina supplements maintained their neuron function much better, and showed far less free radical damage than those of rats fed cucumbers. Bickford's research is significant because it indicates that spirulina may be helpful in warding off the free radical damage related to aging diseases like Parkinson's and Huntington's disease.
- ☆ Kansas State University's International Food Safety Network has published a 2005 article from the Decca, India Herald reporting that Spirulina is approved by both the FDA and the World Health Organization for use as a health food, and mentions it as a source of five essential amino acids which the human body cannot manufacture. The article also refers to spirulina's high mineral content, in particular the amount of iron it contains, its abilities to reduce gastro enteric distress by promoting the growth of intestinal bacteria, and to help prevent gastro enteric infection; and its cholesterol-reducing effects.
- ☆ The abstract of a 2002 study by the Department of Animal Sciences at India's University of Hyderabad from the US National Institutes of Health's research data publication site, also cited by the University of Maryland Medical Center, explains the process by which spirulina is able to inhibit the elevated levels of Cycloxygenase-2 associated with inflammation and cancer.
- ☆ A second 2002 study at Havana's Ozone International Center showed that phycocyanin from spirulina acted as an anti-inflammatory in protecting live mice which had been injected with arthritis-inducing Zymosan. The

mice showed no cartilage damage and an inhibited inflammatory response after receiving spirulina for eight days following their Zymosan injections. The animal nutritionals maker Pharma Chemie of Syracuse, NY has since gone on to patent a compound containing phycocyanin, which gives spirulina its blue-green color, as an anti-inflammatory for use in animals. The UMM site also goes on to mention a Kerala, India study of eighty-seven human subjects in which 45 per cent of those exhibiting oral cancer lesions caused from chewing tobacco experienced complete regression of their symptoms after receiving one gram of *Spirulina fusiformus* daily for twelve months.

- ✰ The NYU Medical Center Hospital for Joint Diseases has, on its Center for Children website, a review of some of the research and claims made concerning the health benefits of spirulina. The NYU site mentions that spirulina has shown preliminary promise as in fighting HIV; and its potential against HIV and other viruses was demonstrated in a 1998 Dana-Farber Cancer Institute and Harvard Medical School study. The NYU site also references other numerous but preliminary test tube, animal and human studies which suggest that spirulina might help lower cholesterol and reduce hypertension; prevent cancer; strengthen the liver's defense against toxins; and alleviate allergic reactions.
- ✰ Spirulina's effectiveness in fighting allergies was shown in 2005 UC Davis research in which thirty-six rhinitis sufferers were treated with spirulina supplementation, and those who took two grams daily experienced a significant reduction in their allergic symptoms. That research followed a 2000 UC Davis study which showed that when incubated spirulina dilutions, cultured human mononuclear blood cells, including macrophages and lymphocytes, are able to mount a stronger immune response.

But the NYU site clearly states that Spirulina has not been proven effective in treating any medical condition and that its use has yet to be well-documented. And both the National Institute of Health and the U.S. National Medical Library, on their Medline Plus site, graded the existing research on spirulina in August of 2006. Their findings were that there was unclear scientific evidence to support the claims of spirulina's use in treating diabetes, high cholesterol, and oral cancers.

Proponents of spirulina's health benefits certainly have reason for optimism, but it seems clear that many more human studies, on much greater numbers of subjects, will be necessary if spirulina's potential is to receive mainstream recognition. Spirulina is effective for the clinical improvement of melanosis and keratosis due to chronic arsenic poisoning; improves weight-gain and corrects anemia in both HIV-infected and HIV-negative undernourished children; and protects against hay fever.

A 2007 study found that 36 volunteers taking 4.5 grams of spirulina per day, over a six week period, exhibited significant changes in cholesterol and blood pressure: (1) lowered total cholesterol; (2) increased HDL cholesterol; (3) lowered triglycerides; and (4) lowered systolic and diastolic blood pressure. However, as this study did not contain a control group, researchers can not be confident that the changes observed are due totally–or even partially–to the effects of the Spirulina Maxima as opposed to other confounding variables (*i.e.*, history effects, maturation effects, demand characteristics).

11. Advocates

The United Nations World Food Conference in 1974 lauded Spirulina as the 'best food for the future'. Recognizing the inherent potential of Spirulina in the sustainable development agenda, several Member States of the United Nations came together to form an intergovernmental organization by the name of the Intergovernmental Institution for the Use of Micro-algae Spirulina Against Malnutrition. IIMSAM aspires to build a consensus to make Spirulina a key driver to eradicate malnutrition, achieve food security and bridge the health divide throughout the world.

Spirulina has been proposed by both NASA (CELSS)[ˡ] and the European Space Agency (MELISSA) as one of the primary foods to be cultivated during long-term space missions.

12. Precautions

If you have any health problems, consult your doctor before using this product. Liquid preparations of this product may contain sugar and/or alcohol. Caution is advised if you have diabetes, alcohol dependence or liver disease. Ask your doctor or pharmacist about the safe use of this product. Blue green algae should be used only when clearly needed during pregnancy. Discuss the risks and benefits with your doctor.

13. Conclusion

In an evaluation of Spirulina by WHO in 1992, the organizations states that 'Spirulina is a high quality food product. It is rich in iron and protein. It is safe to consume and is an excellent nutrient supplement for children'.

Researchers at NASA found it to be an excellent, compact space food for astronauts (1 kg. of Spirulina is equivalent to 1000 kgs. of assorted vegetables). International health organizations have hailed Spirulina as one of the "Greatest Superfoods on Earth". An ideal food supplement, Spirulina provides the body with all the essential nutrients required daily, leaving you fresh, active and energetic through out the day.

Based on this preliminary research, scientists hope the use of Spirulina and its extracts may reduce or prevent cancers and viral diseases. Bacterial or parasitic infections may be prevented or respond better to treatment and wound healing may improve. Symptoms of anaemia, poisoning and immunodeficiency may be alleviated.

Scientists in the USA, Japan, China, Russia, India and other countries are studying this remarkable food to unlock its potential. More research is needed to determine its usefulness against AIDS and other killer diseases. However, it is already clear this safe and natural food provides concentrated nutritional support for optimum health and wellness.

References

Ayehunie S, Belay A, Baba TW and Ruprecht RM (1998) Inhibition of HIV-1replication by an aqueous extract of *S. platensis* (*A.platensis*).Journal of Acquired Immune Deficiencies Syndromes and Retrovirology.18: 7-12.

Bermejo P, Pinero E, Villar AM (2008) Iron-chelatin ability and antioxidant properties of phycocyanin isolated from a protean extract of *Spirulina platensis*.Food Chem 110: 436–445.

Bhat VB and Madyastha KM (2000) C–phycocyanin: a potent peroxyl radical scavenger in vivo and in vitro. Biochemical and Biophysical Research Communication 275: 20–25.

Demule MCZ, Decaire GZ, and Decano MS (1996) Bioactive substances from *S.platensis* (cyanobacteria) Int J Expt Bot 58: 93-96.

Dillon JC, Phuc AP and Dubacq JP (1995) Nutritional value of the alga Spirulina. World Review of Nutrition and Dietetics 77: 32- 46.

Herna´ndez-Corona A, Nieves I, Meckes M, Chamorro G e Barron BL(2002) Antiviral activity of *Spirulina maxima* against herpes simplex virus type 2.Antiviral Research, 56: 279-285.

Hirahashi T, Matsumoto M, Hazeki K, Saeki Y Ui M and Seya T (2002) Activation of the human innate immune system by Spirulina: augmentation of interferon production and NK cytotoxicity by oral administration of hot water extract of *S.platensis*. International Immunopharmacology 2: 423-434.

Jimenez C, Cossio BR, Labella D and Niel FX (2003) The feasibility of industrial production of Spirulina (Arthrospira) in southern Spain. Aquaculture 217: 179-190.

Khan Z, Bhadouria P, Bisen PS (2005) Nutritional and Therapeutic potential of Spirulina. Curr. Pharm Biotechnol. 6: 73- 379.

Miranda MS, Cintra RG, Barros SB and Manchini FJ (1998) Antioxidant activity of the microalga S. maxima, Brazilian Journal of medical Biology Research 31: 1075 -1079.

Ozdemir G, Karabay NU, Dalay MC, and Pazarbasi B (2004). Antibacterial activity of volatile components and various extracts of Spirulina platensis protein extract II Farmaco 59: 497–500.

Pinero Estrada JE, Bermejo Bescos P, Villar del Fresno AM (2001) Antioxidant activity of different fractions of S.platensis protean extract II.Farmaco 56: 497–500.

Simpore J, Kabore F, Zongo F, Dansou D, Bere A, Pignatelli S, Biondi DM, Ruberto G,Musumeci S (2006) Nutrition rehabilitation of undernourished children utilizing Spirulina Misala Nutr J 5: 3.

Subhashini J, Mahipal SVK, Reddy MC, Reddy MM, Rachamallu A and Reddanna P(2004) Molecular mechanisms in C-phycocyanin induced apoptosis in human chronic myeloid leukemia cell line–K562.Biochemical Pharmacology 68: 453–462.

Algal Biotechnology: New Vistas (2010) *Pages 77–92*
Editor: **Mihir Kumar Das**
Published by: **DAYA PUBLISHING HOUSE, NEW DELHI**

Chapter 5

Application of Seaweeds as Food: A Scenario

P.V. Subba Rao**, K. Ganesan and K. Suresh Kumar

Marine Biotechnology and Ecology Discipline,
Central Salt and Marine Chemicals Research Institute (CSIR),
Bhavnagar – 364 002, India

ABSTRACT

Seaweeds have been used as food besides forming a major source for phytochemicals like agar, alginate and carrageenan. The eating patterns of people all over the world have recently undergone marked changes, due to the globalization of markets along with innovation in food technology. Seaweeds are being consumed because they contain considerable quantity of proteins, fatty acids, vitamins and essential micronutrients like Fe, Zn, Mn and Cu needed to maintain a healthy life. They also form a source of antioxidants. There is an increasing interest in the importance of dietary minerals from seaweeds in the prevention of several diseases like cardiovascular and thrombosis.

Keywords: *Seaweeds, Proteins, Fatty acids, Vitamins, Antioxidant activity.*

1. Introduction

Seaweeds (macroalgae) are eukaryotes and live in aquatic biotopes. They are described as lower plants. They do not have true stems, roots and leaves, and they are normally capable of photosynthesis. Some seaweeds are fast growing marine plants

* E-mail: pvsubbarao@csmcri.org

that can grow to considerable size (up to 60 m in length). Growth rates of marine macroalgae (seaweeds) far exceed those of terrestrial ones, mainly due to no water limitation. Annual primary production rates (gram cm^{-2} yr^{-1}) are higher for the major marine macroalgae than those for most terrestrial ones (Gellenbeck and Chapman, 1983). Seaweeds inhabiting intertidal and subtidal regions of the seas and oceans have been used as a source of food, feed, fodder besides forming a potential source for phytochemicals like agar, alginate and carrageenan (Chapman and Chapman 1980).

2. Historical Perspective

From time immemorial, man had utilized seaweeds for food. In the Chinese *Book of Poetry* written during the time of Confucius (between 800 and 600 B.C.) there was a poem that mentioned a house wife cooking seaweeds. At that time, the seaweed was considered as a food of great delicacy and even a worthy sacrificial offering to the ancestors. They are still formed as an important constituent of food supplies and have not been used merely as appetizers or stabilizers of confections as they are being used in the Western countries. The use of marine macroalgae (seaweeds) as food had been traced back to the fourth century in Japan and sixth century in China (McHugh 2003). They had been exploited for food in Japan to a far greater extent than in any other country and are still said to provide about 25 per cent of the daily diet. The first report on collection of a marine macro alga, "nori", *i.e. Porphyra*, dated back to the year 530. The first known documentation of cultivation of this alga occurred in 1640 (Pulz and Gross 2004).

The brown seaweeds had been incorporated in the flour and were used in almost every household as noodles, toasted and served with rice or in soup. Two species, *Laminaria japonica* and *L. angustata*, were used for sweetening and flavoring because of their high contents of mannitol and glutamic acid. In the Western countries, they were never really accepted in human diet and were only eaten in times of scarcity and also where the standard of living had been low.

3. The Present Scenario

Utilization of seaweeds have increased considerably over the past fifty years, with the consequent increase in applied research in various food related fields. The total volume of seaweeds used in foods is considerably greater than that used in industrial applications, in terms of weight and especially in monetary value. In recent decades there has been an increase in direct consumption of seaweed as food, partly because of the nutritional (Mabeau and Fleurence, 1993; Darcy-Vrillon, 1993) and therapeutic (van Netten *et al.*, 2000) benefits that these seaweeds provide. The mean intake of seaweed in Western countries is far from equaling that of Eastern countries and is estimated for the Japanese population as 3.3 g dry seaweed per day per person (Darcy-Vrillon, 1993), but it must not be forgotten the existence of extreme consumers in all countries, such as those who follow a macrobiotic diet

Seaweeds are traditionally consumed in the orient as part of the daily diet. About 221 seaweeds are utilized commercially world- wide of which 65 per cent are used as human food (Zemke-White and Ohno, 1999). Currently, human consumption of green algae (5 per cent), brown algae (66.5 per cent) and red algae (33 per cent) is high in

Asia, mainly in Japan, China and Korea (Dawes, 1998). Recently, other countries, such as the Republic of Korea, the United States of America, South America, Ireland, Iceland, Canada and France, besides Japan and China have significantly increased the consumption, production and marketing of seaweeds Further the demand for seaweed as food has now also extended to North America, South America and Europe (McHugh, 2003). Currently, approximately 15–20 edible algal strains are being commonly marketed for consumption in Europe. These seaweed varieties differ greatly in their quality, color, consistency, and nutrient content.. Moreover, very few of the world's available seaweed species are used commercially. This may be because, they cannot be harvested or cultivated on a commercially viable scale, or because their composition simply makes them unsuitable. They show great variation in the nutrient contents, which are related to several environmental factors like water temperature, salinity, light and nutrients (Dawes, 1998). In India, there are many species of edible seaweeds such as *Gracilaria edulis, Ulva lactuca, Caulerpa racemosa, Sargassum siliquosum, Eucheuma, Enteromorpha, Porphyra, Hypnea, Gelidiella, Laurencia papillosa, Hydroclathrus clathratus*. Especially people living in the coastal areas extensively consume fresh and dried seaweeds. Depending on the type of species, they are generally suitable for variety of dishes like Ulva jam, agar jelly, salad, vegetable curry and porridge (Chennubhotla *et al.*, 1981; Krishnamurthy 2003). The important edible seaweeds are shown in Figure 5.1.

3.1 Green Seaweeds

The green seaweeds *Monostroma, Enteromorpha, Ulva, Caulerpa*, and *Codium* are commonly utilized exclusively as source of food. In Japan, dried fronds of edible *Monostroma* are used in preparation of "nori-jam" and soup. Edible *Monostroma* and *Enteromorpha* are called "Aonori" in Japanese (Ohno, 1997). *Enteromorpha* is eaten as raw salad, dried, or cooked. (Novaczek, 2001) because of varied quality chemical constituents and concentration of basic nutrients needed for human consumption (Aguilera-Morales *et al.*, 2005). *Ulva* is eaten raw or cooked in Pacific regions (Novaczek, 2001). *Ulva lactuca* known as "rimu miti" or "salt seaweed" is an edible species eaten by ancient Tahitians (Payri *et al.*, 2000). *Codium* or commonly known as "miru" in Japan is eaten raw or cooked (Madlener, 1977; Novaczek, 2001) and *Codium geppiorum* being favorite dish eaten raw or with fish (Payri *et al.*, 2000) and *Caulerpa* known as "sea grapes" in Fiji, and "iwazuta" in Japan is commonly sold in market in many Pacific regions. *Caulerpa lentilifera* is eaten as salad in Philippines and in some parts of Indonesia (Trono and Toma, 1997). *Caulerpa sertularioides, Caulerpa peltata* (Novaczek, 2001) and *Caulerpa bikinensis* (Payri *et al.*, 2000) are also eaten with coconut milk in Pacific Islands.

3.2 Brown Seaweeds

Laminaria "kombu" and *Undaria* "wakame" are edible seaweeds forming an important food source in Japan and are being eaten either raw, boiled or dried and they are also eaten with sweetened green beans, jelly, crushed ice, and coconut milk in Southern Vietnam (Tsutsui *et al.*, 2005). *Cladosiphon okamuranus* known as "Okinawa-mozuku" is eaten as salad in Japan (Toma, 1997). *Sargassum* known as horsetail and "hondawara" in Japan is eaten in Korea as soup or dressed with soybean sauce, or after being processed; or dried (Madlener, 1977) and also eaten in Pacific Islands as raw or cooked (Novaczek, 2001). In Pacific region various seaweeds are used as food. Thus, *Turbinaria* or spiny leaf is eaten as soup, or omlet. *Rosenvingea* or slippery cushion and *Colpomenia* or papery sea bubble are eaten as soup or salad or stew. *Dictyota* or brown ribbon weed is used in dressing foods. *Hydroclatharus* or sea colander and *Padina* or sea fan are being eaten fresh as salad (Novazcek, 2001).

3.3 Red Seaweeds

In Pacific region, *Acanthophora* or spiny sea plant is eaten freshly or cooked and used to make soft pudding. Likewise, *Asparagopsis* or supreme limu is eaten as cooked or raw or as a spicy flavour. *Calophyllis* or large wire weed is consumed in cooked or dried and rehydrated form, while *Hypnea* or maidenhair is used to make jellies. *Halymenia* or red sea lettuce is eaten as freshly, cooked or dried. *Laurencia* or flower limu is chopped or salted as a spice while *Scinaia* or tender golden weed, "lumi tamana" is eaten as salads or cooked with coconut milk. *Eucheuma* and *Kappaphycus* or thron grass are eaten raw or cooked in Vietnam or cooked with coconut milk and sugar in Indonesia (Novaczek, 2001; Tsutsui *et al.*, 2005). *Gracilaria* or sea moss is consumed freshly as a salad (Madlener, 1977; Novaczek, 2001). *Gelidiella* or little wire weed, is eaten after simmering as Jelly in Japan, Vietnam and Pacific region (Madlener, 1977; Novaczek, 2001; Tanaka and Nakamura, 2004). *Rhodymenia palmata* or dulse is the most common of edible seaweed in Europe and North America. They also use the following seaweeds as food: *Alaria fistula* "Kausam", *Chordaria flagelliformis* "Nebbpett" and *Porphyra umbilicalis* "Nuru". *Porphyra* or "nori" are eaten freshly or dried in Japan, China, Korea, Vietnam, North America and Europe (Maldlener, 1977; Tanaka and Nakamura, 2004; Tsutsui *et al.*, 2005).

4. A Source of Nutrients

Now-a-days, eating patterns of people all over the world have changed due to the globalization of markets along with innovation in food technology. For years the food industry has sought out new and innovative ways of enhancing a healthy diet by using seaweeds as they contain protein, lipids, vitamins and minerals (Norziah and Ching, 2000; Sanchez-Machado *et al.*, 2004; Wong and Cheung, 2000). The nutritional properties of seaweeds are poorly known and normally are evaluated from their chemical composition (Mabeau and Fleurence, 1993). In addition, seaweeds are rich sources of vitamin C, vitamin B-complex, *e.g.*, folic acid and B12, and vitamin A precursors, such as b-carotene (McDermid and Stuercke, 2003; Takenaka *et al.*, 2001; Watanabe *et al.*, 1999; 2002; Yamada, *et al.*, 1999; Yon and Hyun, 2003). The main component of seaweed is high fiber content with the soluble fraction composed principally of sulphated galactans such as agar or carragenans in red seaweeds and alginates, fucans and laminarin in brown seaweeds. Consumption of dietetic fiber (sulphated galactans, agar or carragenans alginates, fucans and laminarin) has a positive influence on several aspects related to health like reducing constipation, hypercholesterolemia, obesity and diabetes. Further, the consumption of this dietary fibre has been related to the following health promoting effects: (1) Promotes the growth and protection of the beneficial intestinal flora (Fujii *et al.*, 1992; Goni *et al.*, 2001; Kuda *et al*, 1997; 1998a; 1998b), (2) the fiber in combination with high glycemic

Figure 5.1: Some Edible Seaweeds

Hijikia fusiformis

Enteromorpha compressa

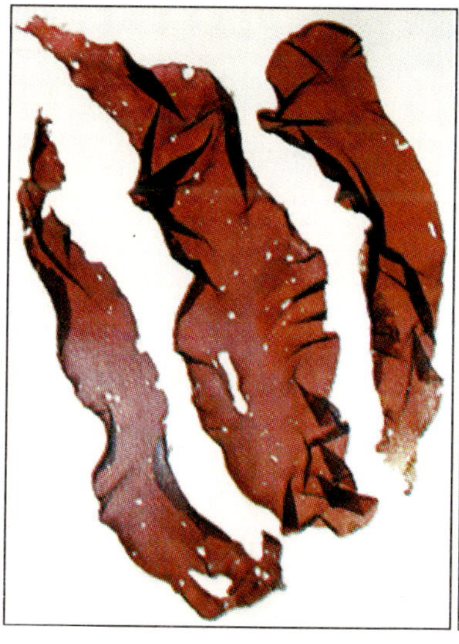

***Porphyra* sp.**

Palmaria palmata

Contd...

Figure 5.1–Contd...

Monostroma **sp.**

Enteromorpha tubulosa

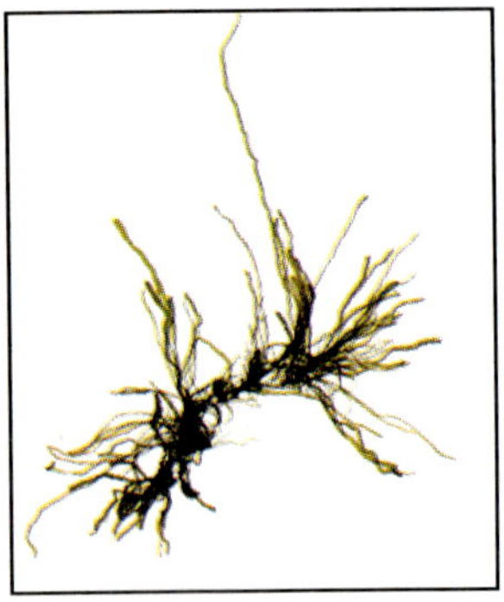

Enteromorpha linza

Ulva faciata

Ulva lactuca

load foods, reduces the overall glycemic response and seaweed fiber acts as a hypoglycemic (Goni *et al.*, 2000), (3) greatly increases stool volume (Jime´nez- Escrig and Sa´nchez-Muniz, 2000) and (4) reduces the risk of colon cancer (Guidel-Urbano and Goni, 2002).

4.1 A Source of Proteins, Amino and Fatty Acids

The plant protein concentrates (PCs) are of growing interest to the food industry with potential for human and animal nutrition (*e.g.* as functional food and fish feed) (Fleurence, 1999). However, only a few studies have been undertaken on the quality of the seaweed proteins (Amano and Noda, 1990; Dam *et al.*, 1986; Ito and Hori, 1989). The protein content of seaweeds differs according to species. Generally, the protein fraction of brown seaweeds is low (3- 15 per cent of the dry weight) compared to that of the green or red seaweeds (10- 47 per cent of the dry weight) (c.f. Fleurence 1999; Arasaki and Arasaki, 1983). Except for the species *Undaria pinnatifida* (wakame) which has a protein level between 11 and 24 per cent (dry weight), most brown seaweeds (*Laminaria digitata, Ascophyllum nodosum, Fucus vesiculosus* and *Himanthalia elongata*) that are industrially exploited have a protein content lower than 15 per cent (d wt).

In some green seaweeds like *Ulva* species the protein content can represent between 10 and 26 per cent (d wt) of the plant. For instance, the species *Ulva pertusa*, which is frequently consumed under the name of "ao-nori" by the Japanese people, has a high protein level between 20 and 26 per cent (dry product) (Indegaard and Minsaas, 1991). Higher protein levels are recorded for the red seaweeds such as *Porphyra tenera* (47 per cent of d wt) or *Palmaria palmata* (35 per cent of d wt) (Arasaki, and Arasaki, 1983; Morgan *et al.*, 1980). These seaweeds known under the names of "nori" and "dulse", respectively, have protein levels higher than those found in high-protein pulses such as soybean. An annual monitoring of protein level from *Palmaria palmata* (Dulse) collected on the French Atlantic coast has shown that the protein content can vary between 9 and 25 per cent (d wt). Higher protein levels are observed during the end of the winter and spring periods whereas lower amounts are recorded during the summer months (c.f. Fleurence 1999). The seasonal variation of the protein content is also reported for various species like *Laminaria digitata* and *Ulva lactuca* (Augier and Santimoine, 1978; Abdel- Fattah and Sary, 1987). *Kappaphycus alvarezii* has also recorded 16 per cent protein (Fayaz *et al.*, 2005). Extraction of proteins from seaweed is difficult because of the occurrence of phenolic compounds (Ragan and Glombitza, 1986) and large amounts polyanionic cell wall mucilages (Fleurence *et al.*, 1995; Jordan and Vilter, 1991).

Seaweeds are very rich source of amino acids. Some seaweeds like *Ulva pertusa, Undaria pinnatifida* and *Porphyra. tenera* have high level of arginine, besides *Porphyra tenera* also contains considerable amount of glycine. With respect to high protein level and amino acid composition, the red seaweeds appear to be a potential source of food proteins. In general amino acid composition in seaweeds is close to that of traditional proteins from leguminous plants and ovalbumin. The free amino acid fraction of seaweeds is mainly composed of alanine, aminobutyric acid, taurine, ornithine, citrulline and hydroxyproline. Levels of these amino acids differ according to the species (Arasaki and Arasaki 1983). In addition, Rhodophyceaen members contain a particular protein called phycoerythrin used in biotechnology applications (dye in immunosence reaction) (Gantt and Lipschutz 1974). The beneficial effects of *Porphyra* meal used as a food additive in the diet of red sea bream are that the diet appeared to increase body weight gain as well as the triglyceride and protein deposition in muscle (Satoh *et al.*, 1987).

The fat content of marine macroalgae (seaweeds) accounts for 1–6 g/100 g d.wt. (Fleurence *et al.*, 1994; Jurkovic´ *et al.*, 1995; Herbreteau *et al.*, 1997). In some brown algal varieties, such as *Hizikia* species and *Arame* species only 0.7–0.9 g/100 g d.w. of fat content was found (Kolb *et al.*, 1999). Seaweeds are also a magnificent source of polyunsaturated fatty acids, as the eicosapentanoic acid (EPA) described in *Himanthalia elongata, Undaria pinnatifida* and *Porphyra* species. These ω3 fatty acids have demonstrated their effect on the reduction of coronary diseases, thrombosis and arteriosclerosis (Simopoulos, 2004). *Kappaphycus alvarezii* is known to possess higher composition of unsaturated fatty acids (44.5 per cent) when compared to saturated fatty acids (37 per cent) (Fayaz *et al.*, 2005).

With respect to their high protein level, amino and fatty acid composition, the red and green seaweeds appear to be interesting potential sources of food proteins. In Europe, the development of novel foods such as functional foods could be a new possibility for use of seaweeds, especially for the protein-rich species, in human nutrition.

4.2 A Source of Minerals and Trace Elements

A characteristic feature of seaweeds is that they possess high mineral content (8–40 per cent) and moreover the essential minerals and trace elements needed for human nutrition are also present in seaweeds (Mabeau and Fleurence, 1993;Ortega-Calvo *et al.*, 1993). Human and animal studies originally have shown that optimal intake of elements such as sodium, potassium, magnesium, calcium, manganese, copper, zinc and iodine could reduce individual risk factors, including those related to cardiovascular disease (Anke *et al.*, 1984; Mertz, 1982; Sanchez-Castillo *et al.*, 1998). Throughout the world, there is an increasing interest in the importance of dietary minerals in the prevention of several diseases. Major minerals are those required in amounts greater than 100 mg per day and they represent 1 per cent or less of bodyweight. These include calcium, phosphorus, magnesium, sulfur, potassium, chloride and sodium. Trace minerals are essential in much smaller amounts, less than 100 mg per day, and make up less than 0.01 per cent of bodyweight. Essential trace elements are zinc, iron, silicon, manganese, copper, fluoride, iodine and chromium. Among functional food components, minerals play a pivotal role for living a healthy life. The wide range in mineral content, not much found in edible land plants, is related to various factors such as seaweed phylum, geographical origin, seasonal, environmental and physiological variations (Mabeau and Fleurence, 1993). Mineral content also depends on seaweed processing (Nisizawa *et al.*, 1987; Yoshie *et al.*, 1994) and on the mineralization methods used (Fleurence and Le Coeur, 1993). Seaweed mineral content is higher than that of land plants and animal products (Ito and Hori, 1989; Ortega- Calvo *et al.*, 1993). *Kappaphycus alvarezii* is known to contain

Ca, Fe, and Zn in admissible limits (Fayaz *et al.*, 2005). In *Porphyra vietnamensis* from India the selected micronutrients (Fe +Zn+Mn+Cu) are found to be higher (45.5- 309 mg/100 g d wt) (Subba Rao *et al.*, 2007) than any of the land vegetables as well as edible seaweeds like *Caulerpa lentilifera, Enteromorpha flexuoza,Monostroma oxyspermum, Eucheuma denticulatom* and *Gracilaria parvispora* reported from Hawaii (McDermid and Stuercke, 2003), *Undearia pinnatifida, Laminaria digitata, Fucus vesiculosus* and *Chondrous crispus* reported from Spain (Kolb *et al.*, 2004; Reperez, 2002).

Bioavailability is a term used to describe the proportion of a nutrient in food that can be utilized for normal body functions (Watzke, 1998; O'Dell, 1984.). For minerals to be bioavailable they are to be in a soluble form, however, not all soluble minerals are bioavailable (Clydesdale, 1988). The bioavailability of minerals and trace elements in humans is best studied using an *in vivo* model; however, experiments using such a model are often complex, time consuming and costly, and yield non-reproducible results. The first step in the study of mineral bioavailability is to measure the concentration of minerals in foodstuff, in the form in which they exist, as well as their solubility.

4.3 A Source of Bioactive Compounds

Antioxidant activity has become a hot topic and is the subject of intensive investigations due to the ever increasing demand by the food and pharmaceutical industries to develop natural bioactive, anti-aging and anti carcinogenic compounds that demonstrate measurable health benefits. Among the most relevant compounds found in the seaweeds are antioxidants probably the substances that have attracted major interest in the recent past. Seaweeds, as photosynthetic organisms, are exposed to a combination of light and high oxygen concentration that induce the formation of free radicals and other oxidative reagents. The absence of structural damage in the seaweeds leads to consider that these organisms are able to generate the necessary compounds to protect themselves against oxidation. In this respect, seaweed can be considered as an important source of antioxidant compounds that could be suitable also for protecting human bodies against the reactive oxygen species formed as a result of metabolism or induced by external factors like pollution, stress, UV radiation, etc.

Seaweed extracts have been demonstrated to possess strong antioxidant properties (Kuda *et al.*, 2005; Yuan and Walsh, 2006), protective effects against liver injury caused by carbon tetrachloride (Wong *et al.*, 2000), antiproliferative activity towards HeLa cells (Yuan and Walsh, 2006), antimicrobial activity (Valdebenito *et al.*, 1982) and antiviral properties (Chatterji *et al.*, 2004). However there are antioxidant substances of very different nature, among which vitamin E or α-tocopherol and carotenoids can be highlighted within the fat-soluble fraction, whereas the most powerful water soluble antioxidants found in seaweeds are polyphenols, phycobiliproteins and vitamins (vitamin C).

In brown seaweeds the principal tocopherol is α-tocopherol (0.052 mg/g in *Himanthalia elongata*) and this compound is highly stable to heat and acids, and unstable to alkali, UV radiation and oxygen. α-Tocopherol has antioxidant activity because it fixes free radicals. Carotenoids are a family of compounds of importance due to their high antioxidant activity and great structural diversity. There are numerous studies that associate the antioxidant activity of different seaweeds with the carotenoid contents. In this regard, a type of carotenoids called xantophylls obtained from *Undaria pinnatifida* have demonstrated some activity against cerebrovascular diseases (Ikeda *et al.*, 2003). Recent studies on the seaweed *Kappaphyvus alvarezii* have revealed that the various solvent extracts have exhibited antioxidant activity in linoleic acid system with ferrothiocyanate reagent (Suresh Kumar *et al.*, 2008).

Besides, many constituents of the dietetic fiber show antioxidant activity as well as immunological activity (Suzuki *et al.*, 2004). In this sense, *Undaria pinnatifida* (wakame) showed some positive effect on several cardiovascular diseases (hypertension and hypercholesterolemia) (Ikeda *et al.*, 2003); this alga contains basically dietetic fiber, being its principal component alginate. This alginic acid has demonstrated to reduce hypertension (Ikeda *et al.*, 2003). These findings, together with consumers' interests in natural food additives, have reinforced the efforts for the development of alternative antioxidants from natural origins.

5. Seaweed Production and Economics

As demand has increased, natural stocks have been unable to meet market requirements, and hence more than 90 per cent of seaweeds that are used commercially are cultivated (McHugh, 2003). For example, "nori", *Porphyra* species., which is used for making sushi, currently provides an industry in Asia with an yearly turnover of ~US$ 1 x 109 (Pulz and Gross 2004). Production of seaweeds world wide in 1994/1995 was over 2 million ton d wt annually and 90 per cent came from just six countries *viz.*, China, France, UK, Korea, Japan and Chile and also from just five genera *viz.*, *Laminaria*, *Lithothamnion*, *Porphyra*, *Undaria* and *Gracilaria*. Of this 52 per cent of seaweed produced was cultured and 90 per cent of the cultured seaweed was produced in China, Korea and Japan. Of the genera cultured, *Laminaria* was produced in the largest amount (682,581 t d. wt) followed by *Porphyra* (130,614 t d. wt), *Undaria* (101,708 t d. wt) and *Gracilaria* (50,165 t d. wt). These four genera made up 93 per cent of seaweed cultured worldwide (c.f Zemke-White and Ohno, 1999) and form marine crop plants since their production through cultivation exceeded their natural standing crop (Tseng, 1981).

In terms of price paid, the seaweed used for food, particularly in Japan, is the most valuable. *Monostroma* has been worth US $ 20–30 kg^{-1} in Japan for the last ten years, while *Porphyra* in Japan is US $ 1.5 billion per year business equating to US $ 25 kg^{-1} (Ohno and Largo, 1998). Not all seaweed food is so expensive. In 1994, *Hizikia* in Japan was worth approximately US $ 9 kg^{-1}, while *Undaria* was US $ 2.25 kg-1. Extrapolating these prices to the total amount of these four algal genera produced worldwide gave an annual value of approximately US $ 3.6 billion (c.f Zemke-White and Ohno, 1999).

In conclusion, the seaweeds, especially from the Chlorophyceae and the Rhodophyceae, could be a complementary source of food proteins for human beings (Fleurence, 1999). Further seaweeds provide a largely untapped reservoir of novel and valuable compounds, and this seems to be the main application area of future commercial algal transgenic. Current exploitation mainly aims to utilize fatty acids, pigments, vitamins and other bioactive compounds.

Acknowledgements

Authors express their gratitude to the Director, Central Salt and Marine Chemicals Research Institute (CSIR), Bhavnagar for his encouragement and the facilities provided. Author KG is thankful to CSIR for financial assistance in the form of Senior Research Fellow.

References

Abdel-Fattah AF and Sary HH (1987) Selective isolation of glycoprotein materials from the green seaweed *Ulva lactuca.* Pakistan Journal of Biochemistry 20: 61-65.

Aguilera-Morales M, Casas-Valdez M,Carrillo-Domý´nguez S, Gonza´lez-Acosta B, and Pe´rez-Gil F (2005) Chemical composition and microbiological assays of marine algae *Enteromorpha* spp. as a potential food source. Journal of Food Composition and Analysis 18: 79–88.

Amano H and Noda H (1990) Proteins of protoplast from red alga *Porphyra yezoensis.* Bulletin of the Japanese Society of Scientific Fisheries 56: 1859-1864.

Anke M, Groppel B and Kronemann H (1984) Significance of newer essential trace elements (like Si, Ni, As, Li, V,) for the nutrition of man and animals. P. Bratter and P. Schramel (Eds.) In: Trace element-analytical chemistry in medicine and biology. vol. 3. pp.421–464. Berlin.

Arasaki A and Arasaki T (1983) Low calories, High Nutrition Vegetables from the Sea to Help you Look and Feel Better, Japan Publications Inc pp. 39-42.

Augier H and Santimoine M (1978) 'Contribution aÁ l'eÂ tude de la composition en azote total, en proteines et en acides amine s prote iques des differentes parties du thalle de *Laminaria digitata* (Huds.) Lamour. dans le cadre de son exploitationindustrielle et agricole' Bull. Soc. Phycol. de France 23 : 19- 28.

Chapman VJ and Chapman DJ (1980) Sea vegetables (Algae as food for man). In: *Seaweeds and their uses* London. Chapman and Hall, pp 62–97.

Chatterji A, Dhargalkar VK, Sreekumar PK, Parameswaran PS, Rodrigues R and Kotnala S (2004) Anti-influenza activity in the Indian Seaweeds–A preliminary investigation. National Institute of Oceanography, Goa.

Chennubhotla VSK, Kaliaperumal N and Kalimuthu S (1981) Seaweed recipes and other practical uses of seaweed. Seafood Experimental Journal 13: 9-16.

Clydesdale FM (1988) Minerals: Their chemistry and fate in food. In: Smith KT (ed) Trace Minerals in Foods, pp. 57-94. Marcel Dekker Inc, New York.

Dam R, Lee S, Fry PC and Fox H (1986) Utilization of algae as a protein source for humans. Journal of Nutrition 65: 376-382.

Darcy-Vrillon B (1993) Nutritional aspects of the developing use of marine macroalgae for the human food industry. International Journal of Food Science and Nutrition 44: S 23–S 35.

Dawes CJ (1998) Marine Botany. John Willey and Sons, Inc. New York pp 480.

Fayaz M, Namitha K, Chidambara Murthy KN, Mahadevaswamy M, Sarada R, Salma Khanam, Subba Rao, PV and Ravisankar GA (2005). Chemical Composition, Iron Bioavailablity and Antioxidant Activity of *Kappaphycus alvarezzi* (Doty).J Agric Food Chem 53: 792-797.

Fleurence (1999) Seaweed Proteins: Biochemical, nutritional aspects and potential uses. Trends in Food Science and Technology 10: 25–28.

Fleurence J and Le Coeur C (1993) Influence of mineralization methods on the determination of the mineral content of brown seaweed *Undaria pinnatifida* by atomic absorption spectrophotometry. Hydrobiologia 260/261 : 531–534.

Fleurence J, Gutbier G, Mabeau S and Leray C (1994) Fatty acids from 11 marine macroalgae of the French Brittany coast. Journal of Applied Phycology 6: 527–532.

Fleurence J, Le Coeur C, Mabeau S, Maurice M and Landrein A (1995) Comparison of different extractive procedures for proteins from the edible seaweeds *Ulva rigida* and *Ulva rotundata*. Journal of Applied Phycology 7: 577-582.

Fujii T, Kuda T, Saheki K and Okuzumi M (1992) Fermentation of water-soluble polysaccharides of brown algae by human intestinal bacteria *in vitro*. Nippon Suisan Gakk 58 : 149–152.

Gantt E and Lipschutz CA (1974) Phycobilisomes of *Porphyridium cruentum*: pigment analysis. Biochem 13: 2960-2966.

Gellenbeck KW and Chapman DJ (1983) Seaweed uses: The outlook for mariculture. Endeavour 7: 31–37.

Goni I, Guidel-Urbano M, Bravo L and Saura-Calixto F (2001) Dietary modulation of bacterial fermentative capacity by edible seaweeds in rats. Journal of Agriculture and Food Chemistry 49: 2663–2668.

Goni I, Valdivieso L and Garcia-Alonso A (2000) Nori seaweed consumption modifies glycemic response in healthy volunteers. Nutrition Research 20: 1367–1375.

Guidel-Urbano M and Goni I (2002) Effect of edible seaweeds (*Undaria pinnitafida* and *Porphyra tenera*) on the metabolic activities of intestinal microflora in rats. Nutrition Research 22: 323–331.

Herbreteau F, Coiffard LJM, Derrien A and De Roeck-Holtzhauer Y (1997) The fatty acid composition of five species of macroalgae. Botanica Marina 40: 25–27.

Ikeda K, Kitamura A, Machida H, Watanabe M, Negishi H and Hiraoka J (2003) Effect of *Undaria pinnatifida* (Wakame) on the development of cerebrovascular diseases in stroke-prone spontaneously hypertensive rats. Clinical and Experimental Pharmacology and Physiology 30: 44 -48.

Indegaard M and Minsaas J (1991) In Animal and Human Nutrition. Seaweed Resources in Europe. Uses and Potential Eds, Guiry MD and Blunden G, John Wiley and Sons, pp. 21- 64.

Ito K and Hori K (1989) Seaweed: chemical composition and potential food uses. Food Review International 5: 101–144.

Jime´nez-Escrig A and Sa´nchez-Muniz FJ (2000) Dietary fiber from edible Seaweeds: Chemical structure, physicochemical properties and effects on cholesterol metabolism. Nutrition Research 20: 585–598.

Jordan P and Vilter H (1991) Extraction of proteins from material rich in anionic mucilages: partition and fractionation of vanadate-dependent bromoperoxidases from the brown algae *Laminaria digitata* and *L. saccharina* in aqueous polymer two-phase system. Bio-chemical and Biophysical Acta 1073: 98-106.

Jurkovic´ N, Kolb N and Colic´ I (1995) Nutritive value of marine algae *Laminaria japonica* and *Undaria pinnatifida*. Die Nahrung 39: 63–66.

Kaehler S and Kennish R (1996) Summer and winter comparisons in the nutrition value of marine macroalgae from Hong Kong. Botanica Marina 39: 11-17.

Kolb N, Vallorani L and Stocchi V (1999) Chemical composition and evaluation of protein quality by amino acid score method of edible brown marine algae Arame (*Eisenia bicyclis*) and Hijiki (*Hijikia fusiforme*). Acta Alimentaria 28: 213–222.

Kolb N, Vallorani L, Kozlek D and Stocchi V (2004) Evaluation of marine algae Wakame (*Undaria pinnatifida*) and Kombu (*Laminaria digitata joponica*) as food supplements. Food Technology and Biotechnology 42: 57–61.

Krishnamurthy V (2003) Phycology-Principles, Processes and Applications. (Ed) Aluwalia Amrik Sing, Daya Publishing House, New Delhi.

Kuda T, Goto H, Yokoyama M and Fujii T (1998a) Fermentable dietary fiber in dried products of brown algae and their effects on cecal microflora and levels of plasma lipid in rats. Fisheries Science 64: 582–588.

Kuda T, Goto H, Yokoyama M and Fujii T (1998b) Effects of dietary concentration of laminaran and depolymerized alginate on rat cecal microflora and plasma lipids. Fisheries Science 64: 589–593.

Kuda T, Tsunekawaa M, Goto H and Araki Y (2005) Antioxidant properties of four edible algae harvested in the Noto Peninsula, Japan. Journal of Food Composition and Analysis 18: 625–633.

Kuda T, Yokoyama M and Fujii T (1997) Effects of marine algae diets Hijiki, Aonori, and Nori on levels of serum lipid and cecal microflora in rats. Fisheries Science 63: 428–432.

Mabeau S and Fleurence J (1993) Seaweed in food products: biochemical and nutritional aspects. Trends in Food Science and Technology 4: 103-107.

Madlener JC (1977) The Seavegetable Book. Carlkson N. Potter, Inc. New York. pp 288.

Marshram S,Scott, GW and Tobin ML (2007) Comparison of nutritive chemistry of a range of temperate seaweeds. Food Chemistry100: 1331-1336.

McDermid KJ and Stuercke B (2003) Nutritional composition of edible Hawaiian seaweeds. Journal of Applied Phycology 15: 513–524.

McHugh DJ (2003) A guide to seaweed industry. FAO Fisheries Technical paper 441. Rome: Food and Agricultural Organization of the United Nations.

Mertz W (1982) Trace minerals and atherosclerosis. Food Proceedings 41: 2807–2812.

Morgan KC, Wright JLC and Simpson FJ (1980) Review of chemical constituents of the red alga *Palmaria palmata* (Dulse). Economic Botany 34: 27-50.

Nisizawa, K., Noda, H., Kikuchi, R., and Watanabe, T. (1987). The main seaweeds in Japan. *Hydrobiologia. 151/152,* 5-29.

Norziah MH and Ching CY (2000) Nutritional composition of edible seaweed *Gracilaria changgi*. Food Chemistry 68: 69–76.

Novaczek I (2001) A Guide to the Common Edible and Medicinal Sea Plants of the Pacific Island. University of the South Pacific pp 40.

O'Dell BL (1984) Bioavailability of trace elements. Nutrition Review 42: 301-308.

Ohno M (1997) Cultivation of the green algae, *Monostroma* and *Enteromorpha"Aonori"* In Seaweed Cultivation and Marine Ranching. Ohno M and Crtichley A 1997 (Eds). JICA. Yokosuka, pp. 7-15.

Ohno M and Largo DD (1998) The seaweed resources of Japan. In: Critchley AT, Ohno M (eds) Seaweed Resources of the World. Japan International Cooperation Agency, Yokosuka: 1–14.

Ortega-Calvo JJ, Mazuelos C, Hermosý´n B and Sa´iz-Jime´nez C (1993). Chemical composition of *Spirulina* and eucaryotic algae food products marketed in Spain. Journal of Applied Phycology 5: 425–435.

Payri C, N'Yeurt ADR and Orempuller J (2000) Algues de Polynésie Française. Imprimé à Singapour. Au Vent des Îles. pp. 320 (in French).

Pulz O and Gross W (2004) Valuable products from biotechnology of microalgae. Applied Microbiology and Biotechnology 65: 635-648.

Ragan MA and Glombitza KW (1986) Phlorotannins, brown algalpolyphenols. In: Progress in phyco-logical research (ed) Round FE and Chapman DJ Bristol Biopress Ltd. 4: 130-230.

Rupérez R (2002) Mineral content of edible marine seaweeds. Food Chemistry 79: 23-26.

Sanchez- Machado DI, Lopez- Cervantes J, Lopez–Hernandez J and Paseiro–Losada P (2004) Fatty acids, total lipid, protein and ash contents of processed edible seaweeds. Food Chemistry 85: 439–444.

Sanchez-Castillo CP, Dewey PJS, Aguirre A, Lara JJ, Vaca R, Leon de la Barra P, Ortiz M, Escamilla I and James WPT (1998) The mineral content of Mexican fruits and vegetables. Journal of Food Composition and Analysis 11: 340–356.

Satoh KI, Nakagawa H, Kashara S (1987) Effect of *Ulva* meal supplementation on Disease resistance of Red Sea Bream. Nippon Suisan Gakkaishi 53: 1115–1120.

Simopoulos AP (2004) Omega-3 essential fatty acid ratio and chronic diseases.Food Review International 20: 77–90.

Subba Rao PV, Mantri V A and Ganesan K (2007) Mineral composition of edible seaweed *Porphyra vietnamensis*. Food Chemistry 102: 215-218.

Suresh Kumar K, Ganesan K and Subba Rao PV (2008) Antioxidant potential of different solvent extracts of *Kappaphycus alvarezii*- an edible seaweed. Food Chemistry 107: 289- 295.

Suzuki N, Fujimura A, Nagai T, Mizumoto I, Itami T and Hatate H (2004) Antioxidative activity of animal and vegetable dietary fibers. Biofactors 21: 329-333.

Takenaka S, Sugiyama S, Ebara S, Miyamoto E, Abe K and Tamura Y (2001) Feeding dried purple laver (Nori) to vitamin B12-deficient rats significantly improves vitamin B12-status. British Journal of Nutrition 85: 699–703.

Tanaka J and Nakamura T (2004) Japanese Seaweeds. Tokyo. pp.248 (in Japanese).

Toma T (1997) Cultivation of the brown alga, *Cladosiphon okamuranus "Okinawa-mozuku"*. In: Seaweed Cultivation and Marine Ranching. Ohno M and Critchely A (Eds) JICA. Yokosuka, pp.51-56.

Trono G and Toma T (1997) Cultivation of the green alga *Caulerpa lentilifera*. In: Seaweed Cultivation and Marine Ranching. Ohno M and Cricthley A (Eds). JICA. Yokosuka, pp. 17-23.

Tseng CK (1981) Commercial cultivation. In: Biology of Seaweeds. (Ed) Lobban CS and Wynne MJ. Blackwell Scientific Publications, Oxford pp. 680- 725.

Tsutsui I, Huybh QN, Nguyên HD, Arai S and Yoshida T (2005) *The Common. Marine Plants of Southern Vietnam*. Japan Seaweed Association. Kochi. pp.250.

Valdebenito H, Bittner M, Sammes PG, Silva M and Watson WH (1982) A compound with antimicrobial activity isolated from the red seaweed *Laurencia chilensis*. Phytochemistry 21: 1456–1457.

Van Netten SA, Hoption Cann DR, Morley JP and van Netten (2000) Elemental and radioactive analysis of commercially available seaweed. Science of Total Environment 255: 169–175.

Ventura MR and Castanon JIR (1998) The nutritive value of seaweed (*Ulva lactuca*) for goats. Small Ruminant Research 2: 325–327.

Watanabe F, Katsura H, Takenaka S, Fujita T, Abe K and Tamura Y (1999) Pseudovitamin B12 is the predominant cobamide of an algae health food, *Spirulina* tablets. Journal of Agriculture and Food Chemistry 47: 4736–4741.

Watanabe F, Takenaka S, Kittaka-Katsura H, Ebara S and Miyamoto E (2002) Characterization and bioavailability of vitamin B12- compounds from edible algae. Journal of Nutritional Science and Vitaminology 48: 325–331.

Watzke HJ (1998) Impact of processing on bioavailability examples of minerals in foods. Trends in Food Science and Technology 8: 320-327.

Wong CK, Ooi VEC and Ang PO (2000) Protective effects of seaweeds against liver injury caused by carbon tetrachloride in rats. Chemosphere 41: 173–176.

Wong KH and Cheung PCK (2000) Nutritional evaluation of some subtropical red and green seaweeds. Part I. Proximate composition, amino acid profiles and some physico-chemical properties. Food Chemistry 71: 475–482.

Yamada K, Yamada Y, Fukuda M andYamada S (1999) Bioavailability of dried Asakusanori (*Porphyra tenera*) as a source of obalamin (vitamin B12). International Journal of Vitaminology and Nutrition Research 69: 412–418.

Yon M and Hyun TH (2003) Folate content of foods commonly Consumed in Korea measured after trienzyme extraction. Nutrition Research 23: 735–746.

Yoshie Y, Suzuki T, Shirai T and Hirano T. (1994) Changes in the contents of dietary fibers, minerals, free amino acids and fatty acids during processing of dry Nori. Nippon Suisan Gakkaishi 60: 117–123.

Yuan YV and Walsh NA (2006) Antioxidant and antiproliferative activities of extracts from a variety of edible seaweeds. Food and Chemical Toxicology 44: 1144–1150.

Zemke-White LW and Ohno M (1999) World seaweed utilization: End-of-Century Summary. Journal of Applied Phycology 11: 369-376..

Algal Biotechnology: New Vistas (2010) *Pages 93–110*
Editor: Mihir Kumar Das
Published by: DAYA PUBLISHING HOUSE, NEW DELHI

Chapter 6

UV-B Radiation-Induced Stress and Protection Strategies in Cyanobacteria

R.P. Sinha[1], M.B. Tyagi[2], Sushil Kumar[3] and Ashok Kumar[3]*

[1]Centre of Advanced Study in Botany, [2]Mahila Maha Vidyalaya,
[3]School of Biotechnology,
Banaras Hindu University, Varanasi – 221 005, India

ABSTRACT

Depletion of the stratospheric ozone layer due to anthropogenically released chemicals such as chlorofluorocarbons (CFCs) has resulted in an increase in ultraviolet-B (UV-B; 280-315 nm) radiation on the Earth's surface. UV-B is a small (less than 1 per cent of total energy) but highly active component of the solar spectrum that can penetrate to biologically significant depths in lakes, ponds, rivers and oceans. A number of physiological and biochemical processes such as growth, survival, cell differentiation, motility, pigmentation, photosynthesis, nitrogen metabolism, proteins and DNA have been reported to be affected by UV-B radiation. However, cyanobacteria have evolved mechanisms to counteract the damaging effects of UV-B radiation by synthesizing UV-mitigating compounds such as mycosporine-like amino acids (MAAs) and scytonemin. Several efficient repair mechanisms are also operative in cyanobacteria to minimize the damage caused by UV-B radiation. This chapter deals with the effects of UV-B radiation on

* E-mail: kasokbt@rediffmail.com

cyanobacteria and the role of certain compounds in mitigating the damaging effects caused by the increasing level of natural UV-B intensity.

Keywords: *Cyanobacteria, UV-B (280–315 nm) radiation, Nitrogenase inactivation, Mycosporine-like amino acids (MAAs), Scytonemin, DNA damage and repair.*

1. Introduction

Depletion of the stratospheric ozone layer due to anthropogenically released atmospheric pollutants such as chlorofluorocarbons (CFCs) has resulted in an increase in ultraviolet-B (UV-B; 280–315 nm) radiation reaching on to the Earth's surface (Crutzen 1992; Kerr and McElory 1993; Madronich *et al.*, 1998; Weatherhead and Anderson 2006). Such increases in UV-B radiation are most pronounced in Antarctica, but similar trends have also been reported at other latitudes (Kerr and McElroy, 1993).

UV-B radiation is an important component of solar radiation having many deleterious effects on biological systems (Vincent and Roy, 1993; Häder *et al.*, 2007). The damaging effects of UV-B radiation on cyanobacteria are of special concern because they form a prominent microflora of nitrogen-fixing microorganisms in tropical rice-fields and enrich the fertility of soil (Singh, 1961). The deleterious effects are in part due to the direct effects of UV-B radiation on membrane component, growth and survival, motility, pigmentation, nutrient uptake, hormones, DNA, proteins, enzymes, and the photosynthetic apparatus, particularly the reaction center of photosystem II and RuBISCO (Döhler *et al.*, 1986; Vincent and Roy, 1993; Wilson *et al.*, 1995; Quesada *et al.*, 1999; Kumar *et al.*, 2003; Wright *et al.*, 2005; Bhandari and Sharma, 2006; Helbling *et al.*, 2006; Cheng *et al.*, 2007; Bjorn, 2007; Rath and Adhikary, 2007), or to its indirect effect through the formation of reactive oxygen species, such as super oxide (O_2^-), hydroxyl radicals (OH^-) or hydrogen peroxide (H_2O_2) (He and Häder, 2002). However, to overcome the above adverse effects, cyanobacteria have developed efficient repair mechanisms including the synthesis of UV-mitigating compounds such as scytonemin and mycosporine-like amino acids (MAAs). This chapter summarizes some of the effects of UV-B radiation as well as modes of UV-B protection mechanisms operative in cyanobacteria (Table 6.1).

2. Effects of UV-B on Physiological and Biochemical Processes

2.1 Growth and Survival

UV-B radiation has been shown to severely affect the growth and survival of several cyanobacteria (Tyagi *et al.*, 1992; Sinha *et al.*, 1995, 1996; Quesada and Vincent, 1997; Prasad and Zeeshan, 2004). Various species differ with respect to their tolerance to UV-B radiation and even closely related strains show differential sensitivity. Complete killing and loss of survival within 120-180 min of UV-B exposure (2.5 W/m^2) has been reported for several cyanobacteria (Tyagi *et al.*, 1992; Sinha *et al.*, 1995). Strains such as *Nostoc commune* and *Scytonema* sp. whose filaments are embedded in a mucilage sheath are more tolerant to UV-B radiation than the species which do not have such coverings (Sinha *et al.*, 1996). A phycoerythrin rich (brown) strain of *Nostoc* sp. was found to be more tolerant to UV-B radiation than the phycocyanin rich (blue-green) form of this strain (Tyagi *et al.*, 1992). Two Antarctic cyanobacteria, *Phormidium murrayi* and *Oscillatoria priestleyi*, isolated from a common habitat showed differential sensitivity to UV-B radiation. In contrast to complete suppression of growth in *Oscillatoria priestleyi*, only a 62 per cent inhibition of growth of *Phormidium murrayi* was observed following exposure to the same quantum of UV-B intensity (Quesada and Vincent, 1997). Survival and growth has been shown to be decreased with selected doses of UV-B (30 and 90 min) in non-heterocystous cyanobacterium, *Plectonema boryanum* (Prasad and Zeeshan, 2004).

Table 6.1: UV-B-Induced Effects in Cyanobacteria

Process	*Effects*	*References*
Movement and orientation	Inhibition; impairment of phototaxis and photophobic responses	Häder and Worrest (1991) Donkor *et al.* (1993); Häder *et al.* (2007)
Growth and survival	Inhibition and killing	Tyagi *et al.* (1992); Sinha *et al.* (1995); Kumar *et al.* (2003)
Cell differentiation	Delayed heterocyst differentiation and akinete formation; changes in spacing pattern of heterocysts	Blakefield & Harris (1994); Sinha *et al.* (1996)
Cell membrane	Membrane disruption; peroxidation of membrane lipids; opening of membrane bound calcium channels	Döhler *et al.* (1986); Sinha *et al.* (1996); Richter *et al.* (1999)
Photosynthetic pigments	Decrease in chl a and phycobilin content, photobleaching, disintegration of phycobilisomes, increase in carotenoids especially echinenone and myxoxanthophyll	Tyagi *et al.* (1992); Vincent and Roy (1993); Sinha *et al.* (1995); Ehling-Schulz *et al.* (1997)
Photosynthesis	Impairment of reaction centre, degradation of D_1 and D_2 polypeptides, instant decrease of O_2 evolution, inhibition of RuBISCO activity and $^{14}CO_2$ uptake	Newton *et al.* (1979); Häder and Worrest (1991); Tyagi *et al.* (1992); Sinha *et al.* (1995); Sass *et al.* (1997); Campbell *et al.* (1998) Cockell and Rothschild (1999)
Nitrogen metabolism	Stimulation of *in-vivo* NR activity, inhibition of activity and inactivation of nitrogenase, inhibition of GS activity	Tyagi *et al.* (1992); Kumar *et al.* (2003)
General protein profile	Elimination of low molecular weight protein bands	Sinha *et al.* (1995)
Protein synthesis	Inhibition	Kumar *et al.* (2003)
DNA	Thymine-dimer formation	Sinha *et al.* (2001a)

2.2 Cell Differentiation

Differentiation of vegetative cells into heterocysts or akinetes is also affected by UV-B radiation. Blakefield and Harris (1994) observed delayed heterocyst differentiation and akinete formation in *Anabaena aqualis* following exposure to UV-B radiation (Table 6.1). It is presumed that UV-B radiation alters the C:N ratio which in turn affects the spacing pattern of heterocyst in a filament (Sinha *et al.*, 1996). A decrease in the level of major heterocyst polypeptides (26, 54 and 55 kDa) following UV-B treatment has been noted which is presumed to be responsible for the disruption of multi-layered heterocyst wall, the essential component for maintaining the active form of nitrogenase. Role of UV-B radiation in the opening of specific calcium channels in the heterocyst of *Anabaena* sp. has also been demonstrated (Richter *et al.*, 1999).

2.3 Motility and Orientation

Cyanobacteria employ motility and orientation mechanisms to escape harmful UV radiations and high light intensities but it has been reported that these avoidance mechanisms themselves are severely affected by UV-B radiation, resulting in a decrease or loss of their ability to escape from harmful UV conditions (Donkor *et a*l., 1993). Filamentous cyanobacteria such as *Anabaena variabilis, Oscillatoria tenuis* and *Phormidium uncinatum* use gliding motility to optimize their position in their microhabitat (Table 6.1). Donkor *et al.* (1993) found a drastic reduction in the percentage of motile filaments of the above cyanobacteria within 10-30 min of exposure to unfiltered solar radiation at Ghana (4.30°N). Removal of UV-B radiation by an appropriate long-pass or band- pass filters resulted in higher percentage of motility in most cases suggesting the inhibitory effect of UV-B radiation on gliding motility of these cyanobacteria. Similarly, inhibition of phototactic orientation and photophobic responses reduce the ability of the organisms to orient themselves in their photo-environment (Bebout and Garcia-Pichel 1995). Failure to quickly respond to changes in the photo-environment may result in death of the organisms (Häder and Worrest, 1991).

2.4 Pigmentation

It has been demonstrated that UV-B radiation causes photobleaching of photosynthetic pigments, a decrease in phycocyanin/chlorophyll *a* ratio and disintegration of phycobilisome complex leading to the bleaching of phycobiliprotein pigment complex (Häder and Worrest 1991; Quesada and Vincent 1997). Phycobiliprotein seems to be the most sensitive pigment most probably due to its water soluble property. Sinha *et al.* (1995) have reported a decrease in the phycobiliprotein content and disassembly of phycobilisomal complex following UV-B irradiation in a number of cyanobacteria (Table 6.1). The pattern of fluorescence emission spectra of phycobiliproteins observed after UV-B irradiation suggests an impairment of energy transfer from the accessory pigments to the photosynthetic reaction centers (Sinha *et al.*, 1995). Exposure of cyanobacterial cultures to intense UV-B radiation causes bleaching of phycocyanin and phycoerythrin which is in accordance with the disintegration of phycobilisomes. UV-B-induced damage of photosynthetic pigments, phycocyanin, carotenoids and chlorophyll *a* has been reported but phycocyanin appeared to be the main target of UV-B in *Plectonema boryanum* (Prasad and Zeeshan, 2004).

2.5 Photosynthesis

Photosynthetic parameters such as $^{14}CO_2$ uptake, O_2 evolution and ribulose-1,5 bisphosphate carboxylose (RuBISCO) activity are greatly inhibited following exposure to UV-B radiation (Wilson *et al.*, 1995; Quesada and Vincent, 1997). Photosynthetic O_2 evolution of a *Phormidium* strain from Lake Baikal declined within minutes of exposure to solar irradiation (Häder and Worrest, 1991) while Gerber and Häder (1995) had reported reduction in photosynthetic O_2 evolution in five species of phytoplankton after UV-B exposure (Table 6.1). Several workers have reported an inhibition of $^{14}CO_2$ uptake by a number of cyanobacteria following exposure to UV-B irradiation (Newton *et al.*, 1979; Tyagi *et al.*, 1992; Sinha *et al.*, 1996). In a study of diurnal variability of the effect of solar UV-radiation on photosynthesis in an algal mat comprising of *Lyngbya aestuarii*, Cockell and Rothschild (1999) found a mean increase in the total primary productivity (measured in terms of $^{14}CO_2$ uptake) over the day when solar UV-B was screened out. In *Plectonema boryanum*, ^{14}C-fixation was found to be more sensitive to UV-B than photosynthetic oxygen evolution. Photosynthetic electron transport activity decreased in response to UV-B. This inhibition was due to a greater susceptibility of PSII than PSI (Prasad and Zeeshan, 2004). UV radiation inhibits photosynthesis of desert *Nostoc* sp. The effect was dependent on the intensity and exposure time. Exogenous chemicals such as NAC (N-acetylcysteine), ASC (ascorbic acid), and SNP (sodium nitroprusside) produced different levels of protection of photosynthesis activity under UV radiation. NAC addition can provide effective protection for photosynthetic activity of *Nostoc* under UV-B radiation. Protective effect of NAC has been shown to be greater than those of ASC and SNP addition after 2 h of UV treatment (Wang *et al.*, 2008).

It has been demonstrated that RuBISCO may undergo a number of modifications such as photo degradation, fragmentation and denaturation of polypeptide chain, change in active site and increased solubility of membrane proteins. In addition, supply of ATP and $NADPH_2$ may also be inhibited which may cause inhibition of CO_2- fixing ability. The D_1 and D_2 polypeptides that are the major constituents of PS-II reaction centers are also degraded, even after exposure to intermediate levels of UV-B radiation (Sass *et al.*, 1997; Campbell *et al.*, 1998). The D_1 core protein of photosystem II is highly susceptible to damage by UV-B-generated radicals. Two genes, *psbA2* and *psbA3*, encoding identical D_1 polypeptides were induced four to six fold by UV-B irradiation at 20 and 60 microeinsteins m^{-2} s^{-1} and intense white light (Mate *et al.*, 1998). Down-regulation of photosynthetic genes (except *psbA*) suggested a decrease of photosynthetic activity as intensity of UV-B exposure increased from 20 to 60 microeinsteins m^{-2} s^{-1} (Huang *et al.*, 2002).

2.6 Nitrogen Metabolism

The nitrogen-fixing enzyme nitrogenase has been found to be extremely sensitive to UV-B radiation (Newton *et al.*, 1979; Tyagi *et al.*, 1992; Kumar *et al.*, 1996; 2003). In a *Nostoc* sp., nitrogenase activity was lost within 45 min of UV-B exposure whereas nitrate reductase and glutamine synthetase activity remained more or less unaffected (Tyagi *et al.*, 1992). Kumar *et al.* (2003) reported complete inactivation of nitrogenase activity within 25-40 min of UV-B exposure in a number of cyanobacteria including

Anabaena sp., *Nostoc* sp., *Calothrix* sp. and *Scytonema* sp. Employing inhibitors of protein synthesis and PS-II activity they have demonstrated that restoration of nitrogenase activity in UV-B-treated cultures occurs only after fresh synthesis of nitrogenase polypeptide (Table 6.1). The inactivation most probably occurs due to complete damage to the nitrogenase polypeptide. It has been suggested that the aromatic amino acids present or the native structure of nitrogenase are responsible for extreme susceptibility of nitrogenase to UV-B radiation.

2.7 Protein Profile

UV-B radiation is known to cause a subtle change in the protein profile of several cyanobacteria (Sinha *et al.*, 1995). In a *Nostoc* sp., the low molecular weight proteins of around 20 kDa (αβ monomers of phycocyanin) were the most affected one. It is interesting to find that UV-B radiation has a drastic effect specifically on low molecular weight proteins (Table 6.1). In a comparative study on total protein profile of several cyanobacteria, Sinha *et al.* (1995), reported complete loss of protein bands between 14.5 and 45 kDa after 90 and 120 min of UV-B exposure in *Nostoc carmium* and *Anabaena* sp. However protein bands of relatively higher molecular weight *viz.*, 55 and 66 kDa were unaffected even after 120 min. In species such as *N. commune* and *Scytonema* sp., which contain a mucilage covering/sheath, complete elimination of bands occurred only after 150 min of exposure (Sinha *et al.*, 1995). The increased tolerance in certain species may be ascribed to the presence of mucilage covering or photoprotectants such as scytonemin and mycosporine-like amino acids (MAAs) in the cell. Appearance of certain new proteins may be due to UV-B shock and may be similar to the *hsp* group of proteins (Shibata *et al.*, 1991). In *Nostoc commune*, semiquantitative analysis of about 1350 proteins revealed that at least 493 proteins (37 per cent) belong to the UV-B stimulon. A minimum of 188 proteins were positively stimulated, whereas a minimum of 305 proteins were negatively affected (Ehling-Schulz *et al.*, 2002).

2.8 DNA Damage

UV-B-induced lesions in DNA have been reported, that result either due to direct absorption of UV-B radiation by DNA or from free radicals and reactive oxygen species (ROS) formed by various photochemical processes. The main class of UV-induced lesions consist of dimeric photoproducts such as cis-syn cyclobutane pyrimidine dimers (CPDs) and pyrimidine (6-4) pyrimidone photoproducts (6-4 PPs) and their Dewar isomers (Mitchell and Karentz, 1993; Friedberg et al.,1995; Cadet *et al.*, 2005; Kumari *et al.*, 2008). It has been reported that DNA lesions, if not repaired, may inhibit the progress of DNA polymerase, and interferes with DNA synthesis and RNA transcription, ultimately leading to mutation or death of the organism (Britt, 1995).

The damaging effects of UV-B radiation on DNA and formation of various photoproducts are well known in a number of organisms including the Antarctic marine planktonic diatom communities (Karentz *et al.*, 1991a; Buma *et al.*, 1995). However, little information is available regarding the specific effects of UV-B radiation on DNA of cyanobacteria, although few studies have been conducted with UV-C radiation (Levine and Thiel, 1987; Eker *et al.*, 1990). Sinha *et al.* (2001a) have demonstrated that exposure to UV-B radiation could induce the formation of thymine dimers in three rice-field cyanobacteria, *Nostoc* sp., *Anabaena* sp. and *Scytonema* sp. (Table 6.1). By using dot-blot and chemiluminescence method these workers successfully developed a relatively quick, simple and efficient method for the quantitative analysis of thymine dimers in aquatic primary producers including cyanobacteria (Sinha *et al.*, 2001a). Similarly, a PCR- based assay technique to assess the impact of UV-B radiation in causing the damages to the DNA of bacteria and cyanobacteria has been developed (Kumar *et al.*, 2004).

3. UV-Mitigating Compounds

Since cyanobacteria are believed to have originated in the Precambrian era at a time when ozone shield was absent, they presumably faced high fluxes of UV radiation which must have acted as an evolutionary pressure leading to selection for efficient UV-mitigating mechanisms. Tolerance of cyanobacteria to intense sunlight as well as UV radiation might have contributed to their success during early stages of colonization. As a consequence they have developed effective mechanisms to counteract the damaging effects of UV-B radiation. In case of microbes behavioral responses as well as external and internal mechanisms of protection against UV-induced damages have been reported and it has been proposed that several of these mechanisms for UV-protection and/or repair of UV-induced damage might have worked singly or in various combinations (Sinha and Häder, 2008). An important mechanism to prevent UV-induced photodamage, is the synthesis of UV-absorbing compounds (Singh *et al.*, 2008a). In cyanobacteria certain compounds such as scytonemin and mycosporine-like amino acids (MAAs), having strong absorption in the UV-A and/or UV-B region of the spectrum play an important role in allowing these organisms to grow and survive in habitats exposed to strong irradiation. In a terrestrial *Nostoc commune* strain, subjected to regular cycles of desiccation and rewetting, UV-absorbing compounds play a key role as UV-protectants (Ehling-Schulz *et al.*, 1997; Oren and Gunde-Cimerman, 2007). Role of UV-mitigating compounds in UV-B tolerance is highlighted below:

3.1 Scytonemin

Scytonemin is a yellow-brown colored pigment occurring in the extracellular sheath of a number of terrestrial cyanobacteria especially in those species growing in habitats exposed to intense solar radiation (Garcia-Pichel and Castenholz, 1991; 1993). It is a lipid soluble compound having *in vivo* absorption maximum at 370 nm in the UV-A region but shows considerable absorption in the UV-B region also (Figure 6.1). Purified scytonemin shows absorption maximum at 386 nm although there being appreciable absorbance at 252, 278 and 300 nm (Sinha *et al.*, 1999a). It is a novel dimeric pigment composed of indole and phenolic subunits with a molecular mass of 544 Da. Microspectrophotometric measurements of the transmittance of pigmented sheath and the quenching of ultraviolet excitation of phycocyanin fluorescence suggest that the pigment is effective in shielding the cells from incoming UV radiation and is able to reduce the incident UV-A radiation entering the cells to around 90 per cent (Garcia-Pichel and Castenholz, 1991; Brenowitz and Castenholz, 1997).

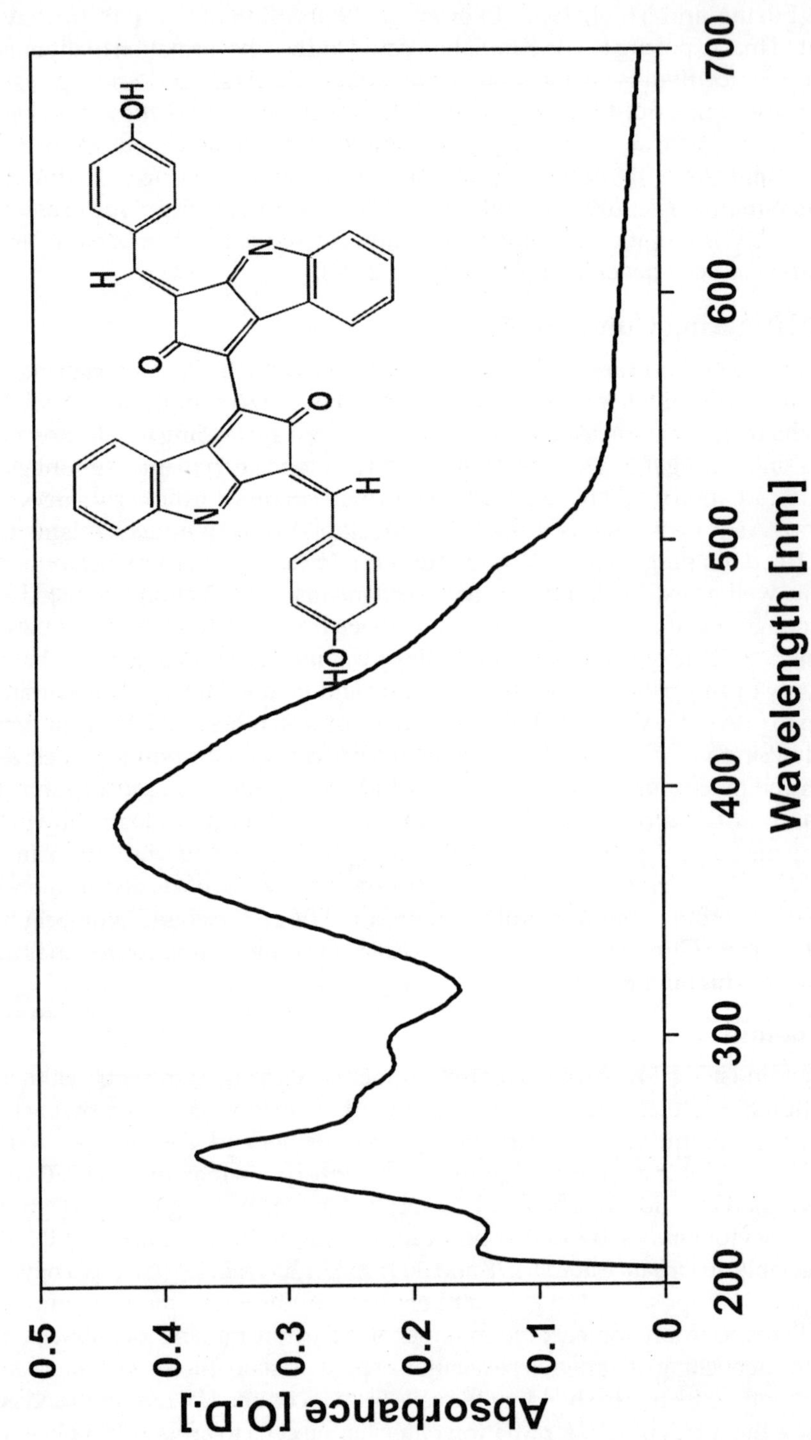

Figure 6.1: Absorption Spectrum and Molecular Structure of the UV-Mitigating Cyanobacterial Sheath Pigment, Scytonemin

Scytonemin may constitute as much as 5 per cent of the total cellular dry weight or even higher in some naturally occurring cyanobacteria (Castenholz, 1997). Since rapid photodegradation of scytonemin does not occur, it is highly stable and able to persist very long in terrestrial cyanobacterial crusts or dried mats (Brenowitz and Castenholz, 1997; Quesada *et al.*, 1999). It has been proposed that appearance of scytonemin during the Precambrian era must have helped the cyanobacteria in the colonization of exposed, shallow water and terrestrial habitats (Quesda *et al.*, 1999). Besides having a role as UV protectant (Fleming and Castenholz, 2007), scytonemin may have additional roles such as protection against pathogenic attack, bacterial decomposition or herbivore grazing (Sinha and Häder, 2008).

3.2 Mycosporine-like Amino Acids (MAAs)

The other class of photoprotectants is colorless, water-soluble MAAs having absorption maxima in the UV range between 310–362 nm (Torres *et al.*, 2006; Callone *et al.*, 2006; Conde *et al.*, 2007; Sinha *et al.* 2007; Sinha and Häder, 2008; Singh *et al.* 2008b) (Figure 6.2). First identified in fungi, these compounds are now reported to be present in a number of taxonomically diverse groups of organisms from eukaryotic and prokaryotic classes. They are present in cyanobacteria (Sinha *et al.*, 1999b; 2001b), eukaryotic algae (Carreto *et al.*, 1990; Karentz *et al.*, 1991b), marine heterotrophic bacteria and marine invertebrates (Karentz *et al.*, 1991b), fish (Dunlap *et al.*, 1989) and in a wide variety of other marine organisms (Karentz *et al.*, 1991b; McClintock and Karentz, 1997; Dunlap and Shick, 1998). Among cyanobacteria, MAAs has been reported from species growing in freshwater, marine or terrestrial habitats (Garcia-Pichel and Castenholz, 1993; Zhang *et al.*, 2007). Their presence has also been reported from strains growing in Antarctica (Quesada and Vincent, 1997) as well as from a community of halophilic species (Oren, 1997).

Structurally these compounds contain cyclohexenone or cyclohexenimine chromophore conjugated with the nitrogen substituent of an amino acid or its imino alcohol and have an average molecular weight of 300 (Cockell and Knowland, 1999). A number of MAAs such as shinorine, mycosporine-glycine, palythene, palythinol, asterina-330 and porphyra-334 has been reported in cyanobacteria (Sinha *et al.*, 2007). The specific MAA content in several isolates has been found to vary between 0.16 to 0.84 per cent of the total dry weight. Sinha *et al.* (2001b) have demonstrated the induction of MAAs by solar ultraviolet-B radiation in three cyanobacteria namely, *Anabaena* sp., *Scytonema* sp. and *Nostoc commune*. They reported that PAR and UV-A were ineffective in causing induction of MAAs. All these cyanobacteria possessed only a single type of MAA which was identified as shinorine. Zhang *et al.* (2007) isolated and characterized three MAAs, M-tau (UV-B sunscreen), dehydroxylusujirene and M-343 in *Synechocystis* sp. PCC 6803 and observed that dehydroxylusujirene and M-343 induction was mainly UV-A dependent but M-tau could be induced by both UV-A and UV-B. Recently, growth media dependent accumulations of MAAs (Singh *et al.*, 2008c) as well as UV-induced biosynthesis of three MAAs, mycosporine-glycine, porphyra-334 and shinorine (Singh *et al.*, 2008d) were for the first time reported in cyanobacteria.

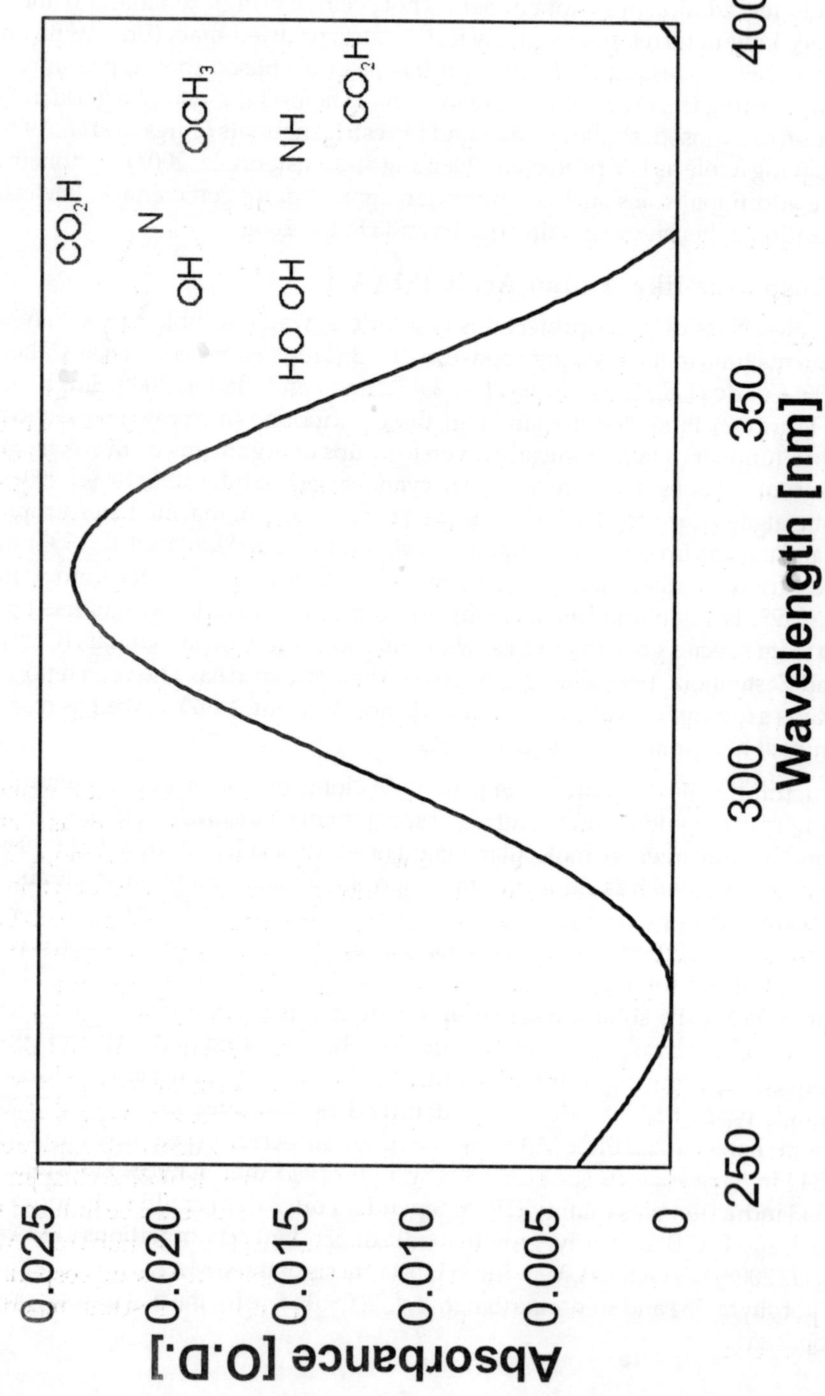

Figure 6.2: Absorption Spectrum and Molecular Structure of the Mycosporine-like Amino Acid (MAA), Shinorine

It has been suggested that the content of MAAs in the cell is regulated by osmotic mechanisms (Oren, 1997). Its role in several processes such as, i) it may transfer energy to the photosynthetic reaction centers (Sivalingam *et al.*, 1976; Sinha *et al.*, 1999b), ii) may have a role in osmotic regulation (Oren, 1997) and, iii) they may act as antioxidants and prevent cellular damage resulting from UV-induced production of highly reactive oxygen species. Interestingly, in laboratory-grown cultures of *Chlorogloeopsis* sp., both UV and osmotic stress are able to induce and regulate the synthesis of MAAs (Portwich and Garcia-Pichel, 1999; 2000). In this species, a specific UV-B photoreceptor, probably pterin may mediate MAA induction (Portwich and Garcia-Pichel, 2000).

A terrestrial species of *Nostoc commune* when exposed to high solar radiation was found to possess a UV-A/B absorbing pigment with absorption maxima at 312 and 325 nm (Scherer *et al.*,1988). In this species, an oligosaccharide MAA (OS-MAA), has been shown to be located in the extracellular glycan sheath (Böhm *et al.*, 1995; Ehling-Schulz *et al.*, 1997). It is the first mycosporine so far known to be covalently linked to the oligosaccharide and located outside the cell (Böhm *et al.*, 1995). This pigment besides providing protection by absorbing UV radiation may also provide additional protection by quenching (Sinha and Hader, 2008).

Besides scytonemin and MAAs, occurrence of a few other UV-absorbing compounds, conferring resistance to UV radiation to the organisms, have been reported (Kumar *et al.*, 1996). The shielding role of a brown-coloured pigment (having major absorption peak at 314 nm) from *Scytonema hofmanii* and a phycoerythrin rich pink extract from *Nostoc spongiaeforme,* against UV-B-induced damage, has been reported (Kumar *et al.*, 1996). A strain of *Nostoc* sp. showing resistance to high light intensity and UV radiation has been reported to produce three different UV-absorbing compounds having absorption maxima at 256, 314 and 400 nm (de Chazal and Smith, 1994).

During the last two decades vast information on UV-B absorbing/sun-screening pigments from diverse organisms have been gathered and this has led to the development of a database on photoprotective compounds (Gröniger *et al.*, 2000; Sinha *et al.*, 2007). Readers may find detailed information on the absorption maxima and extinction coefficients of different compounds together with experimental protocols on the web site (http://www. biologie.uni-erlangen.de/botanikI/ index.html).

3.3 DNA Repair

To conserve the genome integrity, nature has provided several repair pathways for proper functioning and survival of all organisms. DNA repair is a highly complex phenomena, mainly depends on the types and location of lesions in the genome within the cell (Essers *et al.*, 2006). The process of DNA repair proceeds via certain specialized repair proteins that scan the genome continuously and encounter the DNA lesions by triggering different repair mechanisms such as photoreactivation, base excision repair (BER), nucleotide excision repair (NER), mismatch repair (MMR), recombinational repair (RER), and some specialized forms of repair system (Carell and Epple, 1998; Häder and Sinha 2005). It has been demonstrated that cyanobacteria in general are very sensitive to UV-B radiation and death is inevitable in the presence of increased level of UV-B flux. Impact of ultraviolet-B radiation in causing the damages to the DNA of the cyanobacterium, *Anabaena* strain BT2 and *Nostoc muscorum* has been investigated (Kumar *et al.*, 2004). Cyanobacteria protect themselves from photodamage by adopting a number of repair mechanisms such as photoreactivation and light-independent NER (Häder and Sinha, 2005).

4. Conclusions

UV-B radiation has now been unequivocally demonstrated to show detrimental effects on a number of vital physiological and biochemical processes of cyanobacteria. UV-B affects pigmentation, disintegrates phycobilisomes, severely inhibits photosynthetic processes and causes inactivation of nitrogenase. Major alterations in general protein profile have also been reported. Formation of dimers between two adjacent pyrimidine bases, *cis*-syn cyclobutane dimers and pyrimidine (6-4) pyrimidone products following UV-B exposure in certain cyanobacteria has also been demonstrated. Till date most of the studies pertaining to UV-B radiation effects on cyanobacteria are based on short term laboratory experiments employing high UV-B flux. Only a few short-term experiments that utilize solar UV-B radiation obtained by using various long-pass filters are available (Donkor *et al.*, 1993). Conclusions about the long-term ecological impacts of ambient natural UV-B radiation cannot be accurately forecast from results obtained through short-term incubation studies.

Other environmental and biological variables should be taken into account while attempting a realistic assessment of the effects of enhanced levels of UV-B radiation. The existence of various protective/repair strategies provide a high level of natural defense against the potential effects of elevated UV-B radiation. Information related to the ecological significance of sun-screening substances such as scytonemin and MAAs and spatial distribution of MAAs from a wide range of organisms are added every day and data available suggest their great potential. There is need to explore the mitigating role of various natural compounds against UV-B radiation.

Acknowledgements

SK is grateful to University Grants Commission, New Delhi, India, for the award of Junior Research Fellowship. This study was partly supported by a research grant sanctioned to AK by the Department of Biotechnology, Govt. of India, New Delhi (BT/PR 1239/AGR./02/065/98).

References

Bebout BM and Garcia-Pichel F (1995) UV-B induced vertical migrations of cyanobacteria in a microbial mat. App Environ Microbiol 61: 4215-4222.

Bhandari R and Sharma PK (2006) High-light-induced changes on photosynthesis, pigments, sugars, lipids and antioxidant enzymes in freshwater (*Nostoc spongiaeforme*) and marine (*Phormidium corium*) cyanobacteria. Photochem Photobiol 82: 702-710.

Blakefield MK and Harris DO (1994) Delay of cell differentiation in *Anabaena aequalis* caused by UV-B radiation and the role of photoreactivation and excision repair. Photochem Photobiol 59: 204-208.

Böhm GA, Pfleiderer W, Böger P and Scherer S (1995) Structure of a novel oligosaccharide-mycosporine-amino acid ultraviolet -A/B sunscreen pigment from the terrestrial cyanobacterium *Nostoc commune.* J Biol Chem 270: 8536-8539.

Björn LO (2007) Stratospheric ozone, ultraviolet radiation, and cryptogams. Biol Conservat 135: 326-333.

Brenowitz S and Castenholz RW (1997) Long-term effects of UV and visible irradiance on natural populations of scytonemin-containing cyanobacterium (*Calothrix* sp.). FEMS Microbiol Ecol 24: 343-352.

Britt AB (1995) Repair of DNA damage induced by ultraviolet radiation. Plant Physiol 108: 891-896.

Buma AGJ, Van Hannen EJ, Roza L, Veldhuis MJW and Gieskes WWC (1995) Monitoring ultraviolet-B induced DNA damage in individual diatom cells by immunofluorescent thymine dimer detection. J Phycol 51: 314-321.

Cheng K, Zhao Y, Du X, Zhang Y, Lan S and Shi Z (2007) Solar radiation driven decay of cyanophage infectivity, and photoreactivation of the cyanophage by host cyanobacteria. Aq Microb Ecol 48: 13-18.

Cadet J, Sage E and Douki T (2005) Ultraviolet radiation-mediated damage to cellular DNA. Mutat Res 571: 3-17.

Callone AI, Carignan M, Montoya NG and Carreto JI (2006) Biotransformation of mycosporine like amino acids (MAAs) in the toxic dinoflagellate *Alaxandrium tamarense.* J Photochem Photobiol B: Biol 84: 204-212.

Campbell D, Eriksson M-J, Öquist G, Gustafsson P and Clarke AK (1998) The cyanobacterium *Synechococcus* resists UV-B by exchanging photosystem II reaction-centre D1 proteins. Proc Nat Acad Sci (USA) 95: 364-369.

Carreto JI, Carignan MO, Daleo G and de Marco SG (1990) Occurrence of mycosporine-like amino acids in the red tide dinoflagellate *Alexandrium excavatum*: UV-protective compounds? J Plankton Res 12: 909-921.

Carell T and Epple R (1998) Repair of UV light induced DNA lesions: a comparative study with model compounds. Eur J Org Chem 1245-1258.

Castenholz RW (1997) Multiple strategies for UV tolerance in cyanobacteria. Spectrum 10: 10-16.

Cockell CS and Knowland J (1999) Ultraviolet radiation screening compounds. Biol Rev 74: 311-345.

Cockell CS and Rothschild LJ (1999) The effects of UV radiation A and B on diurnal variation in photosynthesis in three taxonomically and ecologically diverse microbial mats. Photochem Photobiol 69: 203-210.

Conde FR, Churio MS and Previtali CM (2007) Experimental study of the excited-state properties and photostability of the mycosporine-like amino acid palythine in aqueous solution. Photochem Photobiol Sci 6: 669-674.

Crutzen PJ (1992) Ultraviolet on the increase. Nature 356: 104-105.

de Chazal NM and Smith GD (1994) Characterization of a brown *Nostoc* sp. from Java that is resistant to high light intensity and UV. Microbiology 140: 3183-3189.

Döhler G, Biermann I and Zink J (1986) Impact of UV-B radiation on photosynthetic assimilation of ^{14}C-bicarbonate and inorganic ^{15}N-compounds by cyanobacteria. Z Naturf 41: 426-432.

Donkor VA, Damian HAK and Häder D-P (1993) Effects of tropical solar radiation on the motility of filamentous cyanobacteria. FEMS Microbiol Ecol 12: 143-148.

Dunlap WC, Williams DM, Chalker BE and Banaszak AT (1989) Biochemical photoadaptations in vision: UV-absorbing pigments in fish eye tissues. Comp Biochem Physiol 93: 601-607.

Dunlap WC and Shick JM (1998) Ultraviolet radiation-absorbing mycosporine-like amino acids in coral reef organisms: a biochemical and environmental perspective. J Phycol 34: 418-430.

Ehling-Schulz M, Bilger W and Scherer S (1997) UV-B-induced synthesis of photoprotective pigments and extracellular polysaccharides in the terrestrial cyanobacterium *Nostoc commune*. J Bacteriol 179: 1940-1945.

Ehling-Schulz M, Schulz S, Wait R, Görg A and Scherer S (2002) The UV-B stimulon of the terrestrial cyanobacterium *Nostoc commune* comprises early shock proteins and late acclimation proteins. Mol Microbiol 46: 827-843.

Eker AP, Kooiman P, Hessels JKC and Yasui A (1990) DNA photoreactivating enzyme from the cyanobacterium *Anacystis nidulans*. J Biol Chem 14: 8009-8015.

Essers J, Vermeulen W and Houtsmuller A B (2006). DNA damage repair: any time, anywhere? Curr Opin Cell Biol 18: 240-246.

Fleming ED and Castenholz RW (2007) Effects of periodic desiccation on the synthesis of the UV-screening compound, scytonemin, in cyanobacteria. Environ Microbiol 9: 1448-1455.

Friedberg EC, Walker GC and Siede W (1995) DNA repair and mutagenesis. American Society of Microbiology Press, Washington, DC.

Garcia-Pichel F and Castenholz RW (1991) Characterization and biological implications of scytonemin, a cyanobacterial sheath pigment. J Phycol 27: 395-409.

Garcia-Pichel F and Castenholz RW (1993) Occurrence of UV-absorbing mycosporine-like compounds among cyanobacterial isolates and an estimation of their screening capacity. Appl Environ Microbiol 59: 163-169.

Gerber S and Häder DP (1995) Effects of enhanced solar irradiation on chlorophyll fluorescence and photosynthetic oxygen production of five species of phytoplankton. FEMS Microbiol Ecol 16: 32–42.

Gröniger A, Sinha RP, Klisch M and Häder D-P (2000) Photoprotective compounds in cyanobacteria, phytoplankton and macroalgae- a database. J Photochem Photobiol B: Biol 58: 115-122.

Häder D-P and Worrest RC (1991) Effects of enhanced solar radiation on aquatic ecosystems. Photochem Photobiol 53: 717-725.

Häder D-P and Sinha RP (2005) Solar ultraviolet radiation-induced DNA damage in aquatic organisms: potential environmental impact. Mutat Res 571: 221-233.

Häder D-P, Kumar HD, Smith RC and Worrest RC (2007) Effects of solar UV radiation on aquatic ecosystems and interactions with climate change. Photochem Photobiol Sci 6: 267-285.

He Y-Y and Häder D-P (2002) Involvement of reactive oxygen species in the UV-B damage of the cyanobacterium, *Anabaena* sp. Photochem Photobiol 66: 73-80.

Helbling EW, Gao K, Ai H, Ma Z and Villafañe VE (2006) Differential responses of *Nostoc sphaeroides* and *Arthrospira platensis* to solar ultraviolet radiation exposure. J Appl Phycol 18: 57-66.

Huang L, McCluskey MP, Ni H and LaRossa RA (2002) Global gene expression profiles of the cyanobacterium *Synechocystis* sp. strain PCC 6803 in response to irradiation with UV-B and white light. J Bacteriol 184: 6845-6858.

Karentz D, Cleaver JE and Mitchell DL (1991a) Cell survival characteristics and molecular responses of Antarctic phytoplankton to ultraviolet-B radiation. J Phycol 27: 326-341.

Karentz D, Mceuen FS, Land MC and Dunlap WC (1991b) Survey of mycosporine-like amino acid compounds in Antarctic marine organism: potential protection from ultraviolet exposure. Mar Biol 108: 157-166.

Kerr JB and McElroy CT (1993) Evidence for large upward trends of ultraviolet-B radiation linked to ozone depletion. Science 262: 1032-1034.

Kumar A, Tyagi MB, Srinivas G, Singh N, Kumar HD, Sinha RP and Häder D-P (1996) UV-B shielding role of $FeCl_3$ and certain cyanobacterial pigments. Photochem Photobiol 64: 321-325.

Kumar A, Tyagi MB, Jha PN, Srinivas G and Singh A (2003) Inactivation of cyanobacterial nitrogenase after exposure to ultraviolet-B radiation. Curr Microbiol 46: 380-384.

Kumar A, Tyagi MB and Jha PN (2004) Evidences showing ultraviolet-B radiation-induced damage of DNA in cyanobacteria and its detection by PCR assay. Biochem Biophy Res Commun 318: 1025-1030.

Kumari S, Rastogi RP, Singh KL, Singh SP and Sinha RP (2008) DNA damage: detection strategies. EXCLI J 7: 44-62

Levine E and Thiel T (1987) UV-inducible DNA repair in the cyanobacteria *Anabaena* sp. J Bacteriol 169: 3988-3993.

Madronich S, Mckenzie RL, Björn LO and Caldwell MM (1998) Changes in biologically active ultraviolet radiation reaching the Earth's surface. J Photochem Photobiol B: Biol 46: 5-19.

Mate Z, Sass L, Szekeres M, Vass I and Nagy F (1998) UV-B-induced differential transcription of *psbA* genes encoding the D1 protein of photosystem II in the cyanobacterium *Synechocystis* 6803. J Biol Chem 273: 17439-17444.

McClintock JB and Karentz D (1997) Mycosporine-like amino acids in 38 species of subtidal marine organisms from McMurdo Sound, Antarctica. Antarc Sci 9: 392-398.

Mitchell DL and Karentz D (1993) The induction and repair of DNA photodamage in the environment. In *Environmental UV Photobiology* (eds. Young AR, BjÖrn L, Moan J and Nultsch W), Plenum Press, London, pp. 345-377.

Newton JW, Tyler DD and Slodki ME (1979) Effect of ultraviolet-B (280-320 nm) radiation on blue-green algae (cyanobacteria): possible biological indicators of stratospheric ozone depletion. Appl Environ Microbiol 37: 1137-1141.

Oren, A. (1997) Mycosporine-like amino acids as osmotic solutes in a community of halophilic cyanobacteria. *Geomicro. J.* 14: 231-240.

Oren A and Gunde-Cimerman N (2007) Mycosporines and mycosporine-like amino acids: UV protectants or multipurpose secondary metabolites? FEMS Microbiol Lett 269: 1-10.

Portwich A and Garcia-Pichel F (1999) Ultraviolet and osmotic stresses induce and regulate the synthesis of mycosporines in the cyanobacterium *Chlorogloeopsis* PCC 6912. Arch Microbiol 172: 187-192.

Portwich A. and Garcia-Pichel F (2000) A novel prokaryotic UV-B photoreceptor in the cyanobacterium *Chlorogloeopsis* PCC 6912. Photochem Photobiol 71: 493-498.

Prasad SM and Zeeshan M (2004) Effect of UV-B and monocrotophos, singly and in combination, on photosynthetic activity and growth of non-heterocystous cyanobacterium *Plectonema boryanum*. Environ Exp Bot 52: 175-184.

Quesada A and Vincent WF (1997) Strategies of adaptation by Antarctic cyanobacteria to ultraviolet radiation. Eur J Phycol 32: 335-342.

Quesada A, Vincent WF and Lean-David RS (1999) Community and pigment structure of arctic cyanobacterial assemblages: the occurrence and distribution of UV-absorbing compounds. FEMS Microbiol Ecol 28: 315-323.

Rath J and Adhikary SP (2007) Response of the estuarine cyanobacterium *Lyngbya aestuarii* to UV-B radiation. J Appl Phycol 19: 529-536.

Richter P, Krywult M, Sinha RP and Häder D-P (1999) Calcium signals from heterocysts of *Anabaena* sp. after UV irradiation. J Plant Physiol 154: 137-139.

Sass L, Spetea C, Mate Z, Nagy F and Vass I (1997) Repair of UV-B induced damage of photosystem II via *de novo* synthesis of the D_1 and D_2 reaction centre subunits in *Synechocystis* sp. PCC6803. Photosyn Res 54: 55-62.

Scherer S, Chen TW and Böger P (1988) A new UV-A/B protecting pigment in the terrestrial cyanobacterium *Nostoc commune.* Plant Physiol 88: 1055-1057.

Shibata H, Baba K and Ochiai H (1991) Near UV irradiation induces shock proteins in *Anacystis nidulans* R-2; possible role of oxygen. Plant Cell Physiol 32: 771-776.

Singh RN (1961) Role of blue-algae in nitrogen economy of Indian agriculture, Indian Council of Agricultural Research, New Delhi, India.

Singh SP, Kumari S, Rastogi RP, Singh KL and Sinha RP (2008a) Mycosporine-like amino acids (MAAs): chemical structure, biosynthesis and significance as UV-absorbing/screening compounds. Indian J Exp Biol 46: 7-17.

Singh SP, Klisch M, Sinha RP and Häder D-P (2008b) Effects of abiotic stressors on synthesis of the mycosporine-like amino acid shinorine in the cyanobacterium *Anabaena variabilis* PCC 7937. Photochem Photobiol 84: 1500-1505.

Singh SP, Klisch M, Häder D-P and Sinha RP (2008c) Role of various growth media on shinorine (mycosporine-like amino acid) concentration and photosynthetic yield in *Anabaena variabilis* PCC 7937. World J Microbiol Biotechnol 24: 3111-3115.

Singh SP, Sinha RP, Klisch M and Häder D-P (2008d) Mycosporine-like amino acids (MAAs) profile of a rice-field cyanobacterium *Anabaena doliolum* as influenced by PAR and UV. Planta, 229: 225-233.

Sinha RP, Kumar HD, Kumar A and Häder D-P (1995) Effects of UV-B irradiation on growth, survival, pigmentation and nitrogen metabolism enzymes in cyanobacteria. Acta Protozool 34: 187-192.

Sinha RP, Singh N, Kumar A, Kumar HD, Häder M and Häder D-P (1996) Effect of UV radiation on certain physiological and biochemical processes in cyanobacteria. J Photochem Photobiol B: Biol 32: 107-113.

Sinha RP, Klisch M, Vaishampayan A and Häder D-P (1999a) Biochemical and spectroscopic characterization of the cyanobacterium *Lyngbya* sp. inhabiting mango (*Mangifera indica*) trees: presence of an ultraviolet-absorbing pigment, scytonemin. Acta Protozool 38: 291-298.

Sinha RP, Klisch M and Häder D-P (1999b) Induction of a mycosporine-like amino acid (MAA) in the rice-field cyanobacterium *Anabaena* sp. by UV irradiation. J Photochem Photobiol B: Biol 52: 59-64.

Sinha RP, Dautz M and Häder D-P (2001a) A simple and efficient method for the quantitative analysis of thymine dimers in cyanobacteria, phytoplankton and macroalgae. Acta Protozool 40: 187-195.

Sinha RP, Klisch M, Helbling EW and Häder D-P (2001b) Induction of mycosporine-like amino acids (MAAs) in cyanobacteria by solar ultraviolet-B radiation. J Photochem Photobiol B: Biol 60: 129-135.

Sinha RP, Singh SP and Häder D-P (2007) Database on mycosporines and mycosporine-like amino acids (MAAs) in fungi, cyanobacteria, macroalgae, phytoplankton and animals. J Photochem Photobiol B: Biol 89: 29-35.

Sinha RP and Häder D-P (2008) UV-protectants in cyanobacteria. Plant Sci 174: 278-289.

Sivalingam PM, Ikawa T and Nisizawa K (1976) Physiological roles of a substance 334 in algae. Bot Mar 19: 9-21.

Torres A, Enk CD, Hochberg M and Srebnik M (2006) Porphyra-334, a potential natural source for UV-A protective sunscreens. Photochem Photobiol Sci 5: 432-443.

Tyagi R, Srinivas G, Vyas D, Kumar A and Kumar HD (1992). Differential effect of ultraviolet-B radiation on certain metabolic processes in a chromatically adapting *Nostoc*. Photochem Photobiol 55: 401-407.

Vincent WF and Roy S (1993) Solar ultraviolet-B radiation and aquatic primary production: damage, protection and recovery. Environ Rev 1: 1-12.

Wang G, Chen K, Chen L, Hu C, Zhang D and Liu Y. (2008) The involvement of the antioxidant system in protection of desert cyanobacterium *Nostoc* sp. against UV-B radiation and the effects of exogenous antioxidants. Ecotoxicol Environ Safety 69: 150-157.

Weatherhead EC and Anderson SB (2006) The search for signs of recovery of the ozone layer. Nature 444: 39-45.

Wilson MI, Ghosh S, Gerhardt KE, Holland N, Sudhakar Babu T, Edeeman M, Dumbroff EB.and Greenberg BM (1995) *In vivo* photomodification of ribulose-1,5-bisphosphate carboxylase/oxygenase holoenzyme by ultraviolet-B radiation. Plant Physiol 109: 221-229.

Wright DJ, Smith SC, Joardar V, Scherer S, Jervis J, Warren A, Helm RF and Potts M (2005) UV irradiation and desiccation modulate the three dimensional extracellular matrix of *Nostoc commune* (cyanobacteria). J Biol Chem 280: 40271-40281.

Zhang L, Li L and Wu Q (2007) Protective effects of mycosporine-like amino acids of *Synechocystis* sp. PCC 6803 and their partial characterization. J Photochem Photobiol B: Biol 86: 240-245.

Algal Biotechnology: New Vistas (2010) *Pages* **111–132**
Editor: **Mihir Kumar Das**
Published by: **DAYA PUBLISHING HOUSE, NEW DELHI**

Chapter 7

Growth Response of Cyanobacteria from Sandy Soil and Mine Waste Burdened Soil to Different Environmental Variables

Pramila Tripathy and S.P. Adhikary*

P.G. Department of Botany, Utkal University, Bhubaneswar – 751 004, Orissa, India

ABSTRACT

The soils of Ramchandi coastal area and the mine waste burdened soils of Damonjodi showed wide variation in their nutrient content, soil pH and salinity levels. Hence the influence of these factors including the effect of exogenous carbon sources on the growth of four selected cyanobacteria isolated from these sites *e.g. Plectonema puteale* UU 62246, *Westiellopsis prolifica* UU 62247, *Microcolus vaginatus* UU 61243 and *Nostoc paludosum* UU 61240 was investigated under light and dark conditions. *M. vaginatus* and *N. paludosum* were most tolerant to a wider pH range, *W. prolifica* preferred alkaline pH where as *P. puteale* was very sensitive to changes of the pH to acidic or alkaline range. In general they fall into three groups with regard to their response to pH; the organisms from sandy soils of

* *Present Address*: Centre for Biotechnology, Visva-Bharati, Santiniketan – 731 235, West Bengal, India; E-mail: adhikarysp@viswa-bharti.ac.in

Ramchandi preferred alkaline pH, and those isolated from mine waste burdended soils tolerated a wider pH range. However, upon their growth they changed the pH of the culture medium towards alkaline range, and those from alkaline side decreased the pH close to 9 suggesting their influence on soil conditioning making it to buffer between 7 and 9 at which it is conducive for cyanobacterial colonization. Further it was observed that the organism which occurred more prominently penetrating several centimeters below the top soil of the coastal zone and those having distinct sheath layer around their trichome tend to be comparatively higher tolerant to NaCl.

All the four organisms showed wide variation in their capability to grow mixotrophically and/or heterotrophically in the dark. *Plectonema puteale* which did not penetrate deep into the soil, and was always grew on the surface of the top soil did not utilize exogenous organic carbon sources for its photoheterotrophic or chemoheterotrophic growth, however, was a mixotrophic form. *M. vaginatus* also did not grow photo and/or chemoheterotrophically utilizing all the sugars tested, but when a specific sugar (15 mM sucrose) was supplemented, it could grow in the dark. However, this mat forming organism grew prominently mixotrophically utilizing almost all the exogenous sugars. These results suggest that the cyanobacteria isolated from carbon sparse sandy soils as well as soils from mine waste burdened areas can colonise once little organic carbon is available, bind the soil with their stratified mucilaginous sheath layers, aggregate the soil making it productive.

Keywords: *Cyanobacteria, Growth response, Heterotrophy, Light intensity, pH, Salinity*

1. Introduction

Soil is a dynamic system comprising of physical, chemical and biological components. While physical and chemical status of soil has a bearing on the fertility level of the soil, productivity of soil largely depends on the microbial population occur therein. The microorganisms occur in soil in specific microenvironments as it is found in a particular niche. Such specialized micro-environments vary with the type of soil, and those occur in different climatic regimes. Usually the upper surface of soil is a photic layer and aphotic zone lies below the oxidized layer, which could be a plough layer or sub-soil.

Several factors influence the occurrence of cyanobacteria in soils and sub-soil layers which include principally the chemical composition and the light regimes. The soils from Damonjodi mine waste dumped areas are in near alkaline range, contain adequate quantity of nutrients including N and P. To the contrary, the top layer of soils of Ramchandi is principally sandy with low nutrient, lesser water holding capacity and the pH is in near neutral range. Further, at both the sites, cyanobacteria forms occur at the top surface layer as well as could penetrate deep into the sub-surface layer where hardly any light is available. Also almost all the study sites at Ramchandi and Damonjodi mine sites are poor in water availability; hence the organisms occurring there are principally drought tolerant forms. Thus, the growth response of four different cyanobacteria of which *Microcoleus vaginatus* principally grow in Damonjodi mine waste dumped soils, *Plectonema puteale* and

Westiellopsis prolifica occur at the study sites of Ramchandi, and *Nostoc paludosum* occur at both the locations were selected for further experiments, and their growth response to different pH, salinity levels, growth pattern at different light intensities and in darkness, and also their capability to utilize exogenous organic carbon sources for mixotrophic and chemoheterotrophic growth was analyzed.

Light plays an important role in determining cyanobacterial growth in the nature. Cyanobacteria are principally photoautotrophic microorganisms. Very few are able to live facultative heterotrophically in the dark, however, many are able to live photoheterotrophically in dim-light. Cyanobacteria also show definite responses to the quality and quantity of light. In the heavily shaded soils, cyanobacteria were found to be abundant as encountered in enrichment cultures (Singh, 1961; Reynaud and Roger, 1978).

pH determines the solubility of CO_2 and minerals in the culture and directly or indirectly influences the metabolism of the cyanobacteria. The branched cyanobacterium *Mastigocladus laminosus* exhibited maximum branching at pH 9.5 and maximal growth at pH 8.5 (Sahu and Adhikary, 1987). There are reports that in the culture of cyanobacterial species, pH of the culture medium always tend to change to the alkaline range (Singh, 1974; Swain and Adhikary, 1991). Such a change is attributed to the influence of extracellular products of cyanobacteria which have been shown to be composed of amino acids, polypeptides, carbohydrates, hormones and vitamins.

Soil salinity, which is on the rise globally, poses a major hindrance to the realization of actual yield potential of most crop plants (Boyer, 1972). Apart from the natural causes of salinization like seawater, inefficient water management practices also result in secondary salinization of farm lands (Flowers *et al.*, 1977). Sodium (Na^+) is the predominant soluble cation in most saline soils and water, particularly in coastal areas. Many cyanobacteria exhibit considerable tolerance to salt (NaCl). Therefore the occurrence of marine and brackish water forms is by no means rare (Kumar and Kaushik, 1994). A number of earlier reports indicate several salt triggered events, *e.g.* (*i*) stimulation of photosynthesis, (*ii*) accumulation of sugars, (*iii*) reorganization of photosynthetic apparatus, and (*iv*) modification of cell surface volume (Blumwald and Tel-Or, 1982a,b).

Several mechanisms have been put forth for salinity tolerance in cyanobacteria. These include the curtailment of Na^+ influx (Apte *et al.*, 1987) or synthesis and accumulation of osmoprotective compounds to achieve equilibrium of osmotic potential (Borowitzka *et al.*, 1980; Reed *et al.*, 1986; Waar *et al.*, 1984). Salinity stresses in cyanobacteria are also relieved by a variety of exogenous saccharides and alcohols. A large number of amino acids protect *Calothrix brevissima* against salt stress (Kumar and Kaushik, 1994). Extracellular saccharides and amino acids secreted to the medium also known to provide protection to rice field cyanobacteria to NaCl stress (Padhi *et al.*, 1998).

Earlier reports have shown that supplementation of 0.4M NaCl decreased biomass accumulation, chlorophyll-*a* content and cell protein levels of *Calothrix brevissima* isolated from saline sodic soils where as *Anabaena* sp. of freshwater tolerated only up to 0.1 M of NaCl (Kumar and Kaushik, 1994). The unicellular cyanobacterium *Microcystis firma* tolerated up to 1M NaCl but another unicellular species *Synechocystis aquatilis* showed salt tolerance up to 0.3 M NaCl. The growth of N_2-fixing cyanobacterium *Nostoc muscorum* was completely inhibited by 0.4M NaCl (Blumwald and Tel-Or, 1982a). All these results showed the existence of a varying degree of tolerance among certain species towards their tolerance to NaCl.

Many workers have found out that most cyanobacteria are obligate phototrophs and apparently cannot grow heterotrophically in the dark (Hoare *et al.*, 1967; Smith *et al.*, 1967; Holm-Hansen, 1968). However, reports of several published work have showed that a variety of cyanobacteria can grow heterotrophically at the expense of various exogenous carbon sources (Smith, 1973, 1982; Wolk, 1973; Khoja and Whitton, 1975; White and Shilo, 1975; Adhikary and Pattnaik, 1979; Sahu and Adhikary, 1981a,b, 1982; Banerjee and Kumar, 1987; Barnum and Gendel, 1987; Bastia *et al.*, 1993).

Four different modes of growth of cyanobacteria have been reported *e.g.* (1) autotrophy *i.e.* growth solely from CO_2 and inorganic matters in light, (ii) mixotrophy in which CO_2 and exogenous organic compounds are assimilated simultaneously, (iii) Photoheterotrophy *i.e.* the growth in the light on an organic substrate in the absence of net CO_2 fixation and (iv) Chemoheterotrophy *i.e.* the growth on an organic substrate in complete darkness. Photoheterotrophic growth condition was achieved in the laboratory by exclusion of CO_2 from the medium (Ingram *et al.*, 1973) by blocking assimilation of CO_2 using herbicide DCMU (3-3(3-4-dichlorophyl 1-1-dimethyl urea) which inhibits PSII, noncyclic electron flow and production of NADPH (Rippka, 1972; Stanier, 1973) or by reducing net CO_2 fixation by lowering the incident light intensity (Van Baalen *et al.*, 1971; Sahu and Adhikary, 1981a, b). In the two modes of heterotrophic growth *i.e.* Photoheterotrophy and chemoheterotrophy, increase in the culture density is totally dependent on exogenous organic compounds which is distinct from mixotrophic growth in which CO_2 and organic carbon are assimilated simultaneously.

In the present work, growth response of 4 different species of cyanobacteria, *e.g. Nastoc paludosum, Microcoleus vaginatus, Plectonema puteale* and *Westiellopsis prolifica* at different pH levels of the culture, to different salinities (different concentration of NaCl), grown at different light intensity and darkness, and also their capability to grow mixotrophically, photoheterotrophically and heterotrophically in the dark in presence of several exogenous carbon sources was studied.

2. Materials and Methods

Four species of cyanobacteria *e.g. Nostoc paludosum* UU1240, *Microcoleus vaginatus* UU61243, *Plectonema puteale* UU62246 and *Westiellopsis prolifica* UU62247 isolated from the Ramchandi coastal area and Damonjodi mine waste burdened site were selected for experiments basing on their colonization potential in the soils of both the areas. Of these *Microcoleus vaginatus* was the dominant organism occurred in almost all the sites in the soils up to 20 cm depth. *Plectonema puteale* and *Westiellopsis prolifica* were the widely occurring organisms at the sandy soils of Ramchandi and its surrounding area. The cyanobacteria *Nostoc paludosum* however, occurred in the soils of all the study sites at both the sandy soils and mine waste burdened areas.

Experiments were conducted in test tubes stoppered with non-absorbent cotton wool plug containing 10 ml of BG11 medium (Rippka *et al.*, 1979). KNO_3 was omitted from the medium for culturing the heterocystous forms, *e.g.*, *Nostoc paludosum* and *Westiellopsis prolifica*. Different pH levels (5, 6, 7, 8,9, 10 and11) of the culture was obtained by addition of 0.1N HCl and 0.1N NaOH to the medium aseptically. The absorbance of the culture suspension at 750 nm of each organism soon after inoculation into the experimental tubes was 0.2. The cultures were incubated for 10 days under continuous light of 7.5 W/m^2 (except for experiment requiring continuous darkness) and then harvested. On the 10th day (harvest day), pH of the culture medium was also recorded.

To study the response of cyanobacteria to different concentrations of NaCl, stock solution of the salt was prepared in distilled water, sterilized separately and then added to the culture medium to obtain the desired concentration (0.025, 0.05, 0.075, 0.1, 0.2, 0.3 M). Equal amount of each organism was inoculated (absorbance of the culture suspension at 750 nm soon after inoculation was 0.2), incubated up to 10 days and then harvested.

The organisms were also grown at different light intensities *e.g.* at 5 W/m^2, 7.5 W/m^2, 12.5 W/m^2 and in the dark up to 10 days and then harvested for assessing their growth. For studying the autotrophic, photoheterotrophic, mixotrophic and chemoheterotrophic growth capability, various sugars, *viz.* glucose, fructose, sucrose, galactose, mannose (15 mM each) and DCMU (10^{-5}M) were prepared, autoclaved separately and subsequently added aseptically as per the design of the experiment to the culture medium to obtain the final concentration of 15 mM and 10^{-5}M, for the exogenous sugars and DCMU respectively. Cultures were incubated under continuous light or in complete darkness and harvested on the 10th day. For all these experiments the inoculum size of the organisms was kept constant.

Growth was estimated by measuring the absorbance of the homogenized suspension of the culture at 750 nm in a Systronics 105 spectrophotometer. Chlorophyll-*a* was extracted in 90 per cent (v/v) methanol and the quantity was estimated following Mackinney (1941).

3. Results and Discussion

3.1 Growth Response of Cyanobacteria from Different Soils to Various pH

Growth response (absorbance of the culture at 750 nm and chlorophyll-*a* content, µg/10 ml culture) of 4 different cyanobacteria at different pH levels of the culture was studied. The organisms were grown at pH 5, 6, 7, 8, 9, 10 and 11 of the culture for a period of 10 days under continuous light and then harvested. Growth pattern of the organisms showed wide variations among themselves with response to different pH. *Microcoleus vaginatus* was found to be tolerating over a wider pH range from 5-11. The decrease of growth was only 16 per cent at pH 5 and 18 per cent at pH 11 in comparison to the control (neutral pH) (Table 7.1). The decrease of chlorophyll-*a* content of the

Table 7.1: Growth Response (Absorbance of culture suspension at 750 nm) of 4 Different Cyanobacteria at Different pH Levels of the Culture

Organism	pH Levels						
	5	*6*	*7 (Control)*	*8*	*9*	*10*	*11*
Nostoc paludosum UU61240	0.29 ± 0.023 (–42.0)	0.32 ± 0.025 (–36.0)	0.50 ±.036	0.46 ± 0.041 (–8.0)	0.42 ± 0.035 (–16.0)	0.40 ± 0.036 (–20.0)	0.30 ± 0.023 (–40.0)
Microcoleus vaginatus UU61243	0.60 ± 0.046 (–16.7)	0.63 ± 0.051 (–12.5)	0.72 ±.041	0.65 ± 0.055 (–9.8)	0.62 ± 0.051 (–13.9)	0.60 ± 0.058 (–16.7)	0.59 ± 0.051 (–18.1)
Plectonema puteale UU62246	0.26 ± 0.021 (–50.0)	0.40 ± 0.03 (–23.1)	0.52 ±.046	0.42 ± 0.032 (–19.3)	0.37 ± 0.028 (–28.8)	0.29 ± 0.022 (–44.2)	0.28 ± 0.019 (–46.2)
Westiellopsis prolifica UU62247	0.25 ± 0.018 (–52.0)	0.36 ± 0.028 (–30.8)	0.52 ±.061	0.46 ± 0.038 (–11.6)	0.45±0.032 (–13.5)	0.43 ± 0.031 (–17.3)	0.42 ± 0.026 (–19.2)

Values represent mean of three determinations ± S.D.

Absorbance of culture suspension at 750 nm soon after inoculation of the organisms was 0.2.

Values in parenthesis indicate percent decrease over control. Cultures were incubated at 25±1°C under continuous light at an intensity of 7.5 w/m² for 10 days and then harvested.

organisms at pH 5 and 11 was less than 50 per cent in comparison to control (Table 7.2). *Plectonema puteale,* the dominant organism in the sandy soils of Ramchandi though occurred at a pH range of 5-11, with increase of the acidity and alkalinity of the culture, the growth as well as the chlorophyll-*content* was decreased proportionately with the increase or decrease of pH levels than the control. To the contrary *Westiellopsis prolifica* isolated from the soils of same site grew well tolerating the alkaline pH up to pH 11; the decrease of growth was 19 per cent and of chlorophyll-*a* was 21 per cent at pH 11 in comparison to control. However, adverse effect of acidic pH on the growth of this organism was more pronounced (Tables 7.1 and 7.2). *Nostoc paludosum* which occurred at all the study sites of both the areas with different soil composition tolerated a wider range of pH next to *Microcoleus vaginatus.* In general, it was observed that *Plectonema puteale* was the most sensitive organism to fluctuation in pH, *Westiellopsis prolifica* preferred alkaline pH to grow, whereas *Microcoleus vaginatus* was the most tolerant of all occurring in a wider pH range (Tables 7.1 and 7.2). Thus, the organisms can be categorized in to three groups with regard to their response to pH; the organisms from sandy soils of Ramchandi either preferred alkaline pH or were less tolerant, whereas those from Damonjodi mine waste soils tolerated a wider pH range. Previous studies have shown that among the several soil properties, pH is the most important in determining composition of cyanobacterial flora in natural ecosystems. Positive correlation between pH and occurrence of cyanobacteria has been reported by several authors (Roger and Reynaud, 1977; Roger *et al.*, 1987).

In culture media the optimal pH for cyanobacterial growth ranges from 7-10 and the lower limit was about 6.5 (Holm-Hansen, 1968). Under natural conditions cyanobacterial growth was higher in neutral to alkaline soils, although their occurrence was reported in a wide range of acidic and alkaline pH. There are reports on the presence of certain cyanobacteria in soils with pH values between 5 and 6 (Prasad *et al.*, 1978) and lower than 5 (Aiyer, 1965). *Aulosira fertilissima* and *Calothrix brevissima* have been found to be ubiquitous in rice fields of Kerala showing pH values varying from 3.5 to 6.5. Presence of *Nostoc muscorum* and *Anabaena torulosa* has been reported in soils with pH ranging from 5-7. However, in general cyanobacteria occur in neutral to alkaline soil in preference, and further, upon their growth in soil they secrete extracellular substances which leads to the increase in pH towards the alkaline range. Such a result was also encountered in the present work where pH of the medium of all the four organisms after 10 days of culture changed to alkaline side (Table 7.3). More significantly in *Microcoleus vaginatus* and *Plectonema puteale,* isolated from 2 different soils showed that the pH of acidic media even from 5 was increased to 9 and above within 10 days of culture. Similarly, the pH of the culture in the alkaline range from 10 to 11 was decreased to a value close to 9 after 10 days of culture of all the organisms except *Nostoc paludosum*. These results showed that cyanobacterial species which occur in the soils of the study sites change the physico-chemical properties of the soil through their metabolic activities.

3.2 Growth Response of Cyanobacteria from Different Soils to Salinity

Growth response in terms of absorption of culture at 750 nm and chlorophyll-*a* content of 4 different cyanobacteria to different NaCl levels (0.025, 0.05, 0.075, 0.1, 0.2,

Table 7.2: Chlorophyll–*a* Content (mg/10 ml culture) of 4 Different Cyanobacteria at Different pH Levels of the Culture

Organism	*pH Levels*						
	5	*6*	*7 (Control)*	*8*	*9*	*10*	*11*
Nostoc paludosum	8.57 ± 0.63	10.71 ± 0.98	18.56±1.08	17.13 ±0.98	15.7 ±1.03	10.71 ±0.95	9.28 ±0.66
UU61240	(–53.8)	(–42.3)		(–7.7)	(–15.41)	(–42.3)	(–50.0)
Microcoleus vaginatus	21.42 ± 0.94	29.99 ± 1.11	37.13±1.69	30.7 ±2.31	28.56 ±1.36	25.7 ±1.21	24.27 ±2.14
UU61243	(–42.3)	(–9.23)		(–17.32)	(–23.1)	(–30.8)	(–34.64)
Plectonema puteale	5.71 ± 0.52	11.42 ± 0.96	17.14±1.01	13.57 ±1.12	12.14 ±0.94	8.57±0.63	7.85 ±0.91
UU62246	(–66.7)	(–33.38)		(–20.83)	(–29.18)	(–50.0)	(–54.2)
Westiellopsis prolifica	5.71 ± 0.42	9.28 ± 0.64	12.85±0.97	12.14 ±1.63	11.42 ±0.88	10.0 ±0.87	9.28 ±0.74
UU62247	(–55.6)	(–17.8)		(–5.5)	(–11.13)	(–22.18)	(–27.8)

Values represent mean of three determinations ± S.D.

Chlorophyll–*a* content of the initial inoculum of *Nostoc paludosum* = 7.14 mg; *Microcoleus vaginatus* = 5.0 mg;

Plectonema puteale = 9.28 mg; *Westiellopsis prolifica* = 4.28 mg

Values in parenthesis indicate percent in increase (+) or decrease (–) over control. Cultures were incubated at 25±1°C under continuous light at an intensity of 7.5 w/m^2 for 10 days and then harvested.

Table 7.3: Change in the pH of the Culture Medium After Cultivation of Different Organisms at Different pH for 10 Days at 7.5 w/m² Light Intensity

Organism	*Strain No.*	*Initial pH of Medium*	*PH of the medium After 10 Days of Culture*
Nostoc paludosum	UU61240	5	6.56 ± 0.63
		6	7.70 ± 0.79
		7	9.14 ± 0.47
		8	9.44 ± 0.62
		9	9.41 ± 0.68
		10	9.88 ± 0.79
		11	10.15 ± 0.91
Microcoleus vaginatus	UU61243	5	9.60 ± 0.71
		6	9.44 ± 0.82
		7	9.86 ± 0.94
		8	9.80 ± 0.91
		9	9.76 ± 0.86
		10	9.64 ± 0.63
		11	9.51 ± 0.51
Plectonema puteale	UU62246	5	9.38 ± 0.69
		6	9.27 ± 0.73
		7	9.26 ± 0.75
		8	9.36 ± 0.69
		9	9.44 ± 0.72
		10	9.95 ± 0.54
		11	9.72 ± 0.58
Westiellopsis prolifica	UU62247	5	6.90 ± 0.43
		6	7.98 ± 0.48
		7	8.99 ± 0.51
		8	8.87 ± 0.54
		9	8.96 ± 0.48
		10	8.98 ± 0.55
		11	9.23 ± 0.46

Values represent mean of three determinations ± S.D.

0.3 M) in culture was studied. Low concentration of NaCl up to 0.025 M showed increased growth of *Westiellopsis prolifica* in culture but with further increase of NaCl in the medium growth as well as chlorophyll-*a* content of the organism was gradually decreased, and at 0.3 M, growth and chlorophyll-*a* was decreased to 57.4 and 92.3 per cent respectively (Table 7.4 and 7.5). Since the organism was isolated from the sandy soils of coastal area, possibly it has been adopted to grow at low concentrations

Table 7.4: Growth Response (Absorbance of culture suspension at 750 nm) of 4 Different Cyanobacteria at Different Salinity Levels in Culture

Organism	Control	NaCl levels (M)					
		0.025	0.05	0.075	0.1	0.2	0.3
Nostoc paludosum	0.64 ± 0.055	0.46 ± 0.051	0.31 ±0.028	0.2 ± 0.031	0.18 ± 0.016	0.15 ± 0.011	0.10 ± 0.004
UU61240		(–28.1)	(–51.5)	(–68.7)	(–71.8)	(–76.6)	(–84.4)
Microcoleus vaginatus	0.56 ± 0.046	0.56 ± 0.047	0.49 ±0.047	0.44 ± 0.042	0.42 ± 0.046	0.36 ±0.031	0.31 ± 0.003
UU61243		(±0)	(–12.5)	(–21.5)	(–25.0)	(–35.7)	(–44.7)
Plectonema puteale	0.44 ± 0.039	0.44 ± 0.028	0.36 ±0.026	0.32 ± 0.031	0.2 ± 0.022	0.15 ± 0.012	0.11 ± 0.005
UU62246		(±0)	(–18.2)	(–27.3)	(–45.5)	(–65.9)	(–75.0)
Westiellopsis prolifica	0.54 ± 0.041	0.60 ± 0.056	0.48 ±0.052	0.46 ± 0.048	0.44 ± 0.044	0.28 ± 0.029	0.23 ± 0.016
UU62247		(+11.0)	(–11.2)	(–14.8)	(–18.5)	(–48.2)	(–57.4)

Values represent mean of three determinations ± S.D.

Absorbance of culture suspension at 750 nm soon after inoculation of the organisms was 0.2.

Values in parenthesis indicate percent decrease (–) or increase (+) over control.

Cultures were incubated at 25±1°C under continuous light at an intensity of 7.5 w/m^2 for 10 days and then harvested.

Table 7.5: Chlorophyll–*a* Content (mg/10 ml culture) of 4 Different Cyanobacteria at Different Salinity Levels of the Culture

Organism	*Control*	*NaCl levels (M)*					
		0.025	*0.05*	*0.075*	*0.1*	*0.2*	*0.3*
Nostoc paludosum	15.7 ± 0.11	10.71 ±0.014	7.14 ±0.093	5.71 ± 0.062	2.86 ±0.026	0.71 ±0.004	0
UU61240		(–31.8)	(–54.53)	(–63.64)	(–81.8)	(–95.5)	(–100)
Microcoleus vaginatus	14.28 ±0.13	14.28 ±0.16	9.97 ±0.12	8.57 ± 0.074	7.14 ±0.055	4.28 ±0.036	2.14 ±0.026
UU61243		(±0)	(–30.9)	(–39.0)	(–50.0)	(–70.3)	(–85.0)
Plectonema puteale	15.7 ± 0.19	15.7 ±0.18	8.57 ±0.062	7.14 ± 0.056	5.4 ±0.043	1.42 ±0.011	0.71 ±0.004
UU62246		(±0)	(–45.42)	(–54.53)	(–65.6)	(–90.96)	(–95.5)
Westiellopsis prolifica	9.28 ± 0.08	11.42 ±0.11	8.57 ±0.067	7.85 ± 0.052	6.43 ±0.048	1.42 ±0.022	0.71 ±0.003
UU62247		(+23.0)	(–7.66)	(–15.41)	(–30.72)	(–84.7)	(–92.3)

Values represent mean of three determinations ± S.D.

Chlorophyll–*a* content of the initial inoculum of *Nostoc paludosum* = 6.43 mg; *Microcoleus vaginatus* = 4.28 mg;

Plectonema puteale = 9.28 mg; *Westiellopsis prolifica* = 4.28 mg

Values in parenthesis indicate percent in increase (+) or decrease (–) over control.

Cultures were incubated at 25±1°C under continuous light at an intensity of 7.5 w/m^2 for 10 days and then harvested.

of the salt, which might be always present in the soils of that area. Growth as well as chlorophyll-*a* content of *Plectonema puteale* and *Microcoleus vaginatus*, both of which have well-defined sheath layer around their trichome, and the later species being isolated from alumina mine waste burdened soil tolerated up to 0.025 M NaCl. With further increase in the concentration of NaCl in the medium from 0.05 to 0.3M, growth as well as chlorophyll-*a* content of both organisms was decreased; decrease was more pronounced in *Plectonema puteale* whose sheath layer was less pronounced. To the contrary, *Nostoc paludosum* was more sensitive to the salt. Growth as well as chlorophyll-*a* content was decreased to 28 and 31 per cent respectively even at 0.25 M NaCl. At 0.05M NaCl the decrease was more than 50 per cent and the organism was almost killed at 0.3 M NaCl. Analysis of these results showed that the organisms which occur more prominently penetrating to several centimeter below top soil in the sandy soils of the coast were more tolerant to salinity, and those organisms having distinct sheath layer around their trichome tend to be comparatively higher tolerant to the salt.

Earlier reports have shown that supplementation of 0.4 M NaCl decreased the biomass accumulation and chlorophyll-*a* of *Calothrix bivissima* isolated from saline sodic soils whereas *Anabaena* sp. of freshwaters tolerated only up to 0.1 M NaCl (Kumar and Kaushik, 1994). The unicellular cyanobacterium *Microcystis firma* tolerated up to 1M NaCl but another unicellular species *Synichocystis aqutilis* showed salt tolerance up to 0.3 M NaCl. All these results showed the existence of varying degree of tolerance among species of cyanobacteria towards their tolerance to NaCl, and this is dependent on from where a species has been isolated, and whether the organisms do posses external envelope layers which might be enabling them to tolerate higher salt concentrations. The sheathe of cyanobacteria have been found to be composed of principally polysaccharides (Adhikary, 1998). There are reports that salinity stress in cyanobacteria is relieved due to accumulation of osmoprotective compounds in their cells to achieve equilibrium of osmotic potential (Reed *et al.*, 1986; Waar *et al.*, 1994). Cyanobacterial species also known to absorb and mobilise Na^+ intracellularly resulting in decrease in the salinity of soils (Kaushik *et al.*, 1981). Improvement of soil property was recorded in saline alkaline soils upon inoculation with cyanobacteria which lead to reduction in electrical conductivity and sodium in the soils. Apte and Thomas (1997) reported that in more than 90 per cent of the cell Na^+ remain extracellularly trapped in the sheath/slimy layer of cyanobacteria resulting in decrease in soil salinity by up to 38 per cent. Hence results of the present finding showed that due to occurrence of cyanobacteria in the sandy soils of the shore as well as in the mine waste areas they might be playing a significant role in amelioration of soil salinity.

3.3 Growth Response of Cyanobacteria from Different Soils to Light and Dark and Under Mixotrophic and Heterotrophic Growth Conditions

The four experimental organisms have been isolated from two different soils at several locations, either confined to the top soil or even penetrating up to several centimeters below the surface. Further the barren soils at both the areas receive light of various intensities due to shading by plant canopy, hence growth response of these organisms at various light intensities (5, 7.5 and 12.5 W/m^2) and in darkness was studied. *Westiellopsis prolifica* isolated from the sandy soils of the coast preferred higher light intensity of 12.5 W/m^2 to grow; growth and chlorophyll-a of the organism was increased to 11 and 7 per cent respectively in comparison to control (7.5 W/m^2) (Table 7.6). To the contrary, *Plectonema puteale* isolated from similar soils at Ramchandi preferred low light of 5 W/m^2 where its growth and chlorophyll-a content was increased by 12.5 and 44.8 per cent over control. The organism did not grow in dark in pure mineral medium (Table 7.7). Hence, possibly is the reason why it is confined to the top soil layers. Even at higher light intensity of 12.5 W/m^2 the growth was decreased. The other 2 organisms, *Microcoleus vaginatus* and *Nostoc paludosum* tolerated without change in their growth and chlorophyll-a content up to 10 days in darkness. These organisms preferred moderate light intensity of 7.5 W/m^2, and with increase or decrease of light intensity, above and below this intensity, the growth and chlorophyll-a decreased. This showed that the capability of an organism to grow at a particular light intensity, or its survivability in darkness in pure inorganic medium was a specific physiological characteristic of the organism making it successful at a defined micro-climate of the soil surface (Tables 7.6 and 7.7).

Table 7.6: Growth Response (Absorbance of culture suspension at 750 nm) of 4 Different Cyanobacteria at Different Light Intensities and in Darkness

Organism	*Light intensity*			
	5 W/m²	*7.5 W/m² (Control)*	*12.5 W/m²*	*Darkness*
Nostoc paludosum	0.52 ± 0.064	0.65 ± 0.052	0.50 ± 0.032	0.2 ±0.002
UU61240	(−20.0)		(−23.1)	(±0)
Microcoleus vaginatus	0.56 ± 0.051	0.60 ± 0.053	0.48 ± 0.029	0.2 ±0.004
UU61243	(−6.7)		(−20.0)	(±0)
Plectonema puteale	0.45 ± 0.038	0.40 ± 0.034	0.38 ± 0.026	0
UU62246	(+12.5)		(−5.0)	(−100)
Westiellopsis prolifica	0.46 ± 0.041	0.54 ± 0.038	0.6 ± 0.034	0.18 ±0.007
UU62247	(−14.82)		(+11.1)	(−10.0)

Values represent mean of three determinations ± S.D.

Culture were grown at 25 ± 1ºC up to 10 days. Values in parenthesis indicate percent increase (+) or decrease (−) over control except in darkness where the percentage decrease was calculated with respect to the initial inoculum.

Absorbance of the culture suspension at 750 nm on the day of inoculation was = 0.2.

In order to find out whether these organisms do possess the capability to utilise exogenous organic substrate for their growth in light and dark certain experiments were carried out. Usually in a soil surface where microorganisms colonise, they tend to secrete extracellular polymeric substances which in turn are utilized as a substrate for growth of other microorganisms, co-inhabiting on the soil surface. The disaccharide

Table 7.7: Chlorophyll–*a* Content (mg/10 ml culture) of 4 Different Cyanobacteria at Different Light Intensities and in Darkness

Organism	*Light intensity*			
	5 W/m²	*7.5 W/m² (Control)*	*12.5 W/m²*	*Darkness*
Nostoc paludosum	13.57 ±1.14	15.7 ± 1.39	12.85 ±1.48	6.43 ±0.36
UU61240	(–13.57)		(–18.6)	(±0)
Microcoleus vaginatus	15.7 ±1.24	18.56 ± 1.95	10.0 ± 0.94	4.28 ± 0.25
UU61243	(–15.41)		(–46.13)	(±0)
Plectonema puteale	22.85 ±2.06	15.78 ± 1.67	10.71 ± 0.69	0
UU62246	(+44.8)		(–32.1)	(–100)
Westiellopsis prolifica	9.28 ±0.98	10.0 ± 1.15	10.71 ± 0.74	2.16 ±0.06
UU62247	(–7.2)		(+7.1)	(–50)

Values represent mean of three determinations ± S.D.

Culture were grown at 7.5 W/m² light intensity at 25 ± 1ºC up to 10 days. Values in parenthesis indicate percent in increase (+) or decrease (–) over control except in darkness where the percentage decrease was calculated with respect to the initial inoculum.

Chlorophyll–*a* content of the initial inoculum of *Nostoc paludosum* = 6.43μg; *Microcoleus vaginatus* = 4.28μg;

Plectonema puteale =9.28 μg; *Westiellopsis prolifica* = 4.28 μg.

and the monosaccharides, *e.g.* hexoses: glucose, galactose and fructose, and the pentose: xylose was selected as the exogenous carbon sources to find out their capability to support heterotrophic growth of these photoautotrophic cyanobacteria of soils. The conditions where the organisms were grown in organic substrates in light is known as mixotrophic, in presence of organic substrates in light with additional supplementation of 10^{-5} M DCMU to block electron flow from PSII to cytochrome b6/f is called photoheterotrophy, and when grown in exogenous organic substrate in darkness is called chemoheterotrophy or heterotrophic condition. In *Nostoc paludosum* sucrose, fructose and xylose supported mixotrophic growth, however, the growth was not accompanied by increase in the chlorophyll-*a* content, hence was not a true mixotrophic form. In photoheterotrophy and chemoheterotrophy conditions all the sugars increased the growth of these organisms. However, from the chlorophyll-*a* data (Tables 7.8 and 7.9), it was found that only sucrose and fructose, at 15 mM, supported the heterotrophic growth of the organism. In *Microcoleus vaginatus* all the sugars supported mixotrophic growth to a significant extent; galactose, xylose and fructose increased the growth as well as chlorophyll-*a* content of the organism to an appreciable proportion. However, these sugars were not effective to stimulate the photoheterotrophic and chemoheterotrophic growth of *Microcoleus vaginatus*. Sucrose at 15 mM did support increase in growth and chlorophyll-a content in darkness but was only to a little extent (Tables 7.10 and 7.11). It was interesting to note that *Plectonema puteale* did not grow phtoheterotrophically. The organism was bleached and

Table 7.8: Growth Response (Absorbance of culture suspension at 750 mm) of *Nostoc paludosum* UU61240 to Exogenous Organic Substrates in Light, Light + 10^{-5} M DCMU and in the Darkness

Sugar	*Concentration*	*Light*	*Light +10^{-5} M DCMU*	*Darkness*
Control		*0.60 ± 0.041*	*0.1 ± 0*	*0.1 ± 0*
Sucrose	15 mM	0.82±0.061	0.25 ± 0.019	0.30 ± 0.003
		(+36.7)	(+150)	(+200)
Glucose	15 mM	0.60 ±0.052	0.40 ± 0.021	0.21 ± 0.002
		(± 0)	(+300)	(+110)
Galactose	15 mM	0.65 ±0.055	0.20 ± 0.011	0.19 ± 0.009
		(+8.0)	(+100)	(+90.0)
Fructose	15 mM	0.80 ±0.63	0.40 ± 0.018	0.25 ± 0.006
		(+33.3)	(+300)	(+150)
Xylose	15 mM	0.80 ±0.058	0.24 ± 0.016	0.20 ± 0.002
		(+33.3)	(+140)	(+100)

Values represent mean of three determinations ± S.D.

Absorbance of the culture suspension at 750 nm on the day of inoculation = 0.1. Values in parenthesis indicate percent increase or decrease over control. For dark and light + 10^{-5} M DCMU cultures, the initial values soon after inoculation was taken as control.

Table 7.9: Chlorophyll–*a* Content (mg/10 ml culture) of *Nostoc paludosum* UU61240 in Presence of Exogenous Organic Substrates in Light, Light +10^{-5} M DCMU and in the Darkness

Sugar	*Concentration*	*Light*	*Light +10^{-5} M DCMU*	*Darkness*
Control		*25.70 ± 0.046*	*4.28 ± 0.008*	*4.28 ± 0.006*
Sucrose	15 mM	6.42 ±0.004	4.99 ± 0.004	4.99 ± 0.009
		(–75.0)	(+16.6)	(+16.6)
Glucose	15 mM	4.28 ± 0.028	2.85 ± 0.003	2.14 ± 0.004
		(–83.3)	(–33.4)	(–50.0)
Galactose	15 mM	23.56 ± 0.062	2.14 ±0.003	3.57 ±0.006
		(–8.3)	(–50.0)	(–16.6)
Fructose	15 mM	10.71 ±0.053	5.71 ± 0.008	4.99 ± 0.011
		(–58.3)	(+33.4)	(+16.6)
Xylose	15 mM	22.13 ± 0.089	2.85 ± 0.002	3.57 ± 0.007
		(–13.9)	(–33.4)	(–16.6)

Values represent mean of three determinations ± S.D.

Chlorophyll–*a*, content of the culture soon after inoculation = 4.28 mg. Values in the parenthesis indicate percent increase or decrease over control. For dark and light + 10^{-5} M DCMU cultures the value soon after inoculation was taken as control.

Table 7.10: Growth Response (Absorbance of culture suspension at 750 nm) of *Microcoleus vaginatus* UU61243 to Exogenous Organic Substrate in Light, Light +10^{-5} M DCMU and in the Darkness

Sugar	*Concentration*	*Light*	*Light +10^{-5} M DCMU*	*Darkness*
Control		*0.46 ± 0.006*	*0.1 ± 0*	*0.1 ± 0*
Sucrose	15 mM	1.80 ± 0.014	0.2 ±0.001	0.2 ±0.001
		(+291.3)	(+100)	(+100.0)
Glucose	15 mM	1.60 ± 0.011	0.08 ±0.0002	0.09 ±0.003
		(+247.8)	(–20.0)	(–10.0)
Galactose	15 mM	2.0 ± 0.019	0.06 ±0.001	0.06 ±0.003
		(+334.8)	(–40.0)	(–40.0)
Fructose	15 mM	2.4 ±0.006	0.17 ±0.006	0.12 ±0.007
		(+412.7)	(+70.0)	(+20.0)
Xylose	15 mM	2.7 ±0.011	0.19 ±0.008	0.17 ±0.006
		(+487.0)	(+90.0)	(+70.0)

Values represent mean of three determinations ± S.D.

Absorbance of the culture suspension at 750 nm on the day of inoculation = 0.1. Values in parenthesis indicate percent increase or decrease over control. For dark and light + 10^{-5} M DCMU cultures, the value soon after inoculation was taken as control.

Table 7.11: Chlorophyll–*a* Content (mg/10 ml culture) of *Microcoleus vaginatus* UU61243 in Presence of Exogenous Organic Substrates in Light, Light +10^{-5} M DCMU and in the Darkness

Sugar	*Concentration*	*Light*	*Light +10^{-5} M DCMU*	*Darkness*
Control		*17.85 ± 0.69*	*2.86 ± 0.051*	*2.86 ± 0.042*
Sucrose	15 mM	24.99 ± 0.94	5.71 ± 0.083	4.99 ± 0.009
		(+40.0)	(+99.7)	(+74.5)
Glucose	15 mM	24.98 ±1.11	2.85 ± 0.022	2.85 ± 0.006
		(+40.0)	(± 0)	(± 0)
Galactose	15 mM	47.83 ± 2.36	2.14 ± 0.016	2.14 ± 0.002
		(+168.0)	(–25.2)	(–25.2)
Fructose	15 mM	35.70 ± 2.14	2.85 ± 0.024	2.14 ± 0.001
		(+100.0)	(± 0)	(–25.2)
Xylose	15 mM	46.41 ± 2.59	0.71 ± 0.001	2.85 ± 0.004
		(+160.0)	(–75.2)	(± 0)

Values represent mean of three determinations ± S.D.

Chlorophyll–*a*, content of the culture soon after inoculation = 2.86 mg. Values in the parenthesis indicate percent increase or decrease over control. For dark and light + 10^{-5} M DCMU cultures, the value soon after inoculation was taken as control.

disintegrated within 10 days of incubation in darkness in pure inorganic medium. Upon supplementing with exogenous sugars the organism grew as well as synthesized chlorophyll-*a* in the dark. Similarly under mixotrophic conditions also the sugars supported growth as well as the increase of chlorophyll-*a* content. Xylose, fructose and sucrose were found to be efficient in supporting mixotrophic growth of the organism (Tables 7.12 and 7.13). These results showed that the organism flourish at the top layer of the soil and cannot grow at different depths in greater abundance. Since these organisms can utilise certain organic substrate for dark growth, if organic carbon is available in the deeper layer of the soil, can also colonize in those habitats sparcely utilizing the available organic carbon. *Westiellopsis prolifica* which also occurred in the Ramchandi soils at the top soil as well as to deeper layers could survive in the mineral medium in darkness to some extent, and upon supplementing with an organic substrate especially sucrose, its growth as well as chlorophyll-*a* content was increased. Similar growth response to sucrose and to other substrates was also observed in the same organism in photoheterotrophic culture in presence of DCMU (Tables 7.14 and 7.15).

Many workers have found that most cyanobacteria are obligate phototrophs and apparently cannot grow heterotrophically in the dark (Hoare *et al.*, 1967; Smith *et al.*, 1967). However, several published work later have shown that a variety of cyanobacteria can grow heterotrophically at the expense of exogenous carbon sources (Smith, 1982; Khoja and Whitton, 1975; Adhikary and Pattnaik, 1979; Sahu and Adhikary, 1981a,b, 1982; Banerjee and Kumar, 1987; Barnum and Gendel, 1987; Bastia *et al.*, 1993). In the present work on the four cyanobacteria from the soil crusts of Damonjodi and Ramchandi sites showed different capability with regard to their mixotrophic, photoheterotrophic and chemoheterotrophic growth in presence of 15 mM of exogenous organic substrates, and the substrate specificity to support such growth varies from species to species. Further, where a specific substrate supported higher growth of an organism in light did not posses the same ability to support its photo–and/or chemohetrotrophic growth. The exogenous substrates used were quite specific and their extent of specificity varies with different genera, species and even growth conditions. This suggests that substrate specificity and efficiency of substrate utilization are manifestations of growth conditions. It is difficult to give any precise explanation for their differential growth response of cyanobacteria to exogenous substrates supporting growth. It is conceivable that light and dark treatments may bring alternations in the membrane permeability and transport properties for assimilation of organic substrates.

Several sugars, *e.g.* pentoses, as well as hexoses have been tested previously as the possible substrates for support of dark growth of various cyanobacterial species. Results have shown that though many species were capable of utilizing a wide spectrum of exogenous organic compounds both in light and dark, only few substrates, mostly monosaccharides and the disaccharide, sucrose that can act as the substrate for the support of dark growth in addition to their normal autotrophic mode of growth in light (Zajic and Chiu, 1970; Smith, 1982; Barnum and Gendel, 1987). The finding of the present investigation also showed that mostly sucrose, glucose or fructose were the efficient substrates to support mixotrophic and/or heterotrophic growth of the

Table 7.12: Growth Response (Absorbance of Culture Suspension at 750 nm) of *Plectonema puteale* UU62246 to Exogenous Organic Substrate in Light, Light +10^{-5} M DCMU and in the Darkness

Sugar	*Concentration*	*Light*	*Light +10^{-5} M DCMU*	*Darkness*
Control		*0.4 ± 0.002*	*0*	*0*
Sucrose	15 mM	1.8 ± 0.016	0	0.6 ±0.001
		(+350)		(+500)
Glucose	15 mM	1.0 ± 0.004	0	0.45 ±0.001
		2.0 (+150)		(+350)
Galactose	15 mM	1.5 ± 0.003	0	0.65 ±0.001
		(+275)		(+550)
Fructose	15 mM	1.6 ± 0.006	0	0.56 ±0.002
		(+300)		(+460)
Xylose	15 mM	1.6 ± 0.006	0	0.55 ±0.004
		(+300)		(+450)

Values represent mean of three determinations ± S.D.

Absorbance of the culture suspension at 750 nm on the day of inoculation = 0.1. Values in parenthesis indicate percent increase or decrease over control. For dark and light + 10^{-5} M DCMU cultures, the value soon after inoculation was taken as control.

Table 7.13: Chlorophyll–*a* Content (mg/10 ml culture) of *Plectonema puteale* UU62246 in Presence of Exogenous Organic Substrates in Light, Light +10^{-5} M DCMU and in the Darkness

Sugar	*Concentration*	*Light*	*Light +10^{-5} M DCMU*	*Darkness*
Control		*19.99 ± 0.92*	*0*	*0*
Sucrose	15 mM	42.12 ±0.99	0	13.56 ±0.15
		(+110.7)		(+58.2)
Glucose	15 mM	32.13 ±1.16	0	12.85 ±0.21
		(+60.7)		(+49.94)
Galactose	15 mM	24.99 ±1.11	0	13.56 ±0.23
		(+25.0)		(+58.2)
Fructose	15 mM	43.55 ±2.36	0	12.85 ±0.098
		(+117.9)		(+49.94)
Xylose	15 mM	88.53 ±0.58	0	14.28 ±0.14
		(+342.9)		(+66.62)

Values represent mean of three determinations ± S.D.

Chlorophyll–*a* content of the culture soon after inoculation = 8.57 mg. Values in the parenthesis indicate percent increase or decrease over control. For dark and light + 10^{-5} M DCMU cultures, the value soon after inoculation was taken as control.

Table 7.14: Growth Response (Absorbance of culture suspension at 750 nm) of *Westiellopsis prolifica* UU62247 to Exogenous Organic Substrate in Light, Light +10^{-5} M DCMU and in the Darkness

Sugar	*Concentration*	*Light*	*Light +10^{-5} M DCMU*	*Darkness*
Control		*0.61 ± 0.006*	*0.09 ± 0.001*	*0.09 ± 0.001*
Sucrose	15 mM	1.0 ± 0.012	0.29 ± 0.002	0.20 ± 0.003
		2.0 (+63.93)	(+190)	(+100)
Glucose	15 mM	0.82 ±0.008	0.23 ± 0.002	0.18 ± 0.002
		0.83 (+39.34)	(+130)	(+80)
Galactose	15 mM	0.70 ± 0.004	0.19 ± 0.001	0.14 ± 0.002
		(+14.75)	(+90)	(+40)
Fructose	15 mM	0.70 ± 0.006	0.20 ± 0.002	0.14 ± 0.002
		(+14.75)	(+100)	(+40)
Xylose	15 mM	0.65 ± 0.009	0.22 ± 0.004	0.14 ± 0.002
		(+6.55)	(+120)	(+40)

Values represent mean of three determinations ± S.D.

Absorbance of the culture suspension at 750 nm on the day of inoculation = 0.1. Values in parenthesis indicate percent increase or decrease over control. For dark and light + 10^{-5} M DCMU cultures, the value soon after inoculation was taken as control.

Table 7.15: Chlorophyll–*a* Content (mg/10 ml culture) of *Westiellopsis prolifica* UU62247 in Presence of Exogenous Organic Substrates in Light, Light +10^{-5} M DCMU and in the Darkness

Sugar	*Concentration*	*Light*	*Light +10^{-5} M DCMU*	*Darkness*
Control		*16.42 ± 0.048*	*2.9 ± 0.012*	*1.8 ± 0.006*
Sucrose	15 mM	35.70 ± 0.091	4.28 ± 0.011	4.99 ± 0.003
		(+117.42)	(+19.89)	(+39.77)
Glucose	15 mM	24.99 ± 0.12	3.57 ± 0.009	3.57 ± 0.006
		(+52.19)	(± 0)	(± 0)
Galactose	15 mM	19.99 ± 0.14	2.85 ± 0.006	3.57 ± 0.004
		(+21.74)	(–20.17)	(± 0)
Fructose	15 mM	25.70 ± 0.18	2.14 ± 0.004	2.85 ± 0.002
		(+56.51)	(–40.06)	(–20.17)
Xylose	15 mM	23.56 ± 0.22	2.14 ± 0.007	2.85 ± 0.002
		(+43.48)	(–40.06)	(–20.17)

Values represent mean of three determinations ± S.D.

Chlorophyll–*a* content of the culture soon after inoculation = 3.57 mg. Values in the parenthesis indicate percent increase or decrease over control. For dark and light + 10^{-5} M DCMU cultures, the value soon after inoculation was taken as control.

experimental organisms. Such a growth capability of drought tolerant cyanobacteria of soils in the coastal area of Ramchandi as well as in the mine waste burden soils of Damonjodi are of immense importance because they can grow utilizing the scarce organic carbon present in their respective habitats, and colonize making the soil productive.

Acknowledgements

The authors thank the Head of the P.G. Department of Botany, Utkal University for providing laboratory facilities. One of us (P.T.) thank the authorities of Higher Education Department, of Govt. of Orissa, for grant of study leave to carry out this work at Utkal University.

References

Adhikary SP (1998) Polysaccharides from mucilaginous envelope layers of cyanobacteria and their ecological significance. J Sci Ind Res 57: 454-466.

Adhikary SP and Pattanaik H (1979) Growth response of *Westiellopsis prolifica* Janet to organic substrates in light and dark. Hydrobiologia 67: 241-248.

Apte SK and Thomas J (1997) Possible amelioration of coastal soil salinity using halotolerant nitrogen fixing cyanobacteria. Plant and Soil 189: 205-211.

Apte SK, Reddy BR and Thomas J (1987) Relationship between Na^+ influx and salt tolerance of nitrogen fixing cyanobacteria. Appl Environ Microbiol 53: 1934-1939.

Banerjee M and Kumar HD (1987) Effect of light quality, anaerobiosis and glucose on nitrogen fixation by the rice field cyanobacterium *Aulosira fertilissima* Ghose. Acta Microbiol Hungarica 34: 57-63.

Barnum SR and Gendel SM (1987) Heterotrophic growth of nine strains of filamentous cyanobacteria. Iowa State J Res 62: 147-159.

Bastia AK, Satapathy DP and Adhikary SP (1993) Heterotrophic growth of several filamentous blue-green algae. Arch Hydrobiol suppl Algolog Stud 70: 65-70.

Blumwald E and Tel-Or E (1982a) Structural aspects of the adaptation of *Nostoc muscorum* to salt. Arch Microbiol 132: 163-167.

Blumwald E and Tel-Or E (1982b) Osmoregulation and cell composition in salt adaptation of *Nostoc muscurum*. Arch Microbiol 132: 168-172.

Borowitzka LJ, Demmerle S, Mackay MA and Norton RS (1980) Carbon-13 NMR study of osmoregulation in blue-green algae. Science 210: 650-651.

Boyer JS (1972) Plant productivity and environment. Science 218: 443-448.

Flowers TJ, Troke PF and Yeo AR (1977) The mechanism of salt tolerance in haplophytes. Ann Rev Plant Physiol 28: 89-121.

Hoare DS, Hoare SL and Moore RB (1967) The photassimilation of organic compounds by autotrophic blue-green algae. J Gen Microbiol 49: 351-370.

Holm-Hansen O (1968) Ecology, physiology and biochemistry of blue-green algae. Ann Rev Microbiol 22: 47-70.

Ingram LO, Clader JA and Van Baalen C (1973) Role of reduced exogenous organic compounds in the physiology of blue-green bacteria (algae): Photoheterotrophic growth of a heterotrophic blue-green bacterium. J Bact 114: 695-700.

Kaushik BD, Krishnamurti GSR and Venkataraman GS (1981) Influence of blue-green algal application on the properties of saline alkaline rice soils. Sci Cult 47: 169-170.

Khoja T and Whitton BA(1975). Heterotrophic growth of filamentous blue-green algae. Br Phycol J 10: 139-148.

Kumar H and Kaushik BD (1994) Response of *Calothrix brevissima* to salt. Ind J Microbiol 34: 37-41.

Mackinney G (1941) Absorption of light by chlorophyII solutions. J Biol Chem 140: 315-322.

Padhi H, Rath B and Adhikary SP (1998) Tolerance of nitrogen fixing cyanobacteria to NaCl. Biol Plant 40: 261-268.

Reed RH, Borowitzka LJ, Mackay MA, Chudek JA, Foster R, Warr SRC, Moore DJ and Stewart WDP (1986) Organic solute accumulation in osmotically stressed cyanobacteria. FEMS Microbial Lett 39: 51-56

Reynaud PA and Roger PA (1978) N_2-fixing algal biomass in Senegal rice fields. Ecol Bull Stockholm 26: 148-157.

Rippka R (1972) Photoheterotrophy and chemoheterotrophy among unicellular blue-green algae. Arch Microbiol 87: 303-322.

Rippka R, Deruelles J, Waterbury JB, Herdman M and Stainer RY (1979) Generic assignments, strain histories and properties of pure cultures of cyanobacteria. J Gen Microbiol 111: 1-61.

Roger PA and Reynaud PA (1977) Algal biomass in rice field of Senegal: relative importance of cyanobacteria that fix nitrogen. Rev Ecol Biol Soil 14: 519-530.

Roger PA, Santiago-Ardales S, Reddy PM and Watabane I (1987) The abundance of heterocystous blue-green algae in rice soil and inocula used for application in rice fields. Biol Fertil Soil 5: 98-105.

Sahu J and Adhikary SP (1981a) Heterotrophic growth of the blue-green algae *Anabaena* sp. Microbios Lett 16: 127-134.

Sahu J and Adhikary SP (1981b) Heterotrophic growth and nitrogen fixation in the filamentous blue-green algae, *Anabaena* sp. Z Allg Microbiol 29: 669-676.

Sahu J and Adhikary SP (1982) Heterotrophic growth and pigment composition of four filamentous blue-green algae. Arch Hydrobiol Algolog Stud 63: 189-199.

Sahu J and Adhikary SP (1987) Branching development of the blue-green alga *Mastigocladus laminosus* Cohn at different pH levels. Curr Sci 56: 308-309.

Singh PK (1974) Effect of pH on growth and nitrogen fixation of unicellular blue-green alga *Aphanothece* sp. Oikos 25: 114-116.

Singh RN (1961) Role of blue-green algae in nitrogen economy of Indian agriculture. Indian Council of Agricultural Research, New Delhi, pp 175.

Smith AJ (1973) Synthesis of metabolic intermediates. In: The Biology of Blue-green Algae. Eds Carr NG and Whitton BA, Blackwell Sci Publ Oxford, pp 1-38.

Smith AJ (1982) Modes of cyanobacterial carbon metabolism. In: The Biology of Cyanobacteria. Eds Carr NG and Whitton BA, Blackwell Sci Publ, Oxford, pp 47-85.

Smith AJ, London J and Stainer RY (1967) Biochemical basis of obligate autotrophy in blue-green algae and *Thiobacilli.* J Bact 94: 972-983.

Stainer RY (1973) Autotrophy and heterotrophy in unicellular blue-green algae. In: The Biology of Blue-green Algae. Eds Carr NG and Whitton BA, Blackwell Sci Publ Oxford, pp 501-518.

Swain N and Adhikary SP (1991) Dependence of growth of *Microcystis acruginosa* (Kutz. Emend. Elenkin) on pH of the culture medium. J Freshwater Biol 3: 131-137.

Van Baalen C, Hoare DS and Brandt E (1971) Heterotrophic growth of blue-green algae in dim light. J Bact 105: 685-689.

Waar SRC, Reed RH and Stewart WDP (1984) Physiological responses of *Nodularia harveyana* to osmotic stress. Mar Biol 79: 21-26.

White AW and Shilo M (1975) Heterotrophic growth of the filamentous blue-green algae *Plectonema boryanum.* Arch Microbiol 102: 123-127.

Wolk CP (1973) Physiology and cytological chemistry of blue-green algae. Bact Rev 37: 32-101.

Zajic JE and Chiu YS (1970) Heterotrophic culture of algae. In: Properties and Products of algae. Ed Zajic JE, Plenum Press, New York and London, pp 1-47.

Algal Biotechnology: New Vistas (2010) *Pages 133–144*
Editor: Mihir Kumar Das
Published by: DAYA PUBLISHING HOUSE, NEW DELHI

Chapter 8

The Impact of Fungicides on Rice Field Cyanobacteria

Anshuman Das*[1] *and Mihir Kumar Das*[2]

[1]*P.G.Department of Microbiology,*
Orissa University of Agriculture and Technology,
Bhubaneswar – 751 003, Orissa, India
[2]*P.G. Department of Botany, G.M. College (Autonomous),*
Sambalpur – 768 004, Orissa, India

ABSTRACT

Cyanobacteria contribute significantly to primary production, nutrient cycling, and decomposition in estuarine ecosystems; therefore, detrimental effects of pesticides on cyanobacterial species may have subsequent impacts on higher trophic levels. Pesticides may affect these microorganisms via spills, runoff, and drift. Both the structure and the function of microbial communities may be impaired by pesticide toxicity. Pesticides may also be metabolized or bioaccumulated by microorganisms. Mechanisms of toxicity vary, depending on the type of pesticide and the microbial species exposed. Fungicides are generally most toxic to phototrophic microorganisms, exhibiting toxicity by disrupting photosynthesis. It is therefore, essential that regular monitoring be undertaken to evaluate the influence of fungicides on cyanobacteria. In the present review, range of fungicide-induced biological responses, induction of biochemical changes in the organisms, the relative tolerance of cyanobacteria to fungicides etc. are highlighted.

Keywords: Cyanobacteria, Fungicides, Toxicity, Rice field.

* E-mail: mihir_gmc@rediffmail.com

1. Introduction

Pesticides continue to be a significant and growing component of modern rice technology. The relative importance of pesticides has increased despite the availability of alternatives to exclusive chemical pest control such as varietals resistance and integrated pest management (IPM). While Asia's elite are becoming increasingly concerned about the adverse long-term effects of pesticides on the environment and human health, little scientific research has been done to address this issue. The few studies that exist are based on speculative and anecdotal paradigms. This chapter provides an overview of a multidisciplinary scientific effort to identify the incidence and magnitude of on-site and off-site environmental effects of pesticide use. Indiscriminate pesticide use can result in one or more of the following: (1)health impairment due to direct or indirect exposure to hazardous chemicals; (2)contamination of ground and surface waters through runoff and seepage; (3) the transmittal of pesticide residues through the food chain to the farm family and urban consumers; (4) an increase in the resistance of pest populations to pesticides, thereby reducing their efficacy and consequently causing pest outbreaks;(5) the reduction of beneficial insects like parasites and predators, thereby reducing the effectiveness of pest control strategies that attempt to minimize pesticide use; and (6) the reduction in the populations of microorganisms in the paddy soil and water that help sustain soil fertility while lowering chemical fertilizer use. The incidence and magnitude of each of these effects depend on the types of chemicals, frequency and quantities applied, and their persistence.

2. Cyanobacteria

Cyanobacteria (Blue-green algae) are a diverse assemblage of prokaryotic photosynthetic microorganisms. The morphology and, to some extent, metabolism of the various genera varies greatly and due to an increasing awareness of the ecological importance of these algae much information on these aspects is now available. Cyanobacteria can grow in many different habitats ranging from sub-zero Arctic ice to geothermal springs above 50°C. Many genera can grow on the surface of temperate soils, where they are often visible as a thin greenish crust on the soil. Such terrestrial cyanobacteria are ecologically important in so far as many species can fix nitrogen *i.e.* reduce molecular atmospheric nitrogen to ammonium which can then be utilized for amino acid and protein biosynthesis. Nitrogenous compounds which can be assimilated by other plants are ultimately released to the soil after death of the algal cells and subsequent mineralization and nitrification. Algal nitrogen fixation, release of the fixed nitrogen and it's assimilation by higher plants may be rapid, especially in specialized symbioses between cyanobacteria and fungi, liverworts, pteridophytes or angiosperms where the algae are in intimate contact with their hosts. Biological nitrogen fixation by free-living or symbiotic cyanobacteria can greatly contribute to the pool of combined nitrogen in a soil when compared to other nitrogen input sources, especially in areas where inorganic nitrogen fertilizers are not used. Over 20 kg N/ha/yr may be fixed by cyanobacteria in temperate soils (Henriksson, 1971).

Although nitrogen fixation by cyanobacteria may often be an important factor in maintaining soil fertility (De, 1939), the algae may directly or indirectly benefit higher plants in other ways. The algae can bind soil particles together, due to the adhesive properties of the copious mucilage excreted by many algal species. Thus they may reduce soil erosion and also evaporative water loss from soils as desiccated algal crusts form a relatively impermeable cover over the soil surface. Some reports indicate that cyanobacteria may release biologically active compounds into the soil and that these compounds may then be assimilated by higher plants, significantly enhancing their growth (Venkataraman and Neelakantan, 1967; Dadhich *et al.*, 1969).

3. Pesticide

The increased use of fertilizers has become an integral part of the cultivation practices. With higher fertilizer inputs, disease and pest problems have become prominent which endangered the yield necessitating the large scale use of pesticides. Until 1850s, the use of pesticides against pests was not in practice. Before 1940, they were used in relatively small amount and considered as 'safe pesticides' which were derivatives of various plants. But from 1940 onward, the development of numerous synthetic organic compounds or so called 'second generation pesticides' completely changed the basis and strategies of pest control. Thus, during 1950s and 1960s, huge amount of organic pesticides were widely used in agriculture and horticulture to facilitate the production and protection of food, feed and fibre in greater quantity. During 1960s, it was realized that any one single pesticide did not completely eliminate all the species of a target pest. This created a number of trends in use and development of new pesticides for effective control. As a result, the number of chemical pesticides has increased during the past 50 years. The pesticides included a variety of insecticides, fungicides, herbicides, nematocides etc.

3.1 Pesticide Hazard

Though pesticides were considered as the miracle drugs for the plants during 1950s, doubts were raised concerning the hazardous effects of these chemicals on the environment from the early 1960s. The DDT used in the field has been detected in the mother's milk. The travel of DDT and other pesticides through various components of the food chain has not only affected the flora and fauna, but also, has reappeared in the consumable food products. Richael Carson was one of the first to become aware of the danger of pesticide pollution and in her famous book entitled ' Silent Spring', she produced evidences to toxic effects of pesticides upon wildlife and potential hazards to man (Misra and Mani, 1994). Despite increasing research efforts towards crop improvement by methods obviating the use of pesticides (Day, 1987), agriculture remains heavily dependent on these agrochemicals (Gadkari, 1988). The soil, being the ultimate sink in the rice fields, harbours many beneficial cyanobacteria, which are potentially susceptible to pesticides. Cyanobacteria are known to contribute for maintenance of soil fertility (Singh, 1961; Venkatarman, 1972). It is therefore, imperative that regular monitoring be undertaken to evaluate the influence of pesticides on these organisms.

The pesticide toxicity is regarded only as the lethal effects of the chemicals upon a target organism. It directly changes the growth rate as well as metabolic activities such as photosynthesis and respiration. It may have a toxic effect upon certain groups of microorganisms (Padhy, 1985). The effect of this kind may alter population equilibrium in indirect manner by changing the competitive efficiency of one or other group. Pesticide may promote either directly or indirectly, the growth of one or more type of organisms, which could be either beneficial or undesirable in paddy field ecosystem. Finally, toxic action of pesticides causing death of certain groups of organisms would result in the release of nutrients by the decay of these organisms, thus inducing a corresponding increase in the number of pesticide tolerant forms.

The research work so far carried out in various laboratories of the world relates to the interaction of pesticides to cyanobacteria through *in vitro* determinations and/ or under *in vivo* conditions (field condition). Most of the works pertain to the determinations of lethal doses of pesticides on different species/strains of cyanobacteria. The more recent studies have been more biochemical and physiological in nature than the earlier ones, departing from examination of pesticide-induced qualitative and quantitative fluctuations in soil microbial populations and concentrating more on factors such as the impairment of nitrogen fixation (Kapoor and Sharma, 1980; Kaushik and Venkataraman, 1983). Although studies of the interaction of cyanobacteria with insecticides continue to be widely conducted, there has been an increase in the amount of attention paid to fungicides and herbicides. Since 1979, emphasis has also been laid on the examination of the response of cyanobacteria to mixtures of pesticides (Stratton, 1983, Tandon *et al.*, 1988). This is encouraging, since in rare cases, would any pesticide be present in isolation in the soil.

4. Fungicides

The extensive use of fungicides has greatly improved many facets of agriculture, but at the same time has created many unintended problems. The effect of fungicides on non-target organisms, either direct or indirect, is a significant problem. Cyanobacteria is very important group of soil biota and interference in their activity by soil incorporated fungicides may be expected to have potentially serious consequences on the overall productivity of soil, Cyanobacteria are more sensitive to fungicides than other groups of micro-organisms. In the present review effect of fungicides on growth and biological nitrogen fixation, mode of action and factors influencing the fungicides and algal interaction have been discussed.

Effect of dichlone on cyanobacteria was first reported by Fitzerald *et al.* (1952). *Microcystis aeruginosa* blooms in ponds and non-bloom forming cyanobacteria were cleared by 30 to 55 ppm and 5 ppm dichlone respectively. Fitzerald and Skoog (1954) conducted large scale field trials in clearing cyanobacteria with dichlone (10 to 100 ppb). Dichlone at 10 ppb prevented the growth of *Anabaena* sp. (Bisiach, 1970). The growth of *Anacystis nidulans* was prevented by 140 ppb dichlone (Whitton and Mc Arthur, 1967).

Dithiocarbamate fungicides are the oldest one and many of the members are still actively used to control the plant diseases with greater effectiveness. The members include zineb, thiram, ziram, maneb, moncozeb etc. They are being used in soil treatment, seed application and foliar application. Zineb (zinc ethylene bisdithiocarbamate) was found highly toxic to *Anabaena* and *Nostoc* at very low concentrations, however, some other strains tolerated even up to 50 ppm. The growth of *Cylindrospermum* sp and *Nostoc muscorum* was inhibited at 1 and 5 ppm of zineb respectively in nitrogen deficient medium, but tolerated higher concentrations in nitrogen supplemented medium (Venkataraman and Rajyalakshmi, 1971). They further reported that zineb at 100 kg/ha did not inhibit the growth of eight strains of *Anabaena,* nine (out of 16) strains of *Nostoc, Aulosira fertilissima, Tolypothrix tenuis* and *Anacystis nidulans.*

Zineb is found to have a sub-lethal range of 1 to 15 ppm with *Cylindrospermum* and 5 to 25 ppm with *Nostoc muscorum* in C-N medium. The complete toxic levels in this medium were 25 and 300 ppm for *Cylindrospermum* and *N. muscorum* respectively. In C+N medium, the sub-lethal range was 5 to 20 ppm and 10 to 30 ppm in *Cylindrospermum* and *N.muscorum* respectively, the complete lethal ranges correspondingly were 30 and 35 ppm zineb (Panigrahy, 1984). Panigrahy (1984) also reported that *Cylindrospermum* and *Nostoc* tolerated zinc *in vitro* at 5 and 15 ppm respectively. Schluter (1964) reported little effect of zineb on the growth of *Scenedesmus quadricauda* and *Kirchneriella lunaris.*

Ziram is another member of dithiocarbamates being used for the control of foliar diseases. Thirty-two species of cyanobacteria were treated with several doses of ziram. Diatoms were more sensitive as compared to cyanobacteria and green-algae (Maloney and Plamer, 1956). But this fungicide had a stimulatory effect on growth of *Cylindrospermum* sp and *Nostoc muscorum* at the 0.01 ppm level in both C+N and C-N media. The sub-lethal range for *Cylindrospermum* was found to be 0.05 to 0.13 ppm, whereas for *N. muscorum,* this range was from 0.1 to 0.4 ppm in C-N medium. The complete lethal dose in C-N medium was 0.1 ppm for *Cylindrospermum* and 0.5 ppm for *N. muscorum.* The sub-lethal range in C+N medium was 0.05 to 0.13 ppm for *Cylindrospermum* and 0.1 to 0.5 ppm for *N. muscorum.* The complete lethal doses in C+N medium were 0.02 and 0.6 ppm for *Cylindrospermum* and *N. muscorum* respectively (Panigrahy, 1984).

Mancozeb (Zinc ethylene bisdithiocarbamate and manganese ions) is still another dithiocarbamate fungicide,highly effective in the control of foliar diseases. It is being intensively used where long term effect is being required. Mancozeb in the range of 5 to 100 ppm was least toxic to 27 strains of cyanobacteria (Venkataraman and Rajyalakshmi, 1972). This fungicide had a sub-lethal range up to 7 ppm for *Cylindrospermum* in both C-N and C+N media and the level of 9 ppm was completely toxic in both the media. The sub-lethal range of mancozeb was up to 10 ppm with *N. muscorum* in both C-N and C+N media and the level of 15 ppm was completely toxic in both media. The highest tolerance level in both the media (C-N and C+N) was 5 ppm mancozeb for both the strains on agar plates (Panigrahy, 1984). Seed treatment fungicide, thiram (tetramethyl thiuramdisulphide) was very toxic as it completely removed *Nostoc* and *Anabaena* at the rate 1.2-12 ppm; however, less than 1.2 ppm, the growth was stimulated by 200 per cent (Cameron and Julian, 1984).

Virtually all data were collected under *in vitro* conditions, except a few, in which the benzimidazole carbendazim at 2.9 ppm, the recommended application rate, was applied to soil held in glass cylinders in the laboratory. The fungicide affected neither the growth nor nitrogenase activity of the soil algal and cyanobacterial flora.Gangawane (1980) also observed that carbendazim(as Bavistin ®) could be tolerated by nitrogen-fixing cyanobacteria. This likewise applied to the chlorophenyl fungicide quintozene (as Brassicol ®). In a separate study, tolerance levels of carbendazim (as MBC) were 500 ppm for *Calothrix* sp. and *Nostoc* sp. and 1,000 ppm for *Aulosira fertilissima* and *Tolypothrix tenuis, Westiellopsis prolifica,* however, was intolerant of concentrations as low as 1 ppm (Gangawane and Saler, 1979). Shukla *et al.* (1990) noted the inhibition of growth of unspecified algae of a sandy loam soil following treatment with carbendazim (as Bavistin ®) at 0.25, 0.5, 0.75 and 1 per cent. Other fungicides (copper oxychloride, as Blitox; EBP, as Kitazin ®, mancozeb, as manzeb and Dithane ®) used in the same study and at the same concentrations also inhibited algal growth. The toxicity of these fungicides can be summerized as follows (most to least toxic: Kitazin® > Dithane® > Blitox® > Bavistin® > Manzeb) (Shukla *et al.*, 1990).

Nabam at 1 ppm affected photoassimilation of acetate of *Nostoc muscorum* (Moore, 1967). Captan at 1 ppm level suppressed development but not growth of *N.muscorum* (Moore, 1967). Growth of *Anacystis nidulans* was inhibited by a low concentrations of captan (Moore and Dorward, 1968). Phthalimide fungicides examined by Gangawane and Saler (1979) were found to be relatively non- toxic to *Calothrix* sp., *Nostoc* sp. *Aulosira fertilissima, Tolypothrix tenuis* and *Westiellopsis prolifica*; capton (as Hexacap®) was tolerated by all five organisms at upto 500 ppm active ingredient and captafol (as Difolatan®) was tolerated at 100, 300, 300, 500 and 1000 ppm active ingredient by *Calothrix sp., Aulosira fertillisima, Westiellopsis prolifica, Tolypothrix tenuis* and *Nostoc* sp. respectively (Gangawane and Saler, 1979). The growth of *Aulosira fertilissima* was actually stimulated by captafol. The authors attributed this to active metabolism and adaptation to the fungicide. The low toxicity of captan to cyanobacteria observed by Gangawane and Saler (1979) was not recorded by Cameron and Julian (1984), in spite of the fact that both studies employed liquid cultures with growth measurement by pigment extraction. In fact, treatments of greater than 15-24 ppm of captan and also other fungicides (benomyl, carboxin and thiram) were lethal to *Anabaena* sp, *Lyngbya* sp., *Nostoc* sp. and *Mastigocladus laminosus.* These concentrations are significantly lower than recommended field application rates (Cameron and Julian, 1984). Of the four fungicides examined in this study, thiram was the most and benomyl, the least toxic. Low concentrations of each of the four stimulated the growth of all test cyanobacteria, due possibly to the control of fungi or other heterotrophs that may have been present as contaminants (Cameron and Julian, 1984).

Hutber *et al.* (1979), while working with thiram and the pyrimidinol fungicide ethirimol (as Milstem®), observed that their toxicity to the cyanobacteria, *Aphanocapsa* sp, *Anabaena variabilis* and *Nostoc* sp. were intermediate in a range of other pesticides tested in the same study. These were completely inhibitory at 100 ppm. Sharma and Gaur (1981) suggested that a reduction in nitrogen fixation in strains adapted to the mercurial seed dressings PMA (Phenylmercury acetate)) and MEMC (2-methoxy ethylmercury chloride, as Agallol®) may have been caused by the inhibition of some step(s) in nitrogen-fixation and/or protein synthesis. Tolerance limits of the test cyanobacteria to these organomercury fungicides were much lower (up to 0.7 ppm

PMA and 2 ppm MEMC) than to the insecticides lindane and leptophos (up to 15 and 300 ppm, respectively) in the same study.

Ceresan did not inhibit growth of 8 strains of *Anabaena* and 11 strains of *Nostoc* and *Aulosira fertilissima* at 20 kg/ha and of *Tolypothrix tenuis* and *Anacystis nidulans* at 200 kg/ha. Thus the latter two genera are more resistant than the former three genera (Venkataraman and Rajyalakshmi, 1971). They also reported that among 10 *Anabaena* strains tested for their resistance to ceresan, nine could tolerate 10 ppm, but one strain was inhibited by levels higher than 1 ppm (Venkataraman and Rajyalakshmi, 1972).

Panacide (dichlorophene) at 20 ppm was toxic to *A. fertilissima* and *Nostoc* (Gupta and Saxena, 1974). *Westiellopsis* sp., *Aulosira* sp., *Nostoc* sp, *Tolypothrix* sp. and *Calothrix* sp. tolerated the highest concentration (1000 ppm) of brassicol. There were 17.1 and 10.2 per cent decrease of growth over control in *Aulosira* sp and *Calothrix* sp. respectively, at 500 ppm brassicol. In *Westiellopsis* sp. and *Nostoc* sp., brassicol at 1000 ppm caused 48.3 and 16.9 per cent reduction in growth over the control. *Tolypothrix* was not affected by 1,000 ppm brassicol (Gangawane, 1980). Bavistin was toxic to *Calothrix* sp. at 1 ppm level. This fungicide at 50 ppm caused 5.8 per cent increase in growth of *Tolypothrix* sp and at 100 ppm 5.6 per cent increase in growth of *Nostoc* sp.Growth of *Aulosira* sp. was decreased by 21.4 per cent at the level of 100 ppm bavistin. The level of 300 ppm bavistin caused 31.6 per cent reduction in growth of *Westiellopsis* sp. Fytolan was not toxic to *Aulosira* sp. at 100 ppm level, where as this concentration caused 58.3 and 39.7 per cent decrease in growth of *Westiellopsis* sp. and *Calothrix* sp. respectively. Growth of *Nostoc* sp. and *Tolypothrix* sp. was reduced by 18.8 per cent and 6.25 per cent by a 500 ppm level of fytolan (Gangawane, 1980).

The introduction of systemic fungicides revolutionized chemical control of plant diseases. They have several attributes, such as they are absorbed by the plant parts, translocate, persist in the plant and control pathogens away from the place of application. Further, they are required in very small quantity as compared to non-systematic fungicides (Vyas and Shroff, 1992). They are used as seed and soil treatment and also foliar and fruit application. Among them, MBC (methyl benzimidazole carbamate) a degradation product of carbendazim and benzimidazoles was found to be highly toxic to *Calothrix* sp., only 1 ppm reduced the growth of algae considerably (Gangawane, 1980). The fungicide at 50 ppm increased the growth of *Tolypothrix* sp by 5.8 per cent and similar stimulation occurred at 100 ppm in *Nostoc*. The growth of *Aulosira* was decreased by 21.4 per cent at 100 ppm of carbendazim and a dose of 300 ppm decreased it by 31.6 per cent in *Westiellopsis* sp. Thiophanate methyl (Topsin-M) is a member of benzimidazole group, which caused 11.5 per cent reduction in the growth of *Calothrix* at 100 ppm and at similar concentration, *Westiellopsis* was reduced by 31.6 per cent (Gangawane and Kulkarni, 1979). They further observed that in *Aulosira* and *Tolypothrix* at 300 ppm thiophanate methyl reduced the growth by 12.8 and 14 per cent, respectively and at 1000 ppm, it reduced the growth of *Nostoc* sp. by 22.5 per cent. MBC decreased the growth of *Nostoc* and *Calothrix* by 7.9 and 1.0 per cent respectively. *Westiellopsis prolifica* could not tolerate MBC while *Aulosira fertilissima* and *T. tenuis* had little effect at 1000 ppm; *Nostoc* and *Calothrix* were unaffected by 500 ppm of MBC (Gangaware and Fater, 1979).

Hosmani (1992) studied the interaction effects of two fungicides (Kitazin and Baytan) on the heterocyst frequency of *Calothrix membranaceae.* Single application of kitazin at various concentrations indicate that heterocyst frequency decreases with increase in the fungicide concentrations. However, the combined effect differs considerably and expected values indicate that the combinations at 0.0001 ppm is synergistic, while at 0.001 ppm, the effect is additive. This indicates that either 0.0001 ppm or 0.001 ppm has a beneficial effect on the heterocyst frequency. In the other two concentrations the antagonism increases with the increase of the concentrations of the fungicides. Dikshit and Tiwari (1992) observed stimulation in *Anabaena variabilis* by benlate at 100 ppm and by dithane at 100-200 ppm concentrations, while in *Aulosira implexa* var. *tenuis* by 13-bavistin at 100-200 ppm concentrations. Lethal effect was observed in *Anabaena variabilis* at 100 ppm of captan, 400 ppm of cyathion; in *Aulosira implexa* var. *tenuis* at 100 ppm of captan, 200 ppm of dithane and cyathion; in *Aulosira fertilissima* var. *tenuis* at 400 ppm of dithane, 200 ppm of cyathion; in *Scytonema chiastum* at 100 ppm of 13-bavistin, 400 ppm of cyathion and in *Scytonema stuposum* at 400 ppm of bavistin, benlate and dithane, 200 ppm concentrations of captan and cyathion.

Anand and Subramanian (1997) studied the physiological response of *Nostoc calcicola* to fungicides, bavistin and dithane. The alga showed better growth at 1 ppm of both the fungicide and chlorophyll-a was more than that obtained for cells grown without any amendment by BG-11-N medium. Nitrogenase activity was not suppressed completely, rather, dithane showed 65 per cent activity. In dithane treated cultures, the photosynthetic activity was more than with bavistin, but the unamended BG-11-N cultures showed highest activity. The CO_2 fixation was 9.4 moles in dithane and 9.0 moles CO_2 fixed/h/mg Chl.-a in bavistin. Total nitrogen, carbohydrate and protein content were maximum in dithane.

Toxicity-studies of two carbamate insecticides carbaryl and carbofuran and three carbamate fungicides ziram, zineb and mancozeb (commercial formulations) with the N_2-fixing filamentous cyanobacterium (blue-green alga) *Cylindrospermum* sp. were studied by Panigrahy and Padhy (2000). Toxicity to vegetative cells, heterocysts and nitrogen fixation, as well as akinete-formation and–germination are recorded. Under N_2-fixing condition and calcium nitrate supplementation, the levels 100 and 120 ppm carbaryl were the respective lethal concentration$_{100}$ or effective concentration$_{100}$, while the levels 20 to 80 ppm (under N_2-fixing condition) and 20 to 100 ppm (with nitrate supplementation) were the partial lethal concentrations (less than LC_{100}). Under N_2-fixing condition and nitrate supplementation, the levels 2000 and 3000 ppm carbofuran were the LC_{100}, respectively, whereas levels 100 to 1500 ppm and 100 to 2000 ppm were respective partial lethal doses. The level of 25 ppm carbofuran had growth enhancement in both conditions, *i.e.*, without and with nitrate supplementation. The fungicide ziram had growth stimulatory effect on the cyanobacterium at 0.01 ppm level in both nitrogen fixing and nitrate supplemented conditions. The partial lethal range for both growth conditions was 0.5 to 0.13 and LC_{100} levels were 0.17 and 0.2 ppm for N_2-fixing and nitrate supplemented media, respectively. The fungicide zineb had no growth stimulation at the lowest concentration used; partial lethal ranges were 1 to 15 ppm in N_2-fixing and 5 to 20 ppm in nitrate supplemented growth media, and the LC_{100}levels were 20 and 30 ppm in respective conditions. The fungicide mancozeb had no growth stimulation at the lowest dose used; partial lethal levels were 1 to 7 ppm mancozeb and LC_{100} levels were 9 ppm in both growth conditions. Average numbers of vegetative cells between two polar heterocysts were affected by all used pesticides. In agar media, highest permissive concentrations (HPC) were 80 ppm for carbaryl and 50 ppm for carbofuran; LC_{100}were 100 and 600 ppm for the respective insecticides. HPC values of ziram, zineb and mancozeb were 0.15, 5 and 5 ppm, respectively and LC_{100} values of these fungicides were 0.17, 10, and 7 ppm in both N_2-fixing and nitrate supplemented agar media. Values of no observed effective concentration (NOEC), minimum inhibitory concentration (MIC) or least observed effective concentration (LOEC), HPC and LC_{100} of the five pesticides were recorded. Dehydrogenase activity of heterocysts (reduction of 2,3,5-triphenyl tetrazolium chloride or TTC) was affected by these pesticides at levels much higher than the respective LC_{100} levels, but N-contents of cultures in N_2-fixing medium in presence of individual pesticides were affected at lower levels of respective LC_{100} levels, as determined by micro-kjeldahl method. Formation and germination of akinetes were impaired at all the levels below the HPC in agar media of each pesticide. It was observed that these three fungicides had damaging rolls on vegetative cells, heterocyst activity and N_2-fixation than these two insecticides; and ziram was most toxic among the fungicides used.

Rajendran *et al.* (2007) studied that Lower concentrations of Bavistin supported good growth of *Tolypothrix scytonemoides* with maximum protein and pigment contents. The rates of photosynthetic oxygen evolution declined in cells grown in the pesticide-amended medium but rates of respiratory oxygen consumption were enhanced in cells grown in lower concentrations of Monocrotophos and Nimbicidin. Activities of nitrogenase and glutamine synthetase were affected in all the pesticide treatments but nitrogenase activity was enhanced in the presence of Bavistin. Release of ammonia and carbohydrates was enhanced in cells grown in pesticide-amended medium except for cells grown in medium containing Bavistin where the release of carbohydrates was reduced. New polypeptides of ~280, 152, and 25 kDa (in 250 ppm Bavistin), ~58 and 28 kDa (in 0.3 and 0.2–0.4 ppm Monocrotophos, respectively) and ~31, 28, and 26 kDa (in 0.5 and 1.0 ppm Nimbicidin) were detected in the treated cells.

5. Mode of Action of Fungicides

Fungicides are *per ce* toxic to the algal cells and filaments. The fungicides reduced the cell number, affect chlorophyll concentration and responsible for their destruction, decrease oxygen production, check carbon dioxide fixation and caused an increase in the proportion of labeled carbon in sugar and glycine, accompanied by a reduction in labelled carbon in lipids and glutamic acid (Zweig *et al.*, 1968). Rapid and irreversible alterations in cell permeability were observed. Ultra structural changes have also been observed. Usually fungicides altered cell membrane permeability and hence the primary site of fungicide attacks is cell membrane which might change the configuration of sulphydryl group (Vyas and Shroff, 1992).

6. Conclusion

Cyanobacteria are unique group of micro-organism of ecosystem and thus application of fungicides drastically altered their activity. It was observed that most of the studies on effect of fungicides on cyanobacteria are carried out in the laboratory. The effect of fungicides on algae in soil is considerably different from effects on the same algae in laboratory condition. Even of the fungicides are only 1 per cent effective in soil as in pure culture, they could still have dramatic effects on soil nutrition. The environmental factors also affect the chemical toxicity. The above findings clearly indicate the involvement of experimental conditions in regulating the intensity of the observed response of the test organism and further provide evidence of the extreme complexity of the interactions between the fungicides and cyanobacteria in the soil system.

Acknowledgement

The authors express their deepest sense of gratitude to the Principal, G.M.College (Autonomous), Sambalpur and Head,P.G.Department of Microbiology,OUAT, Bhubaneswar for their moral support during the preparation of this manuscript.

References

Anand N and Subramanian TD (1997) Effect of certain pesticides on the physiology of *Nostoc calcicola*. Phykos 36 (1&2): 15-20.

Bisiach M (1970) Algal infestation in Italian soils. Riso 19: 129.

Cameron HJ and Julian GR (1984) The effects of four commonly used fungicides on the growth of cyanobacteria. Plant and Soil 78: 409-415.

Dadhich KS, Verma AK and Venkataraman GS (1969) The effect of *Calothrix* inoculation on vegetable crops. Plant and Soil 35: 377-379.

Das MK and Adhikary SP (2006) Toxicity of different insecticides to rice field cyanobacteria. In: Glimpses of Cyanobacteria. Ed. Gupta RK, Kumar M and Paliwal GS. Daya Publishing House New Delhi, PP. 122-134.

Day PR (1987) Crop improvement: constraints and challenges. In: British crop protection Conference on weeds. BCPC Pub. Surrey, UK, PP.3-12.

De PK (1939) The role of Blue-Green algae in nitrogen fixation in rice fields. Proc Roy Soc London 127: 121-134.

Dikshit G and Tiwari GL (1992) effect of pesticides on certain nitrogen-fixing cyanobacteria. In: Proc. NaH Symp Cyanobacterial Nitrogen Fixation. Ed. Kaushik BD; Ass Publ Co, New Delhi, PP 495-500.

Fitzgerald GP and Skoog F (1954) Control of blue green algae blooms with 2,3-dichloronaphthoquinone. Sewage Ind Wastes 26: 1136.

Fitzgerald GP, Gerloff GC and Skoog F (1952) Studies on chemicals with selective toxicity to blue green algae. Sewage Ind Wastes 24: 888.

Gadkari D (1988) Assessment of the effects of photosynthesis inhibiting herbicides diuron, DCMU, metamitron and metribuzin on growth and nitrogenase activity of Nostoc muscorum and a new cyanobacterial isolate strain G4. Bio Fertile Soils 6: 50-54.

Gangawane LV (1980) Tolerance of nitrogen–fixing blue green algae to brassicol, bavistin and fytolan. J Ind Bot Sci 59: 157-160.

Gangawane LV and Kulkarni L (1979) Tolerance of certain fungicides by nitrogen fixing blue green algae and their side effects on rice cultivars. Pesticides 13: 37.

Gangawane LV and Saler RS(1979) Tolerance of certain pesticides by nitrogen- fixing blue green algae. Curr Sci 48: 306-308.

Gupta GS and Saxena PN (1974) Effects of panacid on some green and blue green algae. Curr Sci 43: 492.

Henriksson E (1971) Algal nitrogen fixation in temperate regions.In: Biological nitrogen fixation in natural and agricultural habitats. Plant and Soil Spec Vol. Ed. Lie TA and Mulder EG PP 415-419.

Hosmani SP (1992) Interaction of two fungicides (Kitazin and Baytan) on the heterocyst frequency of *Calothrix membranacea* Schmidle. In: Proc Natl Symp cyanobacterial nitrogen fixation. Ed. Kaushik BD. Ass Publ Co, New Delhi PP 535-537.

Hutber GN. Rogers LJ and Smith AJ (1979) Influence of pesticides on the growth of Cyanobacteria. Z Allg Mikrobiol 19: 397-402.

IRRI (1991) World Rice Statistics, Los Banos, Philippines.

Kapoor K and Sharma VK (1980) Effect of certain herbicides on survival, growth and nitrogen fixation of blue-green alga *Anabaena doliolum* Bharadwaja. Z Allg Mikrobiol 20: 465–469.

Kaushik BD and Venkataraman GS (1983) Response of cyanobacterial nitrogen fixation to insecticides. Curr Sci 52: 321-323.

Maloney TE and Palmer CM (1956) Toxicity of six chemical compounds to thirty cultures of algae. Water Sewage Works 103: 509.

Misra SG and Mani D (1994) Agricultural Pollution Vol. II. Ashish Publ House, New Delhi 11 P.

Moore RB (1967) Algae as biological indicators of pesticides. J Phycol 3: 4.

Moore RB and Dorward DA (1968) Accumulation and metabolism of pesticides by algae. J Phycol 4: 7.

Padhy RN (1985) Cyanobacteria and Pesticides. Residue Reviews. 95: 1–44.

Panigrahy KC (1984) Physiological and genetical effect of pesticides on blue-green algae: effect of carbamate pesticides. Ph.D. Thesis, Berhampur University (Orissa) India.

Panigrahy KC and Padhy RN (2000) Toxicity of carbamate pesticides to cells,heterocysts and akinetes of the cyanobacterium *Cylindrospermum* sp.Archiv fur Hydrobiol 134: 95-115.

Rajendran UM, Elango K and Anand N (2007) Effects of a fungicide, an insecticide and a biopesticide on *Tolypothrix scytonemoides*. Pesticide Biochem and Physiol 87(2): 164-171.

Roger PA (1988) Reconsidering the utilization of blue–green algae in wetland rice cultivation. In: Biological nitrogen fixation associated with rice production. Eds Dutt SK and Sloger C. Oxford and IBH Publ Co Pvt Ltd. New Delhi, PP. 119–137.

Roger PA and Kulasooriya SA (1980) Blue–green algae and Rice IRRI, Manila, Los Banos.

Schluter M (1964 b) Erste Untersuchungsergebnisse Uber die algizide wirkung funf verschiedener Fungizide Chlorophyceen gegenuber. Deut Fischerel–Ztg. 12: 373.

Sharma VK and Gaur YS (1981) Nitrogen fixation by pesticide–adapted strains of paddy field cyanophytes. Int J Ecol Environ Sci 7: 117–122.

Shukla DS, Chaturvedi R and Srivastava SK (1990) Effect of various fungicides on microbial population of soil and some soil factors. J Environ Biol 11: 291–298.

Singh RN (1961) Role of blue-green algae in nitrogen economy of Indian agriculture, ICAR, New Delhi.

Stratton GW (1983) Interaction effects of permethrin and atrazine combinations towards several non-target micro-organisms. Bull Environ Contam Toxicol 31: 297–303.

Tandon RS, Lal R and Rao VVSN (1988) Interaction of endosulfan and malathion with blue-green algae *Anabaena* and *Aulosira fertilissima*. Environ Poll 52: 1–9.

Venkataraman GS (1972) Algal biofertilizer and rice cultivation. Today and Tomorrow's Printers and Publishers, New Delhi.

Venkataraman GS (1981) Blue-green algae: a possible remedy to nitrogen scarcity. Curr Sci 50: 253–256.

Venkataraman GS and Neelakantan S (1967) Effects of the cellular constituents of the nitrogen-fixing blue-green alga *Cylindrospermum muscicola* on the root growth of rice plants. J Gen Appl Microbiol 13: 53-62.

Venkataraman GS and Rajyalakshmi B (1971) Tolerance of blue-green algae to pesticides. Curr Sci 40: 143–144.

Venkataraman GS and Rajyalakshmi B (1972) Relative tolerance of nitrogen–fixing blue-green algae to pesticides. Ind J Agric Sci 42: 119-121.

Vyas SC and Shroff VN (1992) Effect of fungicides biofertilizer agents: 1. Cyanobacteria. In: Proc Natl Symp Cyanobacterial nitrogen fixation. Ed. Kaushik BD. Ass Publ Co, New Delhi, PP 287–292.

Whitton BA and Mc Arthur K (1967) The action of two toxic quinones on *Anacystis nidulans*. Arch Mikrobiol 57: 147.

Zweig G, Hitt JD and Mc Mohan R (1968) Effect of certain quinones, diquat and diuron on *Chlorella pyrenoidosa* Chick (Emerson Strain). Weed Sci 16: 69–73.

Algal Biotechnology: New Vistas (2010) *Pages* **145–158**
Editor: Mihir Kumar Das
Published by: DAYA PUBLISHING HOUSE, NEW DELHI

Chapter 9

Role of Blue Green Algae in Rice Production

Y.V. Singh*

Centre for Conservation and Utilization of Blue Green Algae, Indian Agricultural Research Institute, New Delhi – 110 012, India

ABSTRACT

Rice is the most important crop of India and it provides 31 per cent of calorie requirement of its population. Rice crop provide a livelihood support system to a large population which is mostly poor as the rice system has a major potential to impact, favorably or adversely, on the world's food security and on its politico-economic stability. In a country like India where more than 80 per cent farmers belong to small and marginal category and most of them cannot afford costly chemical fertilizers, the biofertilizers like blue green algae can make the rice production system more viable and reduce the ecological hazards caused due to synthetic fertilizers. Algae are the heterogenous assemblage of plants that includes prokaryotes and eukaryotic organisms. They are the pioneer colonizers both in hydrosphere and xerosphere and occupy the base of the trophic pyramid. These organisms have been found to synthesize huge amount of organic matter and constitute about 40 percent of the total organic matter synthesized annually on this planet. Supplementary effect of BGA with chemical fertilizer in rice nutrition has been widely reported and BGA may replace up to 10-30 per cent of the chemical fertilizers. Techniques of maintenance and scaling up of the algal inoculums have been standardized under unialgal state on agar slants containing appropriate media.

* E-mail: yvsingh63@yahoo.co.in

Therefore, under present biotechnological era we should take into account more realistic, inexpensive and pragmatic programmes to utilize the biofertilizer, which will not only increase the productivity but also increase the long-term soil fertility. The low cost and environmental friendly blue green algal biofertilizer can be used in vast area of rice cultivation in our country and its production and marketing can be an economically viable enterprise for rural poor and unemployed youths. The role of biofertilizers in sustainable agriculture assumes special significance particularly in the present context of increasing cost and subsidy burden due to chemical fertilizers. Biofertilizers can be taken as an essential component of organic rice farming.

Keywords: *Biofertilizers, Blue green algae, Eco-friendly, Nitrogen fixation, Rice production.*

1. Introduction

Rice crop provide a livelihood support system to a large population which is mostly poor as the rice system has a major potential to impact, favorably or adversely, on the world's food security and on its politico-economic stability. Seventeen Asian countries annually harvest at least 600,000 hectares of rice and 11 of these countries have both food deficit and low per capita income. A dozen of these Asian countries have severe prevalence, and depth, of hunger. The annual requirement of rice is 600Mt in Asia, whereas the rest of the world produces only 60Mt rice per year. If during the period 2003-2030, should there be a succession of years in which rice supply is inadequate; the task of substituting it by alternative foods will be immense, if not impossible.

In intensive rice systems, the indigenous N supply is never sufficient, and mineral N fertilizer inputs represent the largest part of the N cycle. In most Asian countries, where 92 per cent of rice is grown under different ecosystems, irrigated rice farmers apply 100–150 kg N ha^{-1} to dry season rice crops and 60–90 kg N ha^{-1} to wet season crops. The cost of N fertilizer usually represents 10–20 per cent of the total variable production costs. More than 20 per cent of N fertilizer produced worldwide is used in the rice fields of Asia, but N recovery efficiency in most farmers' fields is only about 25–40 per cent of applied N. The requirement for mineral fertilizer may be reduced when organic nutrient sources such as farmyard manure, legume green manure, blue green algae(BGA) and azolla are used. Green manuring and organic manure use, however, have decreased in recent years as mineral fertilizer has become a more convenient and cost effective source of N.

Rice (*Oryza sativa* L.)–wheat (*Triticum aestivum* L. emend Fiori and Paol.) is the major cropping system covering 24 million ha area in Asia and out of this 10.0 million ha area is in India. The rice-wheat system is one of the most productive systems in the world. This system covers 32 per cent of total rice area and 42 per cent of total wheat area and account for about one-third the total rice and wheat production of India. However, this cropping system has shown the symptoms of resource degradation and there are indication of stagnation or even decline in productivity of this cropping system due to the over mining and losses of nutrients from soil varying from 600 to 700 kg/ha every year against the application of 400 kg NPK/ha through chemical fertilizers. So in order to get sustainable production from this system Integrated Plant Nutrient Supply System (IPNS) has been advocated since application of organic manures and biofertilizers can counteract the deleterious effect of continuous use of fertilizer on physico-chemical properties of soil.

The growing concern for environment and energy, increasing cost and unavailability of fertilizers have created considerable interest in the use of biofertilizers as source of plant nutrients for rice- wheat cropping system. Production of industrially fixed nitrogen is an energy intensive process and it is presently based on the mineral oil reserves. Biological nitrogen fixation (BNF), an attribute monopolized by some of the prokaryotes, supports sustainable, environmentally sound agricultural production and holds promise to meet at least a part of the nitrogen required for agricultural production. Various biological systems capable of fixing the atmospheric nitrogen, together contribute about 175 million tons of nitrogen every year as compared to only about 47 million tons fixed by the industry, involving establishment of around 600 units throughout the world at a cost of 15 billion US$ and consuming about 2 million barrels of crude oil every day (Goyal, 2002).

In view of the increasing production needs and decreasing fertility status of most of the Indian soils, emphasis is being given to develop sustainable production system that maintains soil fertility status and protect the environment. Traditional agriculture with legume- cereal rotation, green manuring, use of organic inputs etc. not only allowed recuperation of soil but enriched it with desirable microflora having plant growth as well as health promoting nutrient mobilizing and plant disease suppressing activities. In a country like India where more than 80 per cent farmers belong to small and marginal catagory and most of them cannot afford costly inputs, the organic manures and biofertilizers can make the system viable and reduce the ecological hazards. Therefore, a new idea of modern biotechnological era for increasing crop production should take into account more realistic, inexpensive and pragmatic programmes to utilize the biofertilizer, which will not only increase the productivity but also increase the long-term soil fertility. More than half the world population is dependent upon rice that provides 20 per cent of global human per capita energy and 15 per cent of per capita protein.

Biofertilizers are cost effective, ecofriendly and renewable source of plant nutrients to supplement chemical fertilizers in sustainable agricultural system. These are the preparation of live or latent cells that are used for application of seeds, soil or composting areas with the objective of increasing the number of such organisms in that particular system to accelerate their metabolic activity to augment the extent of the availability of nutrients in a form which can be easily assimilated by plants. The role of biofertilizers in sustainable agriculture assumes special significance particularly in the present context of very high cost of chemical fertilizers. The prokaryotic organisms having capability of fixing atmospheric nitrogen and being used as biofertilizers mainly belong to bacteria, fungi, *Azolla* (a water fern) and blue green algae or cyanobacteria. Based on different cropping system and agroclimatic conditions, one or the other kind of biofertilizer is used so as to get maximum benefit from it. For example, legume *Rhizobium* symbiosis has been exploited as an effective means of raising the nitrogen status of soils, the free living *Azotobacter* is effective for wheat and cotton and the associate symbiont *Azospirillum* have proved to be effective in rice and millets. In water logged rice fields, nitrogen-fixing cyanobacteria in free state and in symbiotic association with *Azolla* provide biologically fixed nitrogen. In other terms, the 'biofertilizers' are also described as microorganisms that add, conserve and mobilize the crop nutrients in the soil. This induced efficient management of the crop nutrients leads to long-term sustainability in crop production. Different biofertilizers can be used as a component of organic farming (Singh, 2007 and Singh and Singh, 2006).

2. The Blue green Algae or Cyanobacteria

Algae are the heterogenous assemblage of plants that includes prokaryotes and eukaryotic organisms. They are the pioneer colonizers both in hydrosphere and xerosphere and occupy the base of the trophic pyramid. These organisms have been found to synthesize 0.8×10^{11} tonnes of organic matter, constituting about 40 percent of the total organic matter synthesized annually on this planet (Goyal, 2002).

Blue green algae/cyanobacteria constitute the largest, most diverse and widely distributed group of prokaryotic microscopic organisms that perform oxygenic photosynthesis. These are also known as myxophyceae, cyanophyceae and cyanobacteria. These are ubiquitous in distribution, more common in tropics; and are able to withstand extremes of temperature and drought. Ability to withstand adverse ecological conditions, capacity to thrive well in hostile environments and response to the onset of dry conditions by entering into dormant resistant state, have distinguished this group of organisms as pioneers of plant succession. Nitrogen fixation is carried out in specialized cells known as heterocysts, which have thick walls and hence, physically prevent the entry of oxygen and provide necessary anaerobic conditions for the activity of the enzyme nitrogenase. Heterocysts have a high rate of respiration that scavenges the diffused oxygen, lack photoreaction II of photosynthesis due to which there is no splitting of water and evolution of oxygen during photosynthesis (Thomas, 1970). Therefore, these cells act as ideal site for nitrogen fixation under aerobic conditions. Although, nitrogenase enzyme is reported to be present in vegetative cells also, however, it remains inactive because of the presence of oxygenic photosynthesis.

The basic significance of the ecological observations on the abundance of blue green algae in Indian rice field soils became apparent when it was recognized that heterocystous forms can fix atmospheric nitrogen that is made available to the plants during life cycle and after its death by decomposition of cells which became available to the subsequent crops. Till the findings of Wyatt and Silvey (1969), only forms like *Anabaena. Nostoc, Cylindrospermum, Tolypothrix, Calothrix, Scytonema, Hapalosiphon, Westiellopsis* and *Stigonema* belonging to orders Nostocales and Stigonematales were considered to be capable of fixing atmospheric nitrogen. Since then, many of the non -heterocystous forms are also reported to fix nitrogen under aerobic or micro-aerophillic conditions except *Gloeocapsa* that can fix nitrogen under aerobic conditions (Stewart *et al.*, 1979). In the later case, it is accomplished because of the spatial and temporal separation of the processes of carbon and nitrogen fixation. Stewart and Lex (1970) suggested that all the vegetative cells of trichomes of diazotrophic blue green algae contain nitrogenase enzyme, which gets inactivated by oxygen. Once nitrogen is fixed, many organisms are known to incorporate ammonia into amino acids by the enzyme glutamate dehydrogenase that is either absent or present in low amounts in these organisms (Pearce *et al.*, 1969). Under the conditions of ammonia limitations, glutamine synthetase and glutamate synthase is the major pathway for ammonia assimilation (Thomas *et al.*, 1975; Wolk *et al.*, 1976).

3. Use of BGA Biofertilizer for Rice Crop

De (1939) was the first to suggest a positive role for the blue-greens in the sustenance of the nitrogen status of rice fields of our country. Since then lot of information has been generated in tropics regarding improvement in the fertility status of rice soils to sustain rice yields by utilizing diazotrophic blue-green algae as biological input (De and Sulaiman, 1950; Venkataraman, 1972; Singh, 1961; Singh and Bisoyi, 1989). Better crop yields under pot culture experiments were attributed to enhanced N levels due to BGA inoculation (Prasad, 1949; De and Mandal, 1956). Singh (1939, 1942) in Uttar Pradesh attributed the enhanced nitrogen contributions to the nitrogen fixing capacity of blue-green algae and the nitrogen contribution was calculated as 48 kg N/ha/crop. They also suggested the persistence of residual effect of these organisms in the following season. Venkataraman (1966) initiated the work on the algalization of Indian rice fields and much interest was generated in this aspect. The All India Coordinated Project on the utilization of blue-green algae in rice fields throughout the country revealed that the supplementation of chemical fertilizer input with blue-greens could conserve up to 30 per cent of the commercial chemical fertilizer (Venkataraman, 1972, 1981). Multilocational trials conducted under varying agroclimatic conditions using different rice varieties indicated that the algal inoculation could save 30 kg N/ha that, however, depends upon the agroecological conditions. (Venkataraman and Goyal, 1969). The performance of *Aulosira fertilissima* in water logged rice fields was assessed and a 10-30 per cent increase in soil nitrogen by BGA incorporation was reported (Singh, 1976; Singh; Bisoyi, 1989 and Singh and Mandal, 1987, 2000). Singh *et al.* (1981) also reported that at amending the flooded soils with BGA released ammonium nitrogen besides, enhancing the level of soil carbon. Venkataraman (1980) suggested the use of local or native strains for an efficient utilization and better crop response. Blue-green algae are also considered to minimize the occurrence of weeds in the rice fields (Subramanyan, 1972) besides producing growth-promoting substances. Extracts of *Cylindrospermum* sp. (Mohan and Mukherjee, 1980) or *Hapalosiphon welwitschii* (Bongale and Bharati, 1990) accelerated the germination of seedlings. Gupta and Shukla (1967) found an increase in the shoot and root length and protein content of rice grains by pretreatment of seeds with *Phormidium* extract. Reclamation of the alkaline soils in Uttar Pradesh was demonstrated by Singh (1950, 1961) by introducing blue-greens along with rice and sugarcane crops. These were reported to reduce the pH of soil and improve upon exchangeable calcium and water holding capacity. Under these situations, Na^+ ions are immobilized and the sodium clay is converted to calcium type (Kaushik, 1989; Apte and Thomas 1984). Rice fields in Kerala having acidic soils with low pH support the growth of several blue-green algae (Anand and Shantha Kumar, 1986). Many cyanobacteria exhibit a wide range of pesticide tolerance (Ahmad and Venkataraman, 1973; Kaushik and Venkataraman, 1983; Pabbi and Vaishya, 1992).

4. Algal Biofertilizer versus Chemical Fertilizer

Biofertilizers are renewable sources of plant nutrients to supplement chemical fertilizers in sustainable agricultural system. Chemical fertilizer is the major supplier of nutrients besides organic and green manures. Despite sizeable increase in the use of chemical fertilizers, the gap between the nutrient removal and replenishment is significantly high. The role of biofertilizers in sustainable agriculture assumes special significance particularly in present context of high cost of chemical fertilizers. The effect of biofertilizer is not as instantaneously obvious as in the case of chemical fertilizers. It is really difficult to believe the effectiveness of the 'cover crop' of these 'Microbugs'. These can be hardly seen with the naked eye but one must have the patience to realise the potential of this inexpensive input, which is self-regenerating and has sustained cumulative effect. As the crop plants utilize the products of biological activities of a biofertilizer, there is always a possibility of their population build up in the soil that may exhibit the benefits after a few superimposed applications. These are also known to play an important role in soil conditioning and may also be source of growth regulators and organic matter.

Algal biofertilizers are much cheaper than the chemical fertilizers available in the market. The crop plants are able to utilize more nutrients from the soil in presence of the algal inoculation because of slow release of the fixed and metabolizable nitrogen. The chemical fertilizers provide the nutrients in a cyclic manner but the blue green algal biofertilizers provide the nutrients in a gradual and linear manner. This results in the utilization of the applied nutrients more effectively. The diazotrophic cyanobacteria are more compatible with nitrate nitrogen than ammonium nitrogen. This leads to rapid multiplication of blue green algae in the photic zone of rice field where oxidizing conditions tend to keep the level of nitrate nitrogen high. In presence of combined nitrogen the blue green algae not only continue to grow but show concentration dependent increase in the uptake of added chemical nitrogen. This metabolized nitrogen which is prevented from being lost through denitrification, percolation and run off, is slowly released through excretion and autolysis (Suseela and Goyal, 1995; Singh and Mandal, 1987, 2000).

Further, the algal nitrogenase has 'switch on' mechanism, which is activated when the level of combined nitrogen falls below a threshold value, which seems to be around 40 ppm. This enables the increased algal biomass to contribute more of the biologically fixed nitrogen when the level of fertilizer nitrogen in the ecosystem is reduced due to progressive utilization and loss. (Goyal, 2002; Singh and Mandal, 1987, 2000).

5. Maintenance and Scaling Up of the Algal Inoculum

The selected algal strains are grown and maintained in unialgal state on agar slants containing appropriate media. The unialgal strains are also maintained in liquid media for inoculum scale up. After a small amount of algal cells is transferred to fresh medium, it is placed under optimal conditions in culture under 3000-4000 lux light intensity and 28±2°C temperature. After sufficient growth is observed for one to two weeks, the cultures are transferred to bigger size flasks having appropriate medium. For maintenance of inoculum, after sufficient growth is observed for 1-2 weeks, the culture is transferred to conditions that are just sufficient to keep them in viable state under lower illumination (50-100 Lux) and temperature of 15-20°C. These conditions maintain the cultures in viable state for longer duration without frequent subculturing. Two major forms of algal nutrition can be distinguished *i e* autotrophy and heterotrophy. Most of the algae being photoautotrophic obtain their energy through absorption of light. Main considerations in developing nutrient media for algal cultivation are:

- ☆ Total salt concentration dependent upon the ecological origin of the organisms.
- ☆ The concentration and composition of major ionic components such as potassium, magnesium, sodium, calcium, sulphate and phosphate.
- ☆ Nitrogen sources–nitrate, ammonia and urea
- ☆ Carbon sources
- ☆ pH, trace elements and vitamins etc.

Transferring the unialgal culture from the slant → flasks → aspirator bottles → reagent bottles → carboys will do the laboratory scaling up of the inoculum. Natural transparent polythene bottles and bags can be used in place of the carboys. About 20-25 days old culture from the carboys is used to inoculate the outdoor ponds.

The laboratory scaled up culture with optical density of around 1.0 measured at 560nm is added to the medium contained in the out-door ponds in the ratio of 1:10 so as to obtain the final OD of about 0.1. This culture is used to inoculate the raceways. Up to the aspirator bottle stage, the cultures are grown in sterilized medium under aseptic conditions. At the carboys stage, boiled water is used for preparing the medium. All the cultures under laboratory conditions are maintained in unialgal state. Inoculation of the multiplication ponds or the race ways can be done by using the scaled up culture of the algae grown in culture room conditions.

6. Algal Biofertilizer Production Technology

An inexpensive rural oriented algal biofertilizer production technology or algalization technology for rice was developed for the welfare of small and marginal farmers. This technology was based on the natural ecology of these microorganisms and largely adopted by the farmers due to simultaneous development of the prototypes for mass production of these algae at their own level. The ratio between the volume to be inoculated and the inoculum should be 10:1. Depending on the requirements, four methods, *viz*. Trough or tank method, pit method, field method and nursery-cum algae method were developed (Venkataraman, 1981).

The basic method involves mass production of a mixture of blue green algae mainly constituting nitrogen fixing *Anabaena, Nostoc, Tolypothrix* and *Aulosira* in shallow trays or cement tanks in open air using clean, sieved farm soil as a carrier material. For each meter square, 2 kg of soil, 100 g single super phosphate are added.

5 ml of malathion per unit or equivalent insecticide is used to prevent breeding of mosquitoes and other insects. The contents are thoroughly mixed and allowed to settle down. BGA starter culture (about 100g) is sprinkled on the surface. The algae multiply and under favourable conditions form a thick mat over the surface in 10-12 days. The contents are allowed to dry and dried flakes are collected, packed and used to inoculate (10kg/ha) rice field. The basic advantage of this technology is that farmers after getting the soil based starter culture can produce the algal biofertilizer on their own with minimum inputs.

Despite all the advantages as mentioned and simplicity of the easily manageable technology, it has not become popular with the farmers (Pabbi and Kaushik, 1997). There are a number of technical environmental and socioeconomic constraints that limit wider adoption of this technology and problems are multifaceted (Singh and Pabbi, 2001).

Efforts have been made to improve the technology by developing new economically feasible protocols for the production of quality inoculum so that these organisms can be potentially exploited on large scale. This is possible if the multiplication is carried out under controlled conditions with defined parameters. The production technology has been substantially improved with the introduction of new and cheap carrier materials that support higher microbial load with longer shelf life thus, considerably reducing the quantity of inoculum per unit area. The basic changes the technology has undergone include a) indoor production of algal biomass under almost pure state and may be under controlled conditions b) a suitable and cheap growth medium for faster growth of the organisms and c) mixing with the suitable carrier in desired quantities. Straw has been used successfully as a carrier material. Shanmugasundaram (1996) prepared the biofertilizer inoculum "cyanostraw" by growing the desired algal strains along with the carrier in polythene bags under controlled conditions to get a product of high quality free from fungi, protozoan or contaminating algae.

Another method involves growth of algae in an indoor production unit that may be a polyhouse or glass house. The technology has been refined by growing the cultures in a polyhouse in which growth is faster under semi-controlled conditions for year round production (Pabbi *et al.*, 2000).

The individual unit in the polyhouse can be RCC, brick and mortar or polythene lined pits on ground and dimension may vary so as to utilize the maximum available space with the operational management and easy convenience. The algae are grown individually in a specially formulated medium inoculated with laboratory grown pure cultures. Once fully grown, the culture is harvested and mixed with rice or wheat straw and sun dried (Kaushik, 1996). The dried material is packed in suitable sized polythene packets, sealed and stored for future use. The final product contains 100,000 to 1,000,000 units or propagules per gram of carrier material and therefore, 1kg material is sufficient to inoculate 1 hectare of rice growing area and the quantity is maintained by periodically measuring growth in terms of optical density or chlorophyll content. This modified material has following advantages over the earlier soil based starter inoculum.

- ☆ All the four different blue green algal strains namey *Anabaena variabilis*, *Nostoc muscorum*, *Tolypothrix tenuis* and *Aulosira fertilissima* are grown separately and then mixed with the carrier in equal proportion, thus, ensuring an equal share of each algae in the final product.
- ☆ The modified nutrient medium ensures rapid growth with reduction in harvesting time ensuring economic viability of the product.
- ☆ The carrier (straw) used can support increased algal load thus, reducing the quantity required per unit area. Being of plant origin these carrier materials are easily degradable also.
- ☆ The ready to use material is supplied to farmers with assured quality ensuring establishment and activity of BGA in the field.

7. Multiplication of BGA in Raceways

The protocol envisages growing the algae separately in near unialgal state and then wet mixing with the clay, sun drying and grinding the flakes to the fine powder. Large-scale multiplication of the selected strains of BGA is done in raceways lodged in a covered space. The length of the raceways can be varied depending upon the availability of land but the width should not be more than 3 m. A raceway is a shallow pond of about 35 cm depth with round corners. A central wall, leaving 1m spaces at either end, divides the pond. In one of the channel so formed, a paddle wheel of appropriate size is fitted which is energised by an electric motor. This is used to agitate the culture and does not allow settling of the algal biomass. Lodging of raceways in a covered space ensures round the year production and provides protection against contamination and attack by the parasites and predators. It also accelerates the algal growth and enables complete control on the growth conditions and harvesting of the culture at the desired growth stage. In the raceways, normally, the nutrients are used at half the recommended strength. The culture is allowed to grow till it attains the optical density of 12.0. After this daily harvesting of the algal culture is done by filtering 1/10 of the total volume of the culture in the raceway. The culture is filtered through cheesecloth and the filterate is poured back into the raceway. The growth rate of the culture is regularly monitored as it tends to slow down with the depletion of the nutrients. Whenever needed appropriate quantities of the nutrients are added to the raceway. Under ideal conditions of 30-35°C temperature and 4-5 k. lux light intensity the productivity of the raceway comes to about 100 g fresh algal biomass m^{-2} day^{-1}. The harvested wet algal biomass is mixed with equal weight of the presoaked locally available clay known as 'Multani mitti' which is mainly montmorilinite. It has a very high water holding capacity and practically does not have any microbial population. The clay is soaked in 2.5 times water at least 24 hr. before use. The algae-clay paste is spread over polythene sheet into about 1 cm thick layer and sun dried. The dried flakes are powdered to 200 mesh and packed in polythene bags. The clay based algal inoculum so prepared has been found to carry the titre value of about 10,000 cfu per gram and shelf life of at least 2 years (Goyal *et al.*, 1997). For inoculating one ha area, only one kg of this inoculum is required. The clay-based inoculum is well mixed with about 10-15 kg sieved farm soil and then broadcasted on clear standing water in the field. The inoculum should never be mixed with any chemical fertilizer and applied.

8. Field Application Technique for BGA

The recommended method of application of the algal inoculum is broadcasting on standing water, about 3 to 4 days after transplantation. After the application of algal inoculum, the field should be kept water logged for about a week's time. If required application of the chemical fertilizers and algal inoculum can be done simultaneously but one must allow a gap of 2-3 days with the application of pesticides. Establishment of the algal inoculum can be observed within a week of inoculation in the form of floating algal mats, more prominently seen in the afternoon.

When chemical fertilizers are used to the extent of only 50-60 kg N/ha, BGA inoculum should be used to obtain 10-15 per cent increase in crop yield. When chemical fertilizers are used at high doses (100 kg N or more), the use of chemical fertilizers should be reduced by 25-30 kg N/ha with the use of BGA inoculum to get the same yield as the full dose of chemical fertilizers. Recommended pest control measures and other management practices do not normally interfere with the establishment and activity of the BGA in the field. Heavily fertilized rice fields generally show profuse growth of green algae that act as weeds and also reduce tillering in the rice plants. These can be differentiated from the blue green algae by the grass green colour and fibrous nature. The green algae turn dark violet when treated with iodine but the blue green algae remain unaffected. The green algae can be removed manually and buried in a pit and if their growth is intense, copper sulphate @ 4.0 kg/ha can be used which will selectively kill the green algae. BGA biofertilizer should be used in every rice crop as a kind of insurance to the crop yield as well as to prevent the deterioration of soil physico-chemical properties.

9. Conclusions

There is immense market potential of blue green algal biofertilizer considering the vast area of about 45 mha under rice cultivation in our country. Use of these cheaper and easily manageable inputs can make the agriculture increasingly eco friendly and induce long term sustainability. Commercial production of biofertilizers, which can be achieved with an effective popularization campaign, is attractively profitable and appears to be the appropriate option for popularization and large-scale adoption of the technology. Those who would like to involved in the production, promotion and extensive use in the field must be trained for the purpose. The gap between the researchers, extension workers and farmers need to be bridged through revamped extension system that ensures minimum technology dissemination loss. There is a need to create viable, remunerative and enabling environment for participation of the private sector in order to realize our goal of maximizing blue green algal biofertilizer production to meet the demand and the quality.

References

Ahmed MH and Venkataraman GS (1973) Tolerance of *Aulosira fertilissima* to pesticides. Curr Sci 42: 108.

Anand N and Shanthakumar Hopper (1986) Blue green algae from several rice fields in Kerala state. India. Hydrobiologia 144: 223-227.

Anand N and Revathy G (1992) Physiological response of heterocystous filamentous blue green algae *Nostoc calvicola* Berb ex. Born. et Flah to varying pH. In: Proc Natl Symp Nitrogen fixation (ed) Kaushik BD, IARI, New Delhi,pp. 151-161.

Anand N, Revathy G and Ladha L (1990) Enzymes of ammonia assimilation in two heterocystous filamentous blue green algae (cyanobacteria) *Cylindrospermum musicola* Kutzex. Norn. *et* Flap and *Nostoc calcicola* Breb ex. Born *et* Flah. In: Perspective in Phycology (ed) Raja Rao VN, Today and Tomorrow's Printers and Publishers. New Delhi, pp. 449-452.

Apte SK and Thomas J (1984)Sodium requirement and metabolism in nitrogen fixing cyanobacteria. J Bio Sci 6: 771-794.

Bognale UD and Bharathi SG (1990) Algal flora of cultivated soils of Karnataka state, India, Phykos 19: 95-113.

De PK (1939) The role of blue green algae in nitrogen fixation in rice fields. Proc Roy Soc London, Series B 127: 121-139.

De PK and Mandal LN (1956) Fixation of nitrogen by algae in rice fields. Soil Sci 81: 453.

Goyal S K (2002) A profile on algal biofertilizer. In: Biotechnology of Biofertilizers (ed) Kannaiyan S, TNAU, Coimbatore, T.N. India. pp.250-258.

Goyal SK, Singh BV, Nagpal, Veena and Marwaha TS (1997) Improved method for production of algal biofertilizer. Indian J Agric Sciences 67(7): 314-315.

Gupta AB and Shukla AC (1967) Studies on nature of algal growth promoting substance and their influence on growth, yield and protein content of rice plants. J Sci Tech, Kanpur 5: 162-163.

Kaushik BD and Venkataraman GS (1983) Response of cyanobacterial nitrogen fixation to insecticides. Current Science 52(7): 321-323

Kaushik BD and Krishnamurthi GSR (1981) Effect of blue green algae and gypsum application on the physico-chemical properties of alkali soil. Phykos 28: 91-94.

Kaushik BD (1989) Reclamation potential of cyanobacteria in slat affected soils. Phykos 28: 101-109.

Kaushik BD (1996) Consolidated report, Mission mode project on Technology and Demonstration of algal biofertilizer, IARI, New Delhi, India.

Marathe KV (1972) Role of some blue green algae in soil aggregation. In: Taxonomy and biology of blue green algae (ed) Desikachary TV, University of Madras, India. pp.328-331.

Misra BB, Nanda DR and Misra BN (1985) Reclamation with blue green algae changes in nucleic acid, protein and nitrogen content of algae exposed to solid waste of a chlor–alkali factory. Envir Pollt 37: 97-104.

Mohan M and Mukerji KG (1980) Effect of extra- cellular products of blue green algae (N_2 fixer and Non-N_2 fixer) on the growth of *Oryza sativa* C.V. Pusa 2-21. National Workshop on algal systems. Indian Society of Biotechnology. IIT,New Delhi 32: 31-35.

Pabbi S and Kaushik BD (1997) Algal biofertilizer technology: New concept and opportunities. The Botanica 47: 43-51.

Pabbi S and Vaishya AK (1992) Effect of Insecticides on cyanobacterial growth and nitrogen fixation. Proc Natl Symp Cyanobact Nitrogen fixation. (ed) Kaushik BD, pp.489-94.

Pabbi Sunil, Prasanna Radha, Dhar Dollly Wattal and Singh PK (2000) Algal biofertilization to rice: Potential and Constraints. In: Biofertilizers, Blue green algae and Azolla. (ed) Singh PK, Dhar DW, Pabbi S and Prasanna R and Arora Anju, IARI, New Delhi. pp. 83-99.

Pearce J, Leach CK and Carr NG (1969) The Incomplete tricarboxylic acid cycle in the blue green alga *Anabaena variabilis*. J Gen Microbio 55: 371-378.

Prasad S (1949) Nitrogen recuperation by blue green algae in soils of Bihar and their growth on different soil types. J and Proc Industry Chem 21: 135.

Roychoudhary P, Kaushik BD, Krishnamurthi GSR and Venkataraman GS (1979) Effect of BGA and Azolla application of the aggregation status of the soil. Curr Sci 48: 454-455.

Sardesh Pande JS and Goyal SK (1981) Distributional pattern of blue green algae in rice field soils of Konkan region of Maharashtra state. Phykos 20: 102-106.

Shanmugasundarm SS (1996) Consolidated report, Mission mode project on technology development and demonstration of algal biofertilizer. MKU, Madurai, India.

Singh RN (1950) Reclamation of usar lands in India through blue green algae. Nature 165: 325-326.

Singh PK and Bisoyi RN (1989) Blue green algae in rice fields. Phykos 28: 181-195.

Singh PK, Panigrahi BC and Satapathy KB (1981) Comparative efficiency of Azolla, blue green algae and other organic manures in relation to N and P availability in a flooded rice soil. Plant and Soil 62: 35-44.

Singh PK and Pabbi S (2001) Blue green algae biofertilizer for sustainable agriculture. Status paper on organic farming. Deptt. of Agriculture and Cooperation. (Fert. Division), Govt. of India.

Singh PK (1976) Algal inoculation and its growth in water logged rice fields. Phykos 15: 5-10.

Singh RN (1939) An investigation in to the algal flora of paddy field soils of the united provinces. Indian J Agric Sci 9: 55–77.

Singh RN (1942) The fixation of elementary nitrogen by some of the commonest blue green algae from the paddy soils of the united province and Bihar. Indian J Agric Sci 12: 743.

Singh RN (1961) Role of blue green algae in nitrogen economy of Indian agriculture. Indian Council of Agricultural Research, New Delhi.

Singh YV and Mandal BK (1997) Nutrition of rice (*Oryza sativa*) through Azolla, organic materials and urea. Indian J Agronomy 42(4): 626-633.

Singh YV and Mandal BK (2000) Rate of mineralization of *Azolla,* other organic materials and urea in water logged soils. Tropical Agriculture (Trinidad) 77 (1): 119-122.

Singh YV and Singh BV (2006) Role of cyanobacteria in scented rice farming. In: Glimpses of cyanobacteria.Daya Publishing House, New Delhi,pp.48-66.

Singh YV (2007) Organic farming and biodiversity.In: Advances in applied Phycology. (Ed)Gupta RK and Pandey VD, Daya Publication, New Delhi, pp.229-249.

Stewart DWP, Rowell P, Ladha JK and Sampio MJA (1979) Blue green algae (cyanobacteria)–some aspects related to their role as sources of fixed nitrogen in paddy soils. In: Nitrogen and rice, IARI, Los Banos: pp.263.

Stewart WDP and Lex M (1970) Nitrogenase activity in *Plectonema boryanum* 594. Arch Microbiol 75: 250-260.

Subhashini D and Kaushik BD (1984) Amelioration of salt affected soils with blue green algae. Influence of algalization on the properties of saline alkaline soils. Phykos 23: 273-277.

Subrahmanyan R (1972) Some observation on utilization of blue green algal mixtures in rice cultivation in India. In: Taxonomy and Biology of blue green algae (ed) Desikachary TV, University of Madras, India.

Subrahmanyan R, Relwani LL and Manna GB (1965) Observations on the role of the algae on that of conventional fertilizers. Curr Sci 33: 485-486.

Suseela MR and Goyal SK (1995) Growth and Nitrogen fixing potential of cyanobacteria. Phykos 34 (1&2): 131-134.

Thomas J (1970) Absence of pigments of Photosystem-II of photosysnthesis in heterocysts of blue green algae. Nature 228: 181-183.

Thomas J, Wolk CP, Shaffer PW, Austin SM and Galonsky A (1975) The initial organic products of fixation of 15N–labeled nitrogen gas by the blue green alga *Anabaena cylindrica*. Biochem Biophys Res Commun, 67: 501 -507.

Tripathi Upasana, Pabbi Sunil and Singh PK (2001) Status of blue green algal biofertilizers in India. In: Recent Advances in the exploitation of blue green algae and Azolla. (eds) Singh PK, Dhar Dolly Watttal, Pabbi Sunil, Prasanna Radha and Arora Anju., IARI, New Delhi,pp. 45-57.

Venkataraman GS and Rajalkshmi B (1971) Tolerance of blue green algae to pesticides. Curr Sci 140-143.

Venkataraman GS (1966) Algalization. Phykos 5: 164.

Venkataraman GS (1972) Algal biofertilizers and rice cultivation. Today and Tomorrow Printers and Publishers, India.

Venkataraman GS (1980) Algal biofertilizer: Potential and problems. In: Proc. Natl. Workshop on algal systems (ed) Seshadri CV, Thomas S and Jeeji Bai N, Indian Society of Biotechnology, IIT, New Delhi,pp. 1-10.

Venkataraman GS (1981) Blue green algae: A possible remedy to nitrogen scarcity. Curr Sci, 56: 549-550.

Venkataraman GS and Goyal SK (1969) Influence of blue green algae on high yielding paddy variety IR8 Sci Cult 25: 58.

Walk CP, Thomas J, Shaffer PW, Austin SM and Galonsky A (1976). Pathway of nitrogen metabolism after fixation of 15N–labelled nitrogen gas by the cyanobacterium. J Biol Chem 251: 5027-5234.

Wyatt JT and Silvey JKG(1969) Nitrogen fixation by *Gloeocapsa*.Science 165: 908-909.

Algal Biotechnology: New Vistas (2010) *Pages* **159–169**
Editor: **Mihir Kumar Das**
Published by: **DAYA PUBLISHING HOUSE, NEW DELHI**

Chapter 10

Biotechnological Relevance of Microbes in Agriculture

Rajan Kumar Gupta*

Department of Botany, Pt. L.M.S. Government Post Graduate College, Rishikesh – 249201 Uttarakhand, India

ABSTRACT

The mother Earth is unique in the solar system not only in giving birth to life but also in supporting enormous bio-diversity. It is believed that the first forms of life to originate; about 3-4 billion years ago on the planet were microbes. Though smallest living creatures and invisible to naked eye, microbes play immense role in human welfare and required for the very sustenance of life on the earth. Soil is the habitat as well as the richest source of diverse forms of microorganisms like bacteria, actinomycetes, fungi, protozoa, algae and viruses. A cultivated, fertile field contains more than 100 million microbes per gram of soil. Soil productivity and sustainability depends on the soil microbial activity. Soil microorganisms play several roles beneficial to agriculture such as organic matter decomposition, re-cycling of plant nutrients in nature, biological nitrogen fixation, phosphate solubilization, plant growth promotion, bio control of pathogens, formation of symbiosis, uptake of plant nutrients etc. Though small in size and simple in structure, microbes multiply very fast and are strong metabolically. There is a tremendous scope to exploit the chemical and metabolic potentials of microorganisms in agriculture.

Keywords: *Microbes, Biofertilizers, Biocontrol, Bioenergy, Bioprocess.*

* E-mail: rajankgupta1@rediffmail.com

1. Introduction

Micro-diversity is very important for maintaining the key ecosystem–level functions (*e.g.* decomposition, nutrient mineralization). Decomposition and nutrients are needed for crop productivity. Decomposed matter from dead remains of plant and animals serve as total soil organic matter. If green manures and crop residues are added in soil, the size of microbial community gets increased. The soil microbes have either beneficial or harmful effects on the development of plant. The micro-organism are intimately associated with roots of plant, therefore, any toxic or beneficial substance produced by them has direct effect on plant. The soil microorganisms catalyze the reactions in vicinity of roots in soil and produce CO_2 and form organic acids that in term solubilize the inorganic nutrients of plants. Aerobic bacteria utilize O_2 and produce CO_2, therefore, lower O_2 and increase CO_2 tension that reduces root elongation and nutrient and water uptake. There are many soil microorganism which produce growth stimulating substances and release elements in organic forms through the process of mineralization, many soil organisms including soil fungi and bacteria produce growth stimulating substances such as Indole acetic acid, Gibberellin and Cytokinim in the soil. *Fusarium* species are found to secrete Gibberellin and Gibberellic acid. In the absence of oxygen some soil micro-organism secrete chemicals such as, aldehydes, organic acids, etc. which may show toxic effects on many plants. Toxin secreting microbe may be fungi, bacteria and algae. *Fusarium lini,* which causes wilt of flax secretes, HCN, a deadly poisonous substance and *Fusarium udum* causing wilt of pigeon pea secretes fusaric acid in the roots of the host plants. The soil micro-organism influence phosphorous availability to plant through the process of mineralization and immobilization. However when plant suffers from nutrient scarcity during summer in tropical areas the micro-organisms release the immobilized nutrients. Therefore, they act as sink between soil and plant roots in nutrient poor system. Some soil microbes change the availability or toxicity of sulpher to plants. Many soil microbes improve aeration of soil. Burrowing worms are also helpful in improving the aeration and percolation. Bacteria, blue green algae and some other micro-organisms secrete mucilaginous substances which bind the soil particles into soil aggregator. Now it is clear that soil micro diversity play a critical role in maintaining the productivity of agricultural fields.

Thus soil fertility is a sustainable power to produce good yields of high quality which is directly associated with soil micro diversity.

2. Biotechnological Applications in Agriculture

Biotechnological applications in agriculture can be grouped under four main headings *viz.* (1) Biofertilizers (2) Bio-Control (3) Bio-Energy (4) Bio Processing for value addition to agricultural and agro-industrial wastes.

2.1 Bio-Fertilizers in Agriculture

Bio-Fertilizers are products consisting of selected and beneficial microorganisms which are known to improve plant growth through supply of plant nutrients. Bio-Fertilizers are also known as microbial inoculants. In nature soil contains many beneficial microorganisms, which are used in preparation of biofertilizers to exploit their activities in crop production.

2.2 *Rhizobium* Biofertilizer (Figure 10.1)

This consists of *Rhizobium* bacteria which form root-nodules in leguminous plants and fix atmospheric nitrogen in the nodules in a symbiotic association. *Rhizobium* biofertilizer is usually coated seeds before sowing and a sticker solution is used to facilitate proper coating. There is specificity in legume-rhizobium symbiosis hence specific culture should be used for each legume. The benefits of *Rhizobium* biofertilizer are as follows:

- ☆ The legume crops obtain nitrogen from atmosphere.
- ☆ It saves fertilizer nitrogen.
- ☆ It improves soil fertility.
- ☆ It Improves legume growth and yield.
- ☆ The cereals get fixed nitrogen in mixed cropping and in crop rotations.
- ☆ It is a low cost technology suitable for rained farming.
- ☆ It is environment–friendly technology.

2.3 *Azotobacter*–Biofertilizer

Azotobacters are free living nitrogen fixing bacteria. They also produce several plant growth promoting substances. Seed treatment of *Azotobacters* has given improved and faster seedling growth and higher productivity in vegetables, cotton, sugarcane and paddy. Yield increases of 10-20 per cent are found.

2.4 *Azospirillum* Biofertilizer (Figure 10.2)

Azospirillum are nitrogen fixing bacteria which live in a loose-symbiosis in the root-cortex of several tropical crops, cereals, millets and grasses. They are known to improve plant growth through nitrogen fixation and production of growth substances. Inoculation of sorghum, rage, paddy with *Azospirillum* bacteria has resulted in improved yield to an extent of 10 to 30 per cent.

2.5 Algal Biofertilizers

Blue green algae (BGA) have the ability to carry out both photosynthesis as well as nitrogen fixation. BGA are grown in shallow standing water in paddy field and when the surface is covered with algal growth the water is drained out. The algal culture on surface of the soil is dried into flakes and collected to use as algal biofertilizer. A bag of 40Kg algal flakes is used to inoculate an acre paddy field. Application of BGA to paddy is known to help in saving of 50 per cent fertilizer nitrogen.

2.6 *Azolla* Biofertilizer (Figure 10.4)

Azolla is a small floating water harbours nitrogen fixing cyanobacteria. *Azolla* grows and multiplies very fast and spreads on the surface of shallow layer of water in paddy soils thus providing abundant biomass and nitrogen to soil. Use of *Azolla* to

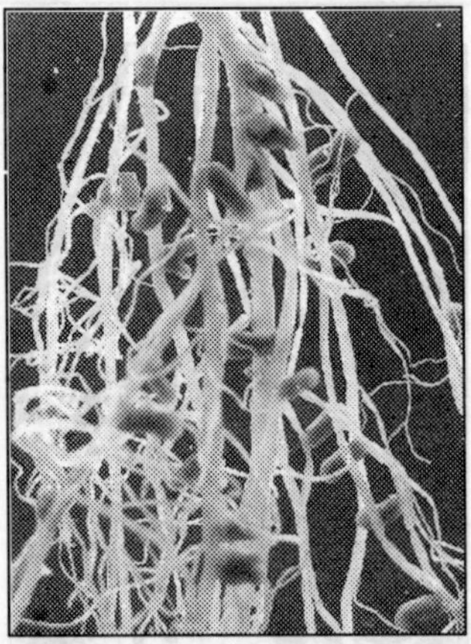

Figure 10.1: Soil Bacterium *Rhizobium leguminosarum*

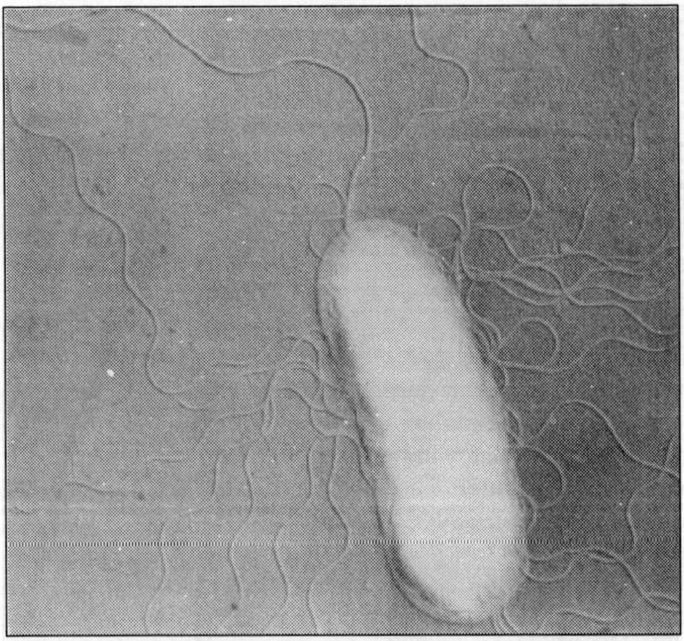

Figure 10.2: *Azospirillum*

paddy cultivation is popular in china and Vietnam. Farmers can themselves maintain a stock culture of *Azolla* in small ponds or cement rings and use for their field application. Application of 15 to 20 tones of *Azolla* can supply 30 to 40 Kg N/ha. Yield increases up to 50 per cent are obtained by use of *Azolla* in paddy.

2.7 Cyanobacterial Biofertilizer

The blue-green algal (cyanobacterial) biofertilizer is an easily manageable, self-generating system, which not only provides valuable nutrients to plant in terms of

Figure 10.3: Cyanobacterial Biofertilizer in Paddy Field

Figure 10.4: *Azolla*

nitrogen, amino acids and growth promoting substances but also improves soil health and texture. A conservative estimate indicates that these BGA contribute 25-30 Kg nitrogen per hectare per season to rice crop.

Amongst the array of biofertilizers developed for different crops, cyanobacteria constitute the most potential inputs in rice field.

2.8 Phosphate Solubilizing Bacteria (PSB)

There are many soil microorganisms which help in solubilization of bound phosphorus in soil. Thus these microorganisms improve phosphate availability to crop plants resulting in improved growth and yield of crops. The bacteria like *Pseudomonas striota, Bacillus megatherium* are important phosphate solubilizing soil microorganisms and are used to prepare biofertilizers.

2.9 Mycorrhiza

There are some fungi in soil which form symbiotic association with plant roots and such fungi are called mycorrhiza. Mycorrhizae fungi are known to improve plant growth by the following mechanisms.

- ✰ Increased up-take of nutrients from the soil particularly the less mobile nutrients like phosphorus, zinc and micronutrients.
- ✰ Increased drought tolerance.
- ✰ Protection against soil-borne plant pathogens biocontrol.
- ✰ Increased hormonal activity biostimulant.
- ✰ Increased nodulation and nitrogen fixation in legumes.

The Vascular Arbuscular Mycorrhizal technology is now used in nursery grown and transplanted horticultural and forest plants (Figure 10.5).

2.10 Plant Growth Promoting Rhizobacteria (PGPR)

There are many microorganisms which colonize the plant roots and it produce growth promoting substances. These are called as Plant Growth Promoting Rhizobacteria or in short P.G.P.R.

2.11 Compost Inoculant and Phosphor-Compost

There are may fungi which can degrade cellulose and lignin. Such fungi help in faster degradation of organic matter and quick compost making. Mushroom fungi are useful to degrade lignin rich organic wastes like coir-pith. The availability of phosphorous from low-grade rock phosphate can be increased by adding rock-phosphate at the time of composting.

3. Bio-Control in Agriculture

There is increasing concern all over the world about the environmental pollutions and health hazards caused indiscriminate use of pesticides in agriculture. Bio-Pesticides are eco-friendly sustainable, cost effective and also help to get premium price to farmers for their produce at international market. Following microorganisms are now widely used in biocontrol of pests and diseases.

3.1 Nucleopolyhedosis Virus

This is a virus pathogenic to several insect pests of crop plants. Bio-pesticide using NPV is found effective against podborers, bollworms, Caterpillars causing damage to pulses, cotton tobacco, vegetable etc., Bio-pesticide made of NPV is produced.

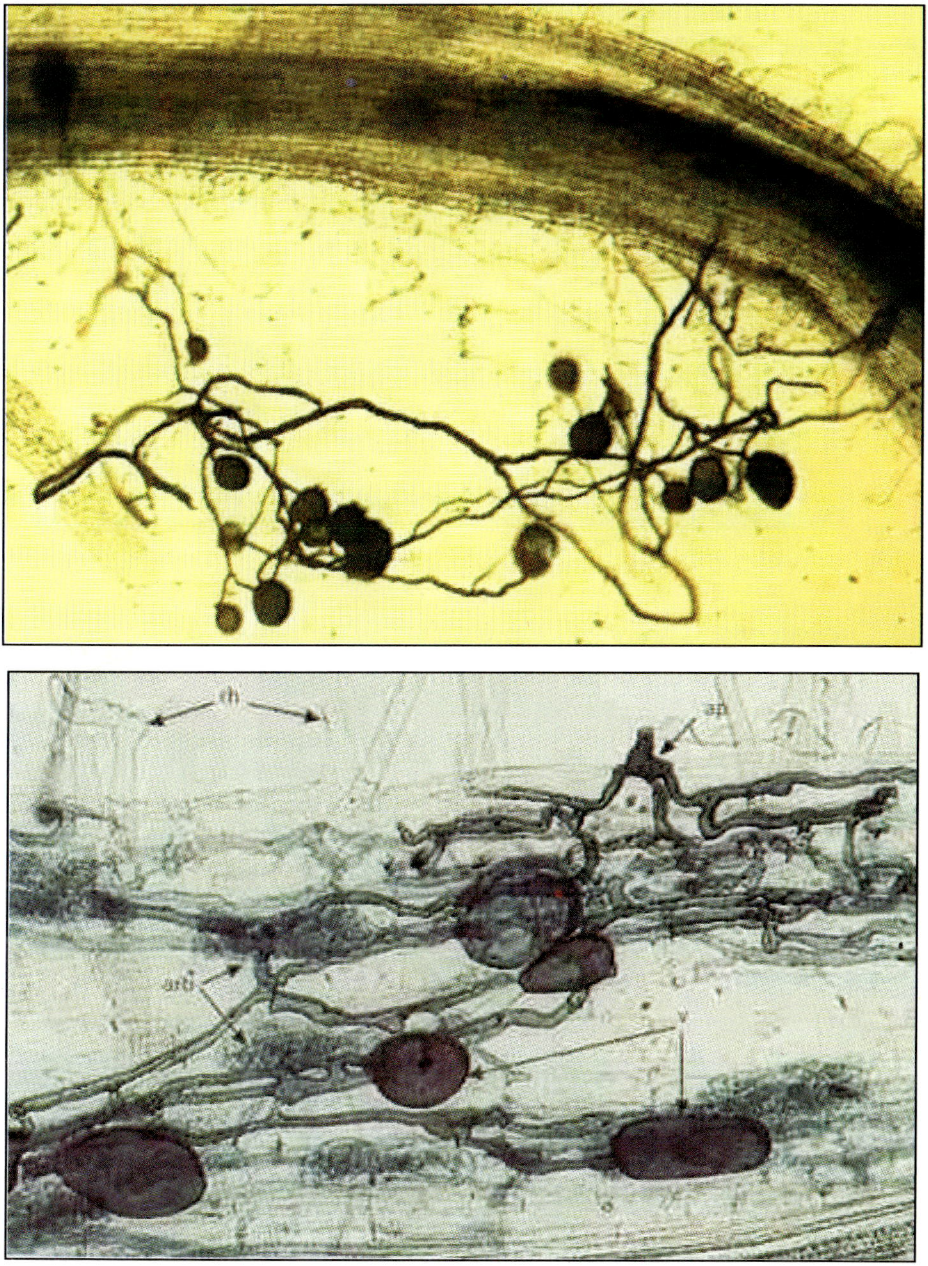

Figure 10.5: Fungal Hyphae with VAM Spores

3.2 *Bacillus thuringiensis* (Figure 10.6)

Bacillus thuringiensis or B.T is a bacterium capable of killing several insect pests of crop plants. It produces a toxin which kills insects. Biopesticide made at BT bacterium available. The BT toxin producing gene is also isolated and transferred to cotton, tobacco etc., and such transgenic Crops become resistant to insect pests.

Figure 10.6: *Bacillus thuringiensis*

3.3 *Trichoderma* (Figures 10.7 and 10.8)

It is a fungus that is useful to control several plant diseases caused by soil-borne plant pathogens. This fungus is also known to stimulate plant growth and solubilize insoluble phosphorous. It is widely used in control of diseases of horticultural crops. *Trichoderma* is commercially produced in India.

3.4 *Pseudomonas* (Figure 10.9)

Pseudomonas fluorescence, P. putida, etc., are known to control plant diseases caused by soil-borne plant pathogens. They also help in phosphate solubilization.

3.5 Bio-Energy

Fuel is as important as food. Bio-methanation of agricultural and agro-industrial wastes is simplest and cost effective way to meet fuel and energy needs for domestic and industrial uses. Bioenergy is renewable energy made available from materials derived from biological sources. In its most narrow sense it is a synonym to biofuel, which is fuel derived from biological sources. In its broader sense it includes biomass, the biological material used as a biofuel, as well as the social, economic, scientific and technical fields associated with using biological sources for energy. This is a common misconception, as bioenergy is the energy extracted from the biomass, as the biomass is the fuel and the bioenergy is the energy contained in the fuel. Anaerobic digestion of organic matter concomitantly provides fuel and compost. Methanogenic bacteria are involved in methane gas production; proper understanding of the physiological and growth requirements of methanogens and suitably manipulating the digester will help in increased gas production per unit substrate. All agricultural and agro-industrial organic wastes if subjected to anaerobic digestion the energy and compost needs can be met to a substantial extent.

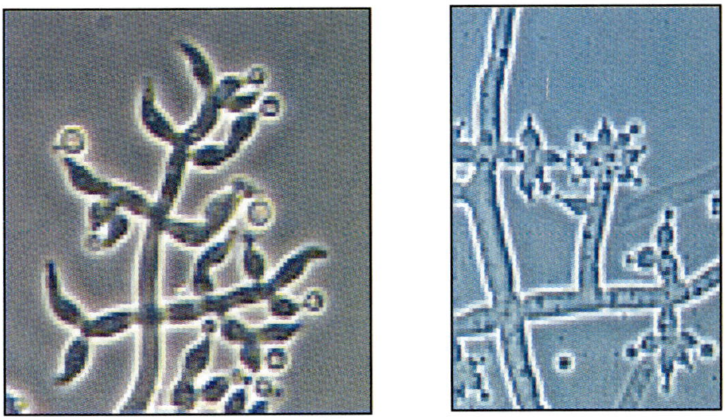

Figure 10.7: ***Trichoderma conidiophore***

Figure 10.8: ***Trichoderma pustules***

3.6 Bio-Processing

Farm wastes and agro-industrial organic wastes can be converted into value added products by using selected microorganisms. Production of mushrooms on the compost prepared from farm wasted is well known. The algae *spirulina* is commercially produced for SCP and medicinal values. Single cell protein can be produced by growing an yeast in waste coconut processing industries, as well as sugar industrial wastes. There are several microorganisms which produce industrially useful products like organic acides, food colours, stabilizers, food and industrial enzymes.

The ever increasing population is posing a tremendous pressure on limited natural resources of cultivable land to fulfill the requirement of food grains. It is estimated that for feeding a population of 1.4 billion by 2025, our country will need to produce 3.1 mt food grain and to achieve this, India will need at least 45 mt plant nutrients, out of which at least 35 mt should be chemical fertilizer sources and rest 10 mt nutrient should come from other sources like biofertilizers or manure. The present day chemical fertilizers based agriculture is unsustainable and slowly leading to

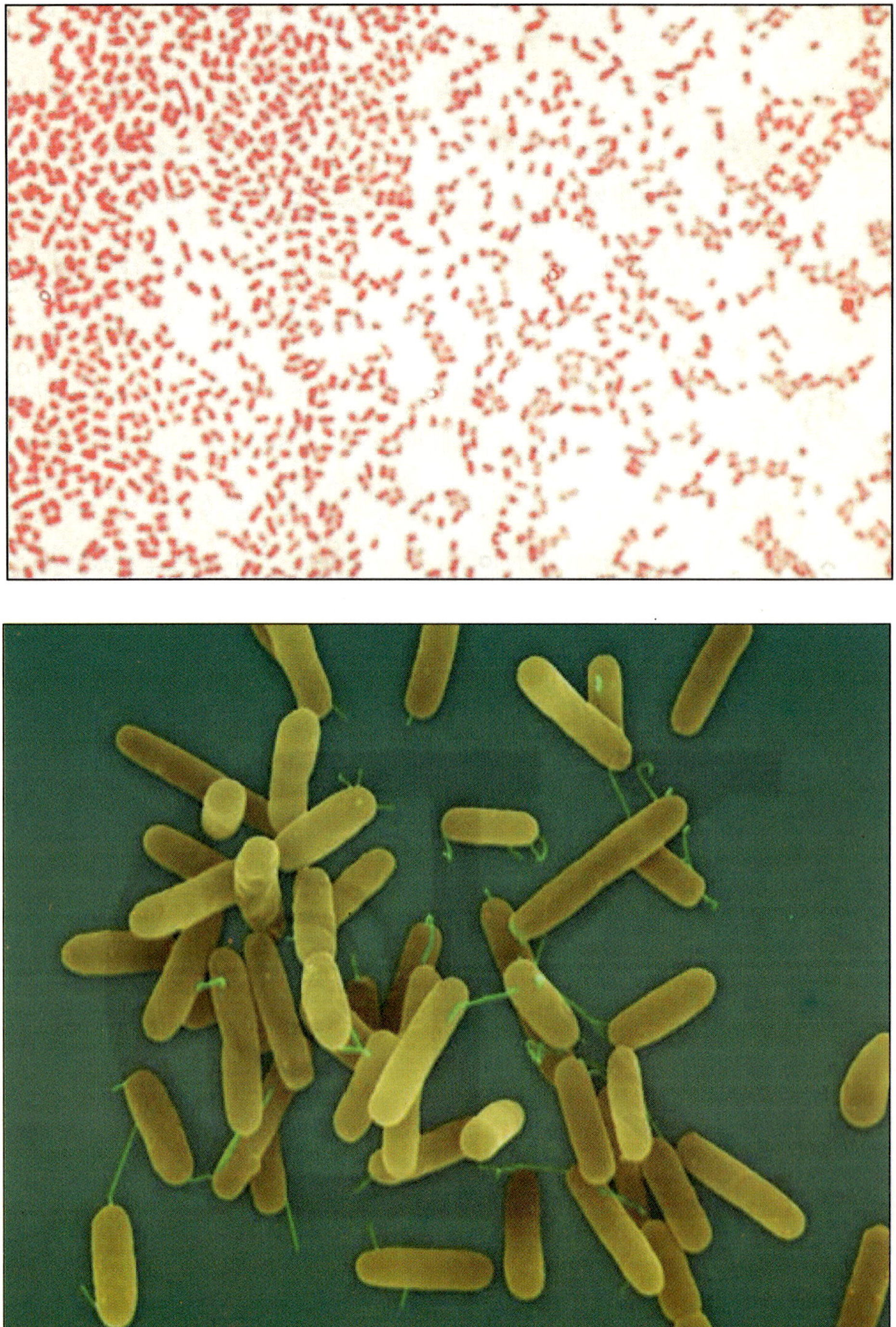

Figure 10.9: *Pseudomonas aeruginosa* in Low and High Power

ecological crisis in terms of decreased soil fertility. The adverse effect of chemical fertilizers acts as stimuli to develop alternate system *i.e.* biological system to cope with the increasing demand of fertilizers to obtain the food grains by maintaining the ecosystem sustainable. In this way, the concept of 'Organic Agriculture' originated advocating minimum use of the chemical fertilizer and increasing dependence on biological inputs like compost, farm yard manure, green manures and biofertilizers.

Acknowledgement

The author is thankful to UGC, Delhi and UCOST, Dehradun for financial help, Principal, Govt. P.G. College, Rishikesh for laboratory facilities.

Algal Biotechnology: New Vistas (2010) *Pages 170–175*
Editor: **Mihir Kumar Das**
Published by: **DAYA PUBLISHING HOUSE, NEW DELHI**

Chapter 11

Lipids and Fatty Acids from Marine Algae: A Potential Biofuel Resource

S. Chakraborty[1]* *and T. Bhattacharya*[2]

[1]*Department of Biological and Environmental Science, N.V. Patel College of Pure and Applied Sciences, V.V. Nagar – 388 120, Gujarat, India*

[2]*Department of Environmental Science and Technology, Institute of Science and Technology for Advanced Studies and Research, V.V. Nagar – 388 120, Gujarat, India*

ABSTRACT

Marine algae contain lipids and fatty acids up to 30 per cent as membrane components, storage products, metabolites and sources of energy. Lipid accumulation in algae typically occurs during periods of environmental stress, including growth under nutrient-deficient conditions. Biochemical studies have suggested that acetyl-CoA carboxylase (ACCase), a biotin-containing enzyme that catalyzes an early step in fatty acid biosynthesis, may be involved in the control of this lipid accumulation process. Therefore, it may be possible to enhance lipid production rates by increasing the activity of this enzyme via genetic engineering. These microalgal strains with high oil or lipid content are of great interest in the search for a sustainable feedstock for the production of biodiesel.

* E-mail: su_kalyanc@yahoo.co.uk

While a number of bio-feedstock is currently being experimented for biodiesel production, algae have emerged as one of the most promising sources for bio diesel production. Though research into algae as a source for bio diesel is not new, the current oil crises and fast depleting fossil oil reserves have made it more imperative for organizations and countries to invest more time and efforts into research on such suitable renewable feedstock. Though some of the countries have already advocated this feedstock and have started large projects on algal biodiesel, but still it has to be implemented at large scale in a country like India where the total coast line, excluding its islands is 5495KM in length. Indian coastline is one of the most rich in terms of macro and micro algae. Therefore we have the potential to harness this magnanimous resource into a less polluting source of renewable energy. However we still need to focus on commercial cultivation and production of bio diesel from marine algae which requires state of the art infrastructure as well as policy support. This can be a step forward to benefit human welfare in the global scenario of fuel crisis and its natural availability would affect the economic rates of the underprivileged population.

Keywords: *Lipids, Fatty acids, Marine algae, Biofuels.*

1. Introduction

The need of energy is increasing continuously, because of increases in industrialization and population. The basic sources of this energy are petroleum, natural gas, coal, hydro and nuclear. The major disadvantage of using petroleum based fuels is atmospheric pollution created by the use of petroleum diesel. Petroleum diesel combustion is a major source of greenhouse gas (GHG). Apart from these emissions, petroleum diesel is also major source of other air contaminants including NOx, SOx, CO, particulate matter and volatile organic compounds. Therefore Biomass has been focused on as an alternative energy source, since it is a renewable resource and it fixes CO_2 in the atmosphere through photosynthesis. If biomass is grown in a sustained way, its combustion has no impact on the CO_2 balance in the atmosphere, because the CO_2 emitted by the burning of biomass is offset by the CO_2 fixed by photosynthesis. Among biomass, algae (macro and microalgae) usually have a higher photosynthetic efficiency than other biomass. On the other hand India has a vast coastline of over 7500 km including the shorelines, estuaries and backwater areas along with a tropical climate that results in abundance of natural populations of varied algal diversities. Phycological researches have also made us aware of the unique chemical composition of marine algae and the potential for developing a wide variety of commercial products from marine algae. As the supply of fossil fuel decreases, mass culture techniques, both in ocean and in controlled coastal facilities have great potential to provide industrially significant fuel alternatives from marine algae. Work has already begun to assess marine and other euryhaline algae as basic resources for biomass, with the expectation that through fermentation biomass can be easily converted to methane gas, ethanol and other useful chemicals. Considerable literature gives the insight that microalgae contain lipids and fatty acids as membrane components, storage products, metabolites and sources of energy. Macroalgae along with diatoms, and cyanobacteria (categorized collectively as "Microalgae") have been found to contain proportionally high levels of lipids (over 30 per cent). According to some estimates, the yield (per acre, say) of oil from algae is over 200 times the yield from the best-performing plant/vegetable oils. While a number of bio-feedstock is currently being experimented for biodiesel production, algae have emerged as one of the most promising sources for biodiesel production. Though research into algae as a source for biodiesel is not new, the current oil crises and fast depleting fossil oil reserves have made it more imperative for organizations and countries to invest more time and efforts into research on suitable renewable feedstock such as algae. Therefore it is the prime time to have a serious discussion and start a national effort to harness the power of algae for the benefit of mankind. (Recently DBT has started similar networking in India).

2. Algal Lipids and Fatty Acids

Marine macro algae, representing a large and diverse group of marine organisms have been point of attraction for the last few decades for their various nutritional and biologically active compounds potentiality. The fatty acid composition has been one of such distinguishing important area of research. Marine algae typically have a high content of Long chain polyunsaturated fatty acids. The chloroplast fatty acids of eukaryotic algae and of vascular plants are largely poly-enoic acids, of which α-linoleic acid (18:3α) almost exclusively found in chloroplasts is a major constituent in the leaf tissues of higher plants and certain group of algae (Nichols, 1970). Lipids represent considerable portion of algal dry matter and show an interesting polyunsaturated fatty acid composition particularly regarding with ω3 and ω6 acids which play a role in the prevention of cardiovascular diseases. Lipid extracts of some edible seaweed showed antioxidant activity and a synergistic effect with the tocopherol (Le Tutour, 1990).

PUFAs consist of two families (ω-3 or n-3 and ω-6 or n-6) of FAs containing 18-22 carbon atoms. The omega designation differentiates the fatty acids in relation to the position of the first double bond, counting from the methyl end of the molecule.

In recent years, fatty acids (FAs) in marine algae have aroused considerable interest among researchers. This is because marine plants can produce C_{18} and C_{20} polyunsaturated fatty acid (Kayama *et al.*, 1989). Red algae are rich in PUFA with 20 carbon atoms, mainly arachidonic acid (AA) and eicosopantenoic acids, which are precursors of prostaglandins (PG). There are also other published reports on algae specially the marine species being rich potential sources for these identified pharmaceutically valuable fatty acids (Cohen, 1990; Cohen and Simon, 1991; Seto *et al.*, 1984. The distribution of lipids in algae is affected by different factors. First of all, it is determined by systematic position of algae. All of the algal divisions are distinctive in their lipid quantity and even quality. They have a specific PUFA composition and especially differ in some usual and less common phospholipids (PL) patterns. The best chemotaxonomical markers are phosphatidylcholine (PC), phosphatidylserine, diacylglyceryl-(N, N, N-trimethyl)–homoserine (DGTS), diacylglyceryl-hydroxymethyl-(N, N, N-trimethyl)-b-alanine (DGTA) and inositolphosphoryl ceramide (IPC)].

Lipid accumulation in algae typically occurs during periods of environmental stress, including growth under nutrient-deficient conditions. Biochemical studies have suggested that acetyl-CoA carboxylase (ACCase), a biotin-containing enzyme that catalyzes an early step in fatty acid biosynthesis, may be involved in the control of this lipid accumulation process. Therefore, it may be possible to enhance lipid production rates by increasing the activity of this enzyme via genetic engineering.

2.1 Bio-fuel Generation from Algae

Shay reported that algae were one of the best sources of biodiesel. In fact algae are the highest yielding feedstock for biodiesel. It can produce up to 250 times the amount of oil per acre as soybeans. In fact, producing biodiesel from algae may be only the way to produce enough automotive fuel to replace current gasoline usage. Algae produce 7 to 31 time greater oil than palm oil. It is very simple to extract oil from algae. The best algae for biodiesel would be microalgae. Microalgae are organisms capable of photosynthesis which are less than 2 mm in diameter. Macroalgae, like seaweed, is not as widely used in the production of biodiesel. microalgae has much more oil than macroalgae and it is much faster and easier to grow. Microalgae can provide several different types of renewable biofuels. These include methane produced by anaerobic digestion of the algal biomass biodiesel derived from microalgal oil and photobiologically produced biohydrogen. The idea of using microalgae as a source of fuel is not new but it is now being taken seriously because of the escalating price of petroleum and, more significantly, the emerging concern about global warming that is associated with burning fossil fuels.

3. Trans-Esterification Technology for Production of Biodiesel

There are three basic routes to biodiesel production from oils and fats:

- ☆ Base catalyzed transesterification of the oil.
- ☆ Direct acid catalyzed transesterification of the oil.
- ☆ Conversion of the oil to its fatty acids and then to biodiesel.

Most of the biodiesel produced today is done with the base catalyzed reaction for several reasons:

- ☆ It occurs in low temperature and pressure.
- ☆ It yields high conversion (98 per cent) with minimal side reactions and reaction time.
- ☆ It is a direct conversion to biodiesel with no intermediate compounds.
- ☆ No exotic materials of construction are needed.

The chemical reaction for base catalyzed biodiesel production is depicted below.

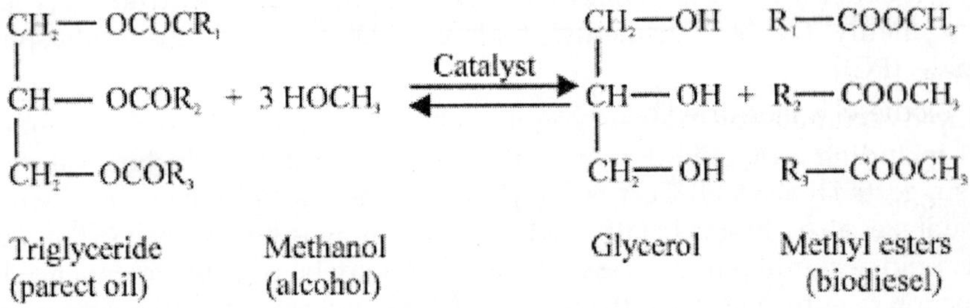

After site selection and sample collection from the prospective places, they are cleaned and brought to the laboratory in insulated containers. Within six hours the samples are washed twice with seawater followed by freshwater and single distilled water to remove the adhering impurities, sand debris, and epiphytes.

Oil Extraction

Algae are ground with mortar and pestle as much as possible. The ground algae are dried for 20 min at 80°C in a incubator for releasing water. Hexane and ether solution (20 and 20 mL) are mixed with the dried ground algae to extract oil. Then the mixture is kept for 24 h for settling.

Biomass Collection

The biomass is collected after filtration and weighed.

Evaporation

The extracted oil is evaporated in Vacuum to release hexane and ether solutions using rotary evaporator.

Mixing of Catalyst and Methanol

0.25 g NaOH is mixed with 24 mL methanol and stirred properly for 20 min.

Biodiesel Production

The mixture of catalyst and methanol is poured into the algal oil in a conical flask. The following reaction and steps are as follows:

Transesterification

The reaction process is called transesterification. The conical flask containing solution is shaken for 3 h by electric shaker at 300rpm.

Settling

After shaking the solution is kept for 16 h to settle the biodiesel and sediment layers clearly.

Separation of Biodiesel

The biodiesel is separated from sedimentation by flask separator carefully. Quantity sediment (glycerine, pigments, etc.) is measured.

Washing

Biodiesel is washed by 5 per cent water until it becomes clean.

Drying

Biodiesel is then dried by using dryer and finally kept under the running fan for 12 h.

Storage

Biodiesel production is measured by using measuring cylinder; pH is measured and stored for analysis.

4. Conclusion

Algae are an economical choice for biodiesel production, because of its availability and low cost. The expected outcome of this concept is some good sources of fatty acids which have promises as future biofuel. This can be a step forward to benefit human welfare in the global scenario of fuel crisis and its natural availability would improve the economic rates of the underprivileged population. Further research should be done having macroalgae and microalgae to compare the ratio of biodiesel production, chemical analysis and statistical significance.

References

Cohen Zvi (1990) J Am Oil Chem Soc 67: 916.

Cohen Zvi and Cohen Simon. 1991. Ibid.68: 16.

Kayama M, Araki S and Sato S (1989) In: Ackman RG (Ed) Marine Biogenic Lipids, Fats and Oils. Vol 2. CRC Press. Florida, pp.3.

Le Tutour B (1990) Antioxidative activities of algal extracts, synergistic effect with vitamin E. Phytochem 29(12): 3759-3765.

Nichols BW and Wood BJB (1970). Comparative lipid biochemistry of photosynthetic organisms. In: Harbourne JB (ed.), Phytochemical Phylogeny. Academic Press. Inc. New York. pp 105–118.

Seto A, Wang HL and Hesseltine CW (1984). Ibid. 61: 892.

Sharif Hossain ABM., Salleh Aishah (2008) Biodiesel Fuel Production from Algae as Renewable Energy. American J Biochem and Biotech 4 (3): 250-254.

Shay EG (1993) Diesel fuel from vegetable oils: Status and Opportunities. Biomass Bioenergy 4: 227-242.

www.biodiesel.org/.,National Biodiesel Board, 2002. USA.

Algal Biotechnology: New Vistas (2010) *Pages* ***176–193***
Editor: **Mihir Kumar Das**
Published by: **DAYA PUBLISHING HOUSE, NEW DELHI**

Chapter 12

Algal Biodiesel: Procedures and Resources for Laboratory Study

*Simrat Kaur[1] *, H.K. Gogoi[1], R.B. Srivastava[1] and M.C. Kalita[2]*

[1]*Bioenergy Division, Defense Research Laboratory (DRDO), Assam*
[2]*Department of Biotechnology, Gauhati University, Guwahati, Assam*

ABSTRACT

In the global trend of increasing primary energy consumption and large CO_2 emissions into the atmosphere, it is necessary to raise production of energy from renewable resources, as well as to promote cleaner alternative fuels. The reserves of fossil fuels will continue to decline over the next 50 years and those that remain will be harder to extract; high oil prices will drive R&D in renewable substitutes for petrochemicals, to ensure energy security. One of the renewable energy resources that have attracted a lot of attention is biodiesel, obtained by converting the fats and oils in agricultural products into a combustible diesel fuel. In light of critical global oil shortages, algae biodiesel currently represents one of the most promising alternative fuel feedstock. Currently most research into efficient algal-oil production for biodiesel purposes is being done in many parts of the world. Two decade long aquatic species program of National Renewable Energy Laboratory US, clearly came out with the finding that a long term R& D would be required for the production of algae-biofuel.Thus making it economically competitive with other renewable energy sources. Microalgal biodiesel is the only renewable biodiesel which can completely displace liquid fuels derived from

* E-mail: kaur.simrat@rediffmail.com

petroleum. Algae could produce almost 100,000 liters of biodiesel a year per hectare of land; compared to 6000 liters a hectare of oil from palm, currently the most productive biofuel.The article provides an overview on various fundamental aspects of obtaining biodiesel from microalgae. A detailed account on practical methods involved in the isolation of microalgal strains from natural waters and maintenance of pure cultures in the laboratory has been discussed. We have also pondered over methods of mass culturing to obtain the microalgal feedstock and biochemical procedures to extract algal lipids and its conversion to biodiesel.

Keywords: *Biodiesel, Energy security, Lipids, Microalgae, Renewable.*

1. Introduction

An over-arching future consideration upon ever-present threat of changing climate and energy security is determining the strategies in developing sustainable renewable resources of biofuel. Algae have gained a considerable boost as a source of biodiesel due to its incredibly fast growth rate and species specific high lipid content. It is also amongst the most photosynthetically efficient plants, which sequesters the carbon di-oxide and form oil as a storage reserve. Thus, commercial scale algal-cultivation systems can also be established near coal power plants, sewage treatment or any industry that emit large quantities of CO_2.

This chapter provides an overview on production of algal biomass, extraction of biomass to obtain algal lipids and its conversion to biodiesel. A decade long extensive research program of the National Renewable Energy Laboratory in the United States provided a detailed findings and prospects of considering algae as a renewable biodiesel resource. The two important approaches that can make algal biodiesel economically competitive with conventional fuel are upgrading the biomass production, and genetic improvement of microalgal strains (Chisti, 2007). Large scale algal biomass production can be achieved either through open pond cultivation under natural sunlight or under controlled conditions of the photobioreactors. The focus of this chapter will be on methods involving isolation of freshwater microalgae, its assessment as biodiesel feedstock and ways of up scaling the biomass production. The aspects of genetic manipulation of algal strains to upgrade lipid production has not been considered, however lipid augmentation through nutrient stress has been discussed.

2. Establishment of Algal-Biodiesel Laboratory

2.1 Equipment and Other Facilities

The research and development in any area, depends on the intellectual minds of the men power as well as the laboratory facilities available at their exposal. Therefore, it is necessary to establish the laboratory before beginning of a new research venture.

2.1.1 Culture Room

The algal-biodiesel laboratory includes a dedicated culture room for isolation and maintenance of algal cultures, and facilities for biochemical work. Cultures are normally maintained in climate-controlled room or incubators that can regulate the temperature and are illuminated by day-light-type, white fluorescent tubes. Fabricated cultural racks and shelves for keeping culture vessels are placed in the culture room. Shelves made of glass or metal are recommended, because they can be cleaned or sterilized easily. Light sources should be above or at the side of shelves to prevent overheating (Andersen, 2005). The temperature and photoperiod are maintained through temperature-controller regulated air-conditioner and photoperiodic timer respectively (Figure 12.1). A laminar flow hood can be placed in the culture room or in a separate inoculation room.

2.1.2 Instruments

The following instruments are required to set up a complete laboratory dedicated for algal-biodiesel production.

1. *Microscopes*: A trinocular compound microscope along with a camera is needed for the isolation, growth studies (cell counting) and documentation of algal strains. Fluorescent microscope is also required for the observation of intracellular lipid globules using fluorescent dye in the live cells of algae.
2. *Autoclave*: This is required to sterilize various glassware and media preparation.
3. *Spectrophotometers*: A UV-VIS spectrophotometer is required for the growth studies of algae and various biochemical procedures. Fluorescence spectrophotometer is required for rapid screening test of lipids as explained later in the following section.
4. *Homogenizer or Tissue lyser*: Extraction of lipids from algae requires a complete breakdown of the cells, therefore an instrument which can mechanically lyses the cells is required.
5. *Rotary Evaporator*: As explained later, the biochemical procedure for gravimetric estimation involves evaporation of chloroform. This instrument is used to evaporate various solvents using vacuum.
6. *GC-MS*: Algal lipids consist of different fatty acids, all of which may not be suitable for biodiesel conversion. Therefore, before the further processing of algal lipids, qualitative analysis for fatty acid characterization using GC-MS is done.

These are the main equipments which are required besides various other instruments such as pH-meter, magnetic stirrer, oven, vortex mixer.

3. Isolation of Freshwater Algae

Desired strain of microalgae is required to obtain the feedstock required for biodiesel. According to Aquatic Species Program of US, there are five groups of microalgae which are of primary importance for biodiesel: diatoms (Bacillariophyceae), green algae (Chlorophyceae), golden brown algae (Chrysophyceae), prymnesiophytes (Prymnesiophyceae) and the eustigmatophytes (Eustigomatophyceae). These microalgae are obtained from the water samples collected from natural sources.

Figure 12.1: Racks with Culture Flasks (Plugged with Non-absorbent Cotton) in a Temperature Regulated Room. The photoperiod of 16:8 is maintained through a photoperiodic timer and illumination is from the fluorescent tubes fixed at the ceiling of each rack (DRL, algal culture laboratory).

In the laboratory, first step is to isolate the dominant algae from the collected water samples. The source of these water samples can be either freshwater or marine. In this chapter, focus will be on the isolation of freshwater algae obtained from the wetlands. Various species of green microalgae have been isolated from the richest treasure trove of the freshwater bodies in Assam (Figure 12.2) in our laboratory. This part of our country is well known as a biodiversity hotspot and has bestowed upon with vivid wetland resources which provide habitat and a conducive environment to support a rich diversity of freshwater algae (Kaur *et al.*, 2009).

Freshwater inhabiting the algal flora displays a wealth of environments. Ability to colonize a particular environment and the selective action of the chemophysical environment affects the distribution of algal species in freshwaters. Therefore, to obtain the desired algae, it is important to mimic the physiochemical environment of the natural habitats which mainly include nutrients, light or photoperiod, pH and temperature. Apart from this, two other factors which play an important role in isolation are, taxonomic knowledge to know the particular nutrient requirement of the target species and the elimination of various contaminants such as bacteria, fungi, zooplanktons and those undesired algae that outcompete the target species.

3.1 Nutrient Requirement of Microalgae

Algal growth in culture depends on an adequate supply of essential macronutrients elements (carbon, nitrogen, phosphorous, silicon), major ions (Na^+, K^+, Ca^{2+}, Cl^-, and $SO4^{2-}$) and a number of micronutrient metals (iron, manganese, zinc, colbalt, copper, and molybdenum) and metalloid selenium. However, silicon is required for diatoms, silicoflagellates, and some chrysophytes (Andersen, 2005).

The macronutrients are used generally as building materials, whereas the micronutrients are commonly metal constituents of enzymes which enter into biological reactions (Mc Elroy and Nason, 1954). The five elements which are required by algae for photosynthesis are manganese, iron, chlorine, zinc and vanadium; five elements required by nitrogen fixing algae are iron, calcium, boron, molybdenum, and cobalt (Eyster, 1964).

3.2 Vitamins

The three vitamins which are generally added to microalgae culture media are vitamin B12 (cyanocobalamin), thiamine, and biotin. However, very few algae need all three vitamins. These are added aseptically through a 0.2μm filter after autoclaving of the media. Autoclaving may cause decomposition of some vitamins, but it is thought that some algae may be able to use some of these decomposition products (Provasoli and Carlucci, 1974).

3.3 Media Selection for Isolation

The first step towards isolation is microscopic examination of the water samples, to select the best possible method and the enrichment media. When species are identified, the selective enrichment medium can be prepared which will favor their growth, else the general medium can be used. The difference between enrichment

Figure 12.2(a,b): Algal Blooms Observed during the Month of December in the Stagnated Waters Near Paddy Field. (c,d,e) Collection of water samples. Various species of algae have been isolated from water sample collected from these sites.

Figure 12.3: (a) Isolation of cells on agar plates, natural or pond water is directly spread on the surface of agar. (b) An agar plate showing growth of colonies of different green algae. (c) Individual colonies are picked from the agar plate and put in the culture flasks.

and general medium is in the presence or absence of particular nutrient or nutrients and pH. The natural water sample can also be used as an enriched media by adding into the nutrients, which support the growth of most algal species. Our personal experience suggests that the use of natural waters for making enriched culture medium supports rich growth as well as maintain the diversity of algae found in the habitat.

Prior to enriching, filter the water samples using Whatman 1 to eliminate the macroscopic impurities and other debris. The filtrate is then added to the selective or general enrichment culture medium which can support the growth of microalgae. Once, the water sample has been enriched, it should be regularly monitored through microscopic examination. Selective medium will usually favor the growth of particular specie, so further purification becomes easy; general medium will support many species, therefore it is preferable to do agar plating (Figure 12.3) as soon as medium shows a slight change in the color. When number of species are more in the enrichment culture, it is serially diluted followed by agar plating and agar streaking to obtain the unialgal culture (Figure 12.3). The bacterial contamination can be avoided using antibiotics but it should be the last step. A consortium of penicillin, streptomycin and gentamycin can be used to obtain the axenic culture. Different algal species tolerate different concentration of antibiotics, so a range of concentration should be used (generally 50-500 mg/L). Sterility of cultures should be checked by microscopic examination and by adding a small amount of sterile bacterial culture medium (0.1 per cent peptone) to a microalgal culture and observing regularly for bacterial growth. Some microalgal cultures may die when made axenic, probably due to the termination of obligate symbiotic relationships with bacteria. The commonly used media for culturing freshwater algae are given in Table 12.1 (Andersen, 2005).

3.3.1 Laboratory Scale Mass Culturing

Analytical experiments to standardize the procedures of biodiesel making requires a sufficient amount of biomass which can be cultivated in large volume glasswares such as 2 litres conical flasks, 10-15 litres large glass bottles or in carboys. For cheaper alternatives, plastic bags and plastic troughs can also be used for mass multiplication. Small circular pits can be dug outside to grow algae as an open cultivation system. This will further be improvised through open ponds for commercial scale mass cultivation as discussed in next section.

3.3.2 Up Scaling Mass Cultivation

The history of mass production of algae dates back to 1953 with the production of *Scenedesmus* species in Washington. A number of devices for cultivating on mass scale has been suggested in many countries like USA, Germany, Japan, Israel, England, Czechoslavakia and others. Algae are first grown in inorganic nutrients and then in the second phase, cultivation is done in wastewaters. Commercial cultivation can be done in different ways such as open cultivation using natural sunlight, closed cultivation using natural light and closed cultivation using artificial light (in photobioreactors). Each system has advantages and disadvantages, and the choice of system depends on the degree of parameter control needed to produce the desired product and on the value of the product (Apt and Behrens, 1999). The final product affects pond choice. If the biomass is used for human consumption, then the system must be capable of cleaner production. If the biomass is used for animal feed or if a specific product is extracted from the algae, then contamination is of less concern. Open pond culture is cheaper than culture in closed photobioreactors (Borowitzka 1999) but it is limited to a relatively small number of algal species.

Table 12.1: Recipes for Preparation of Commonly Used Media for Freshwater Microalgae

Sl.No.	Stock Solutions	Per Litre Distilled Water			
		BG[1]	WC[2]	BBM[3]	DM[4]
1.	$NaNO_3$	15.0g	85.01g	25g	–
2.	$MgSO_4.7H_2O$	7.5g	36.97g	7.5g	24.65g
3.	$CaCl_2.2H_2O$	3.6g	36.76g	2.5g	–
4.	K_2HPO_4	4.0g	8.71g	7.5g	–
5.	KH_2PO_4	–	–	17.5g	54.44g
6.	$NaHCO_3$	–	12.60g	–	–
7.	Na_2CO_3	2.0g	–	–	–
8.	$NaSiO_3.9H_2O$	–	28.42g	–	28.42g
9.	$FeCl_3.6H_2O$	–	3.15g	–	–
10.	Na_2EDTA	–	4.36g	–	–
11.	Citric acid	0.6g	–	–	–
12.	Ferric ammonium citrate	0.6g	–	–	–
13.	NaCl	–	–	2.5g	–
14.	EDTA	0.1g	–	50.0g	–
15.	KOH	–	–	31.0g	–
16.	$FeSO_4.7H_2O$	–	–	4.98g	0.278g
17.	H_2SO_4	–	–	1 mL	–
18.	$Ca(NO_3)_2.4H_2O$	–	–	–	70.85g
19.	H_3BO_3	–	–	11.42g	–
20.	Trace metal solution[a]				
	$ZnSO_4.7H_2O$	0.22g	0.02g	8.82g	–
	$MnCl_2.4H_2O$	1.81g	0.81g	1.44g	0.02g
	$CuSO_4.5H_2O$	0.079g	0.01g	1.57g	–
	$Co(NO_3)2.6H_2O$	0.0049g	–	0.49g	–
	MoO_3	–	–	0.71g	–
	H_3BO_3	2.86g	–	–	–
	$Na_2MoO_4.2H_2O$	0.39g	0.006g	–	–
	$CoCl_2.6H_2O$	–	0.01g	–	–

Contd...

Table 12.1–Contd...

Sl.No.	Stock Solutions	Per Litre Distilled Water			
		BG[1]	WC[2]	BBM[3]	DM[4]
21.	Vitamin stock[b]				
	Cyanocobalamin (Vitamin B12)	–	0.0005mg	–	1mL*
	Thiamine HCl (Vitamin B1)	–	0.10mg	–	1g
	Biotin	–	0.0005mg	–	1g
	Nicotinic acid (niacin)	–	–	–	1g
22.	Tris stock	–	500.0g	–	–
23.	Soil Extract[c]	–	–	–	50Ml
24.	pH	7.4	7.2	6.6	6.75

a,b Dissolve each component separately to 950 mL of distilled water, fully dissolving between additions. Finally make volume up to 1L with distilled water. This is the stock solution.

* Make a separate stock of 1g/L strength and add 1 mL of this stock to the vitamin stock.

c To prepare soil extract, add 1 part soil to 2 parts dH_2O and autoclave for about 2 hours. Allow the particulate matter to settle, and then filter using Whatman 1. The extract is again autoclaved and should be tightly capped and stored at 4°C.

1 Blue green medium for green algae

To prepare 1 L medium add:

Stock 1–100 mL

Stock 19–1mL

Other stocks–10 mL

2 Wright's Cryptophyte medium

To prepare 1 L medium add, 1 mL from each stock to 950 mL, bringing final volume to 1L.

3 Bold Basal medium

To prepare 1L medium add:

Stock 1–5 and 12– 10 mL

Make a separate stock of 14 and 15 as alkaline EDTA solution and add 1 mL of this stock.

Add 1 mL of 17 to make the stock 16 and add 1mL of this acidified iron solution.

Stock 19 and 20– 1mL

4 Diatom medium

Add 1 mL of each stock to 900 mL of distilled water and make final volume to 1L.

Single, rectangular pond with a paddle wheel (raceway ponds) are the most widely used for the production of *Spirulina*, *Dunaliella salina* and *Haematococcus* and represents the most efficient design for the large scale culture of most microalgae. Individual ponds are up to 1 ha in area, with an average depth of about 20- 30 cm (Andersen 2005). The pond depth is often compromised between the need of providing adequate light to the algal cells and the need of maintaining an adequate water depth for mixing. Algae only need about 1/10th of the light amount they receive from direct sunlight. Paddle wheels can be used to circulate the water in a pond and compressed air can be introduced into the bottom of a pond to agitate the water, bringing algae from lower levels upwards. Raceway channels are built in concrete or compacted earth, and may be lined with white plastic. During daylight, the culture is fed continuously in front of the paddlewheel where the flow begins. Broth is harvested behind the paddlewheel, on completion of the circulation loop. The paddlewheel operates all the time to prevent sedimentation (Chisti, 2007). The largest raceway-based biomass production facility occupies an area of 440,000 m^2 (Spolaore *et al.*, 2006). This facility is owned by Earthrise Nutritional (www.earthrise.com), is used to produce cyanobacterial biomass for food. In India, Pary Nutraceuticals, part of the Chennai based Murugappa group has been focusing on microalgal research and development and ' are commercially producing *Spirulina*, *D. salina* and *H. pluvialis* for nutraceutical.

Production of microalgal biomass for making biodiesel has been extensively evaluated in raceway ponds in studies sponsored by the United State Department of Energy (Sheehan *et al.*, 1998). Several different closed systems using natural sunlight have been described (Richmond *et al.*, 1993, Molina Grima *et al.*, 1995, Spektorova *et al.*, 1997). In such systems, the algae are grown in transparent glass or plastic and the vessel are placed under natural illumination. A higher surface to volume ratio is provided, so cell densities are often higher than in the open ponds; however these systems are subject to variations in light intensity and temperature that make cultivation reproducibility problematic. In addition removal of oxygen from the culture and the provision of adequate temperature control pose a major problem with closed systems.

Large scale indoor cultivation using highly controlled photobioreactors or fermentors have also been used successfully. These provide the ability to control and optimize culture parameters, and, coupled with the closure that they provide, photobioreactors are suitable for culturing many different algae. The basic features which must be considered when fabricating a photobioreactor are: source of light, churning of algae (to avoid biomass sedimentation and for uniform availability of nutrients and light), material for construction, CO_2 supply, and removal of O_2, pH and temperature control. Assembly of the photobioreactor system is merely the integration of various monitor and control systems (Andersen 2005).

In nutshell, closed culture systems or photobioreactors are very expensive and many are difficult to scale up; therefore widely used large commercial systems are open-pond systems. However, the number of species that has been successfully cultivated for the production of biodiesel in an outdoor system is relatively small. Open cultivation is also vulnerable to be invaded by other algal species and bacteria. Another noteworthy point is the growth which largely depend on location and apart from tropical areas it is limited to the warmer months with productivity achieved lesser than it is theoretically possible. Therefore, it is necessary to screen out a suitable strain of microalgae that work well at the specific site where technology is to be used. The native strains are most likely to be successful. According to the result of close out report of NREL's Aquatic Specie Program, desert location in New Mexico had sufficient sunlight but low night time temperature limited the ability to achieve consistently high productivities. Hence it can be foresighted that weather conditions in India, in particular North East which is also a bio- diversity hotspot can be very suitable for pond cultivation. However, due to high and frequent rainfall in this region, ponds have to be made under some shed protection. Our laboratory is putting lot of efforts to find a dominant native strain of algae from Assam which has biodiesel potential and which show desired productivity in open system.

4. Algal Lipids

The composition of microalgal lipids is similar to those of vegetable oils. The total lipids include three major classes viz Neutral lipid, Glycolipid and Phospholipid. The lipids of microalgae are generally esters of glycerol and fattyacids with a chain length of C14 to C22, which may be either saturated or unsaturated (Borowitzka and Borowitzka, 1988). The green algae contain primarily C16 and C18 fatty acids with a high degree of unsaturation; non-chlorophyte algae contain high amounts of C20 polyunsaturates. Some of algal lipids are also rich in essential fatty acids such as the C18 linoloeic (18:2ω6) and linolenic (18:3ω3) acids and their derivatives, eicosapentaenoic acid (20:5ω3) and arachidonic acid (20:4ω6).

The microalgal fatty acid composition depends not only on the species and strain used, but also on factors related to culture conditions, including composition of the medium, light intensity, photoperiod, salinity and temperature (Liang *et al.*, 2006).

4.1 Changes in Fatty Acid with Culture Age

Age of culture often has a dramatic effect on the fatty acid composition of microalgae (Fernadez-Reiriz *et al.*, 1989; Siron *et al.*, 1989; Fidalgo *et al.*, 1998). Culture age has been determined as a major determinant of lipid content and composition in two diatoms, *Phaeodactylum tricornatum* and *Chaetoceros muelleri*. The main fatty acids of *P. tricornatum* in all growth phases were 14:0 (6.3 per cent–7.8 per cent), 16:0 (18.3 per cent -25.1 per cent), 16:1 n-7 (24.5 per cent -40 per cent) and EPA (eicosapentaenoic acid) (8.9 per cent -22.8 per cent). For C. *muelleri*, the main fatty acids in all growth phases were 14:0 (10.6 per cent -14.9 per cent), 16:0 (22.3 per cent -29.2 per cent), 16:1 n-7 (26.0 per cent -31.6 per cent) and EPA (6.7 per cent -10.3 per cent).The increases in the 14:0 acid of *C. muelleri* and the increases in the 16:0, 16:1 n-7 acids of *P. tricorntum* has been observed with increasing growth period (Liang *et al.*, 2006) which supports the view that these fatty acids have a role as readily mobilizable reserves (Brown *et al.*, 1996).

Fatty acid composition of various species belonging to classes Bacillariophyceae, Prasinophyceae and Chlorophyceae has been determined at the exponential phase and the stationary phase (Pratoomyot *et al.*, 2005). The major fatty acids of Bacillariophyceae, *Nitzschia* cf. *ovalis* and *Thalassiosira* sp. at the exponential phase were C16:1 n-7 and at the stationary phase were C16:0 and C20:5 n-3. The species also contained highly unsaturated fatty acids (HUFA) at two growth phases as arachidonic acid (C20:4 n-6), EPA and Docosahexaenoic acid (C22:6 n-3). The

Prasinophyceae represented by *Tetraselmis* sp have reported amounts of C16:0, C18:1 n-9, C18:3 n-3 as principle fatty acids. Chlorophyceae represented by genera *Dictyosphaerium pulchelum*, *Stichococcus* sp., *Chlorella* sp., *Scenedesmus falcatus* reported to contain high amount of fatty acids at both growth phases as C18:3 n-3, C16:0, C18:2 n-6.

Some algal species accumulate triglycerides, often intermittently and in inconspicuous quantities. For example, *Chlorella ellipsoidea* dividing in synchrony on a light-dark regime formed triglycerides during the light period and then utilized them extensively in darkness (Thompson Jr, 1996).

4.2 Changes in Fatty Acids upon Nutrient Stress

The proximate chemical composition of eukaryotic algae is generally regarded as species specific and usually regulated by environmental factors (Lewin 1974). Under nitrogen starvation, the cellular composition of *Chlorella* showed a general trend of protein decrease and lipid increase (Spoehr and Milner, 1949). Nitrogen is known to have a strong influence on the metabolism of lipids and fatty acids in various algae. In most algae, enhancement of lipid accumulation occurred under nitrogen deficient conditions. At low concentrations of nitrogen, *Chlorella* and *Euglena* synthesized saturated (16:0) and monounsaturated (18:1) fatty acids, whereas at high nitrogen concentrations the 16:2, 16:3, 16:4 and 18:2 fatty acids predominated (Pohl and Wagner, 1972; Pohl and Zurheide, 1979). In both the species, total lipid coincided with the formation of polyunsaturated fatty acids (Pohl and Zurheide 1979). At low N levels (< 0.001-0.003 per cent NH_4Cl or KNO_3) algae of the classes Chloro-, Rhodo-, Phaeo- and Euglenophyceae have a tendency to synthesize neutral lipids (triacylglycerols, waxesters) and fatty acids with a low degree of unsaturation (especially 16:0, 16:1 and 18:1). At higher N levels (>0.003 per cent NH_4Cl or KNO_3) the algae predominantly synthesize polar lipids such as monogalactosyl diacylglycerol, diagalactosyl diacylglycerol, sulfoquinovosyl diacylglycerol and phosphatidyl glycerol containing polyunsaturated C_{16} and C_{18} fatty acids (Piorreck *et al.*, 1984).

Under nitrogen reduction *Chlorella* strains, *C. emersonii*, *C. minutissima* and *C. vulagris* gained an increase in lipid content of 63 per cent, 56 per cent and 40 per cent biomass by dry weight respectively (Illman *et al.*, 2000).

Apart from nitrogen stress, a high iron concentration also induces considerable lipid accumulation in marine strain *C. vulgaris*. High iron concentration however reduced the final cell density slightly, the onset of stationary phase when the alga began to accumulate lipids is earlier and thus the intervals for harvesting are shortened (Liu *et al.*, 2008). In this regard, both the biomass and lipid accumulation are increased under high iron concentration.

An increase in the environmental concentration of CO_2 (from 1 per cent -30 per cent) over the heterotrophic cultures of *C. fusca* produced a 40 per cent increase in total lipids and nearly a 50 per cent increase in fatty acids. Palmitic acid (16:0) and 18:1 largely account for the fatty acid change by increasing from 12 mg to 33mg/g dry weight and 8 mg to 17 mg/g dry weight respectively (Dickson *et al.*, 1969). The greatest response to a downward shift of CO_2 concentration is the desaturation of 18:2. From an upward shift of CO_2 concentration, the response is represented by the decrease in unsaturation of fatty acids (Tsuzuki *et al.*, 1990). The environmental CO_2 concentration affects on membrane lipid composition in addition to photosynthetic characteristics. Under low CO_2 concentration, *de novo* synthesis of carbonic anhydrase is induced and a pyrenoid with starch sheath is developed in various eukaryotic microalgae (Miyachi *et al.*, 1986; Tsuzuki *et al.*, 1986).

The Algal-derived fuels centered around the green alga *Botryococcus* in which 86 per cent of the dry weight of the cell consists of a mixture of long chain linear alkenes (C27, C29, C31 dienes) and branched isoprenoids, called botryococcenes (Taylor and Otvos, 1998). *B. braunii* is a freshwater green colonial alga of widespread occurrence which occurs in a green exponentially growing stage and a brown resting stage. The brown resting stage often arises as massive rust-colored algal blooms on the surface of lakes (Knight *et al.*, 1970). The only limitation with this specie is its slow growth rate; therefore one has to find a strain with high growth rate and optimal lipid content. This is done through experimental studies with some selected or screened strains of microalgae under varied cultural conditions. The biochemical methods involved in screening of lipids and its extraction from culture cells have been discussed in the next section.

5. Biochemical Methods to Study Lipids in Microalgae

5.1 Strain Selection

As a rough estimate there are over 300,000 algae strains in global algal culture collections. Out of these, very few strains contain substantial amount of oil. The oil per cent in various algae cultures may vary from 2 per cent to 70 per cent. However each strain requires different agro-climatic conditions for growth, selecting right and robust strain for a particular geographic location is the key to success for commercial algae farming projects. To screen appropriate strain of microalgae with desired lipid content is the essential step if one starts with their own isolated cultures. This method needs to be rapid, reproducible and which requires a very less quantity of culture. There is a significant relationship between in vivo fluorescence of cells stained with Nile Red and lipid content. Nile red is a phenoxazone dye that is responsible for the metachromatic coloring of tissue lipids and is used for the detection of intracellular lipid droplets by fluorescence microscopy and quantification by spectrofluorometer. Cooksey *et al.* (1987) used diatom *Amphora coffeiformi* to optimize the Nile red staining procedure. The dye was dissolved in acetone and used at a concentration of 1 mg/mL of cell suspension. The fluorescence level was linearly correlated with an increase in the total lipid content, determined gravimetrically, in a growing culture of algal cells. The resulted protocol involved taking a fixed volume of a diluted algal culture (typically 4 mL), adding 0.04 mL of a nile red solution (0.1 mg/mL in acetone), and determining the fluorescence after 5 min using a fluorimeter equipped with the appropriate excitation and emission filters.

5.1.1 Extraction of Lipids

Lipids can be extracted from the algal mass using non polar solvents such as ethanol, n-hexane and Chloroform. The aim is to separate cellular lipids from the other constituents such as proteins and carbohydrates and to preserve these lipids for further analysis. Neutral lipids are extracted with relatively non-polar solvents such as diethyl ether or chloroform but membrane associated lipids are more polar and require polar solvents such as ethanol or methanol to disrupt hydrogen bonding or electrostatic forces.

The most popular and efficient method adopted for extraction of algal lipids is Bligh and Dyer (1956) method. Total lipids can be extracted from the dry material by modification of this method. Briefly, the algal mass is first separated from the medium either by centrifugation or filtration using GFC filters. The biomass is then dried or lyophilized to improve the efficiency of solvent extraction during homogenization either in a mechanical bead beater/tissue lyser or simply using the pestle mortar. The solvent system used in the modified method is Chloroform:methanol in 2:1 ratio. Phase separation of the biomass-solvent mixtures in the separating funnels is achieved by adding chloroform and water to obtain chloroform:methanol:water ratio of 1:1:0.9. A known portion of each total lipid extract recovered from the lower chloroform phase is used for further analysis.

5.1.2 Transesterification Process

Biodiesel is defined as the mono-alkyl esters of fatty acids derived from vegetable oils or animal fats. In simple terms, biodiesel is the product obtained when a vegetable oil or animal fat is chemically reacted with an alcohol to produce fatty acid alkyl esters. A catalyst such as sodium or potassium hydroxide is required. Glycerol is produced as co-product. Chemical conversion of the oil to its corresponding fatty ester, and thus biodiesel is called transesterification. The biodiesel or Fatty Acids Methyl Ester (FAME) is obtained upon transesterification of various fatty acids namely Palmitic acid/Hexadecanoic acid (R-$(CH_2)_{14}$-CH_3) C16:0 Methyl palmitate/Methyl hexadecanoate; Stearic acid/Octadecanoic acid (R-$(CH_2)_{16}$-CH_3) C18:0 Methyl stearate/ethyl octadecanoate; Oleic acid/9(Z)-octadecenoic acid (R-$(CH_2)_7$-CH=CH-$(CH_2)_7$-CH_3) C18:1 Methyl oleate/Methyl 9(Z)-octadecenoate; Linoleic acid/9(Z),12(Z)-octadecadienoic acid; (R-$(CH_2)_7$-CH=CHCH_2-CH=CH-$(CH_2)_4$-CH_3)C18:2 Methyl linoleate/Methyl 9(Z),12(Z)- octadecadienoate Linolenic acid/9(Z),12(Z),15(Z)-octadecatrienoic acid (R-$(CH_2)_7$-(CH=CHCH_2) 3-CH_3) C18:3 Methyl linolenate/Methyl 9(Z),12(Z),15(Z)-octadecadienoate.

Acknowledgement

One of the authors (Simrat Kaur) thanks Mr. Jayanta Talukdar, Research Scholar, Algal Biotechnology Laboratory, Department of Biotechnology, Gauhati University, who has spent time on many occasions supporting and working with them. They also thank Mr A.K. Das, Technical Officer and Mr. A. Bora, Scientist 'B' of the Defense Research Laboratory, for their unconditional support and help. Thanks are due to Mr. P.J. Gogoi, Technical Officer, for his help in collection of water sample.

References

Andersen RA (2005) Algal Culturing Techniques. Elsevier Academic Press, London, UK.

Apt KE and Behrens PW (1999) Commercial developments in microalgal biotechnology. J Phycol 35: 215-226.

Bligh D and Dyer WJ (1959) A rapid method of total lipid extraction and purification. C J Biochem Physiol 37(8): 911-917.

Borowitzka MA (1999) Pharmaceuticals and agrochemicals from microalgae. In: Chemicals from microalgae. (Ed) Cohen Z, Taylor and Francis. pp: 313-52.

Borowitzka MA and Borowitzka LJ (1988) Micro-algal Biotechnology. Cambridge Univ. Press, Cambridge.

Brown MR, Dunstan GA, Nonwood SJ and Miller KA (1996) Effects of harvest stage and light on the biochemical composition of the diatom *Thalassiosira pseudomonas*. J Phycol 32: 64-73.

Chisti Y (2007) Biodiesel from microalgae. Biotech. Advances 25: 294-306.

Cooksey KE, Guckert JB, Williams SA and Collis PR (1987) Fluorometric determination of the neutral lipid content of microalgal cells using Nile Red. J Microbiol Meth 6: 333-345.

Dickson LG, Galloway RA and Patterson GW (1969) Environmentally-induced changes in the fatty acids of *Chlorella*. Plant Physiol 44: 1413-1416.

Eyster C (1964) Micronutrient requirement for green plants, especially algae. In: Algae and Man.(Ed) Jackson Daniel F, Plenum Press, New York.

Fernandez-Reiriz MJ, Perez-Camacho A, Ferreiro MJ, Blanco MP, Campos MJ and Labarta U (1989) Biomass production and variation in the biochemical profile (total protein, carbohydrates, RNA, lipids and fatty acids) of seven species of marine microalgae. Aquaculture 83: 17-37.

Fidalgo JP, Gid A, Torres E, Sukenik A and Herrero (1998) Effects of nitrogen source and growth phase on proximate biochemical composition, lipid classes and fatty acid profile of the marine microalga *Isochrysis galbana*. Aquaculture 166: 105-116.

Illman AM, Scragg AH and Shales SW (2000) Increase in *Chlorella* strains calorific values when grown in low nitrogen medium. Enzyme Microb Tech 27: 631-635.

Kaur S, Gogoi HK, Srivastava RB and Kalita MC (2009) Algal diversity as a renewable feedback for biodiesel. Current Science 96(2): 182.

Knights BA, Brown AC and Conway E (1970) Hydrocarbons from the green form of the freshwater alga *Botryococcus braunii*. Phytochemistry 9: 1317-1324.

Lewin RA (1974) Biochemical taxonomy. In: Stewart, DWP (Ed) Algal Physiology and Biochemistry, Blackwell, Oxford, pp 1-39.

Liang Y, Beardall J and Heraud P (2006) Changes in growth, chlorophyll fluorescence and fatty acid composition with culture age in batch cultures of *Phaedactylum tricornatum* and *Chaetoceros muelleri* (Bacillariophyceae). Botanica Marina 49: 165-173.

Liu Z, Wang G and Zhou B (2008) Effect of iron on growth and lipid accumulation in *Chlorella vulgaris*. Bioresorce Technology 99: 4717-4722.

McElroy WD and Nason A (1954) Mechanism of action of micronutrient elements in enzyme systems. Ann Rev Plant Physiol 5: 1-30.

Miyachi S, Tsuzuki M, Maruyama I, Gantar M, Miyachi S and Matsushima (1986) Effects of CO2 concentration during growth on the intracellular structure of *Chlorella* and *Scenedesmus*. J Phycol 22: 313-319.

Molina Grima E, Sanchez Perez J A, Garcia Camacho F, Fernandez Sevilla J M, Acien Fernandez F G and Urda Cardona J (1995) Biomass and icosapentanoic acid productivities from an outdoor batch culture of *Phaeodactylum tricornatum* UTEX 640 in an airlift tubular photobioreactor. Appl Microbiol Biotechnol 42: 658-63.

Piorreck M, Baasch K and Peter P (1984) Biomass production, total protein, chlorophylls, lipids and fatty acids of freshwater green and blue green algae under different nitrogen regimes. Phytochemistry 23(2): 207-216.

Pohl P and Wagner H (1972) Control of fatty acid and lipid biosynthesis in *Euglena gracilis* by ammonia, light and DCMU. Z Naturforsch 27(b): 53-61.

Pohl P and Zurheide F (1979) Fatty acids and lipids in marine algae and the control of their biosynthesis by environmental factors. In: Marine Algae in Pharmaceutical Science. (Eds) Hope HA, Levring T and Tanaka Y Walter de Gruzter, Berlin, pp. 433-523.

Pratoomyot J, Srivilas P and Noiraksar T (2005) Fatty acids composition of 10 microalgal species. J Sci Technol 27(6) 1179-1187.

Provosoli L and Carlucci AF (1974) Vitamins and growth regulators. In: Algal Physiology and Biochemistry (Ed) Stewart WDP, Blackwell Scientific,U K, pp. 741-87.

Richmond A, Boussiba S, Vonshak A and Kopel R (1993) A new tubular reactor for mass production of microalgae outdoors. J Appl Phycol 5: 327-32.

Sheehan J, Dunahay T, Benemann J and Roessler P (1998) A look back at the US Department of Energy's Aquatic Species Program- Biodiesel from Algae. National Renewable Laboratory, Golden, CO, Report NREL/TP-580-24190.

Siron R, Giusti G and Berland B (1989) Changes in the fatty acid composition of *Phaeodactylum tricornatum* and *Dunaliella tertiolecta* during growth and under phosphorous deficiency. Mar. Ecol. PROG. Ser. 55: 95-100.

Spektorova L, Creswell R L and Vaughan D (1997) Closed tubular cultivators: an innovative system for commercial culture of microalgae. World Aquaculture 28: 39-48.

Spoehr HA and Milner HW (1949) The chemical composition of Chlorella, Effect of environmental conditions. Plant Physiol 24: 120-149.

Spolaore P, Joannis-Cassan C, Duran E and Isambert A (2006) Commercial applications of microalgae. J Bioscience and Bioengineering 101: 87-96.

Taylor S E and Otvos J (1998) Melvin Calvin: Fuels from Plants. In: Report, Ernest Orlando Lawrence Berkeley National Laboratory. National Technical Information Service U.S. Department of Commerce, Springfield, VA.

Thompson Jr. GA (1996) Lipids and membrane function in green algae. Biochimica et Biophysica Acta 1302: 17-45.

Tsuzuki M, Gantar M, Aizawa K and Miyachi S (1986) Ultrastructure of *Dunaliella tertiolecta* cells grown under low and high CO_2 concentrations. Plant Cell Physiol 27: 737-739.

Tsuzuki M, Ohnuma E, Sato N, Takaku T and Kawaguchi A (1990) Effects of CO_2 concentration during growth on fatty acid composition in microalgae Plant Physiol 93: 851-856.

Algal Biotechnology: New Vistas (2010) *Pages 194–205*
Editor: **Mihir Kumar Das**
Published by: **DAYA PUBLISHING HOUSE, NEW DELHI**

Chapter 13

Industrial Utilization of Algal Fatty Acids

S.B. Padhi*, P.K. Swain, S.K. Behera and G. Behera

Algal Research Laboratory,
Department of Botany, Berhampur University, Berhampur, Orissa, India

ABSTRACT

The n-3 Polyunsaturated fattyacids (PUFAs),n-3 Linolenic acid, Eicosapentaenoic acid (EPA) and Decosahexanoic acid(DHA) have attracted much interest as they have prime role in reducing the risk of heart disease and atherosclerosis as well as in the treatment of cancer,inflammatory problems,rheumatoid arthritis and asthma.These fatty acids are of increasing interest since they have been demonstrated to have several biotechnological and pharmaceutical applications. Interest in fatty acid compositions originates to a large degree from the growing interest in aquaculture, which requires nutrient rich algae and zooplankton for the diets of larval stages of commercially important fishes.The lipids of marine algae play an essential role in the biosynthesis of unsaturated fatty acids.The fatty acid composition was found to be different in different species of algae. The algal isolates from Cyanophyceae, Chlorophyceae, Bacillariophyceae and Rhodophyceae were investigated for fatty acid composition in order to know the suitability of these algae in aquaculture industries. The algal isolates maintained as bacterial free cultures are *Anabaena circinalis, Anabanea* sp., *Nostoc muscorum, Oscillatoria agardhii* and *Microcystis* sps. from Cyanophyceae, *Scenedesmus quadricauda, Chlamydomonas* sps., *Enteromorpha intestinalis, Chaetomorpha linum, Pithophora oedogonia* from Chlorophyceae, *Skeletonema costatum,*

* E-mail: sailabalapadhi@gmail.com

Nitzschia pinnata, Nitzschia acuta, Navicula longa from Bacillariophyceae, *Gracilaria verrucosa, Grateloupia filicina, Ceramium elegans, Polysiphonia sertularioides* and *Chondrus crispus* from Rhodophyceae.

From fatty acid analysis on blue-green algae, certain unicellular forms have very simple fatty acid patterns. The most favourable algal compositions from the human point of view are presence of high proportion of PUFA and a high n3/n6 ratio. Among the algal isolates examined comparatively the diatom *Navicula longa* and *Nitzschia acuta* contained large amounts of EPA and DHA. The blue greens *Oscillatoria and Microcystis* were characterized by relatively high n3/n6 ratios, whereas *Anabaena* and *Nostoc* had only traces of n3 acids. In comparison to Cyanophytes, the Chlorophytes had a higher content of PUFA. The red algal species also contained higher amounts of PUFA and the maximum amount was present in *Grateloupia filicina.*

Keywords: *Aquaculture, Bacillariophyceae, Chlorophyceae, Cyanophyceae, Fatty acids.*

1. Introduction

Aquaculture is a growing industry in the country and has become a major part of the production of fish for human consumption. A major concern today is the distinct possibility that planet earth may not be able to produce enough food for the burgeoning population. Early work on the development of aquaculture centered on finfish, molluscs and lobsters and was later expanded to include other molluscs, other crustaceans and aquatic plants. The present scenario of Indian aquaculture is promising and is showing signs of emerging trends as a major sector in Asia in view of its rich natural water resources. In India, with its abundant brackish water areas, the requirement of our aqua industry is an improved intensive or semi-intensive farming system. Side by side with the progress in aqua farming there is an upsurge in feed companies. The aquaculture of macro and micro-algae is a valuable global industry.

Macroalgae are farmed for their hydrocolloids as well as for food (Abbott, 1996; Bixler, 1996) and microalgae are cultured commercially for use as health food and as a source of valuable chemicals such as betacarotenes. (Belay *et al.*, 1994; Borowitzka, 1994). Microalgae are also an important food source and feed additive in the commercial rearing of many aquatic animals. Algal biomass form an important commercial component of aquaculture, providing for aesthetic requirements and upkeeping of the environment. Algae in general prove to be the best natural feed material for aquaculture and the best supplements in other animal husbandry programmes. The importance of algae in aquaculture is not surprising as algae are the natural food source of these animals. Although several alternative for algae exist such as yeast and micro encapsulated feeds (Jones *et al.*, 1987; Nell, 1993; Heras *et al.*, 1994; Nell *et al.*, 1996). Live algae are still the best and the preferred food sources. The best growing aquaculture industry is dependent on algae from which the aquaculture animals obtain the nutrients. Many algae are cultured for use in aquaculture industries (Day *et al.*, 1991; Johnson and An, 1991; Orus *et al.*, 1991; Bennemann, 1992; Takeyama *et al.*, 1996; Borowitzka, 1997; Lynch, 1998). The growing demand for safe natural products by the nutraceutical and aquacultrual industries has signified the importance of algae as a source of these products and the need to screen more algae for the purpose.

Aquaculture contributes to human nutritions through the production of unicellular algae for use in animal feeds. Unicellular algae, particularly Chlorella, are usually added together with rotifers to rearing tanks when marine fish larvae are reared to fry or even, to metamorphosis (Liao, 1975; Nash and Kuo, 1975; Howell, 1979; Pullin and Kuo, 1981; Juario *et al.*, 1984). In rearing milkfish (*Chanos chanos*) larvae to fry (2, days old), better survival rates were obtained when *Isochrysis galbanes* and *Tetraselmis* sp. as well as *Chlorella* sp. were added to the rearing tanks (Juario *et al.*, 1984).

With the current fish cost of the majority of commercial fish feeds many novel dietary ingredients, particularly protein source, are under consideration for future inclusion in such rations. Algae are receiving increasing attention as a possible protein source for fish diets, particularly in tropical developing countries where algal production rates are high (Venkataraman *et al.*, 1980). Several algae have been shown to have the relatively high protein contents required for their inclusion in balanced fish feeds (50-65 per cent protein as a dry weight basis).

The interest in fatty acids of the n6 family derives from the fact that they are precursors for Prostaglandins, Thromobonanes and Leukotrienes (Holman, 1981). GLA (18:3, n-6) was also shown to alleviate the symptoms of the premenstrual syndrome (Horrobin, 1983). The Cyanobacterium *Spirulina* has found worldwide recognition as health food due to its high content of this rare fatty acid, GLA; Cohen *et al.* (1988) reported the increase in GLA to 30.5 per cent of total fatty acid and 1.3 per cent of the dry weight by varying growth conditions.

Research on the nutritional requirements of zooplankton started in the 1950s and was originally oriented towards marine species (Corner and Coway, 1968). An increasing interest has been put into the lipid fractions and particularly the context of unsaturated fatty acids of n6 and n3 type (two families of polyunsaturated fatty acids, differing by the position of the first double bond in the carbon chain). Interest in fatty acid compositions originates to a large degree from the growing interest in aquaculture, which requires nutrient rich micro algae and zooplankton for the diets of larval stages of commercially important fish and shell fish. The work carried as in the research project "Biotechnological utilization of algae as fish feed" sponsored by DBT, Government of India, in algal Research Laboratories, Berhampur University have shown that the fatty acid composition varied a great deal between the algal species. The diatoms contained long-chained polyunsaturated fatty acids with 20 or 22 carbon and consist of eicosapentaenoic acid (20:5, n-3) and DHA (decosahexaenoic acid (22:6, n-3). The blue-green algae instead consist of shorter n-3 acid, 18:3 n-3 (α-linolenic acid) Corner, E.D.S. ed and Coway, C.B.(1968).

The lipids of marine algae would also play an essential role in the biosynthesis of these unsaturated fatty acids. Fatty acids constituting the fats may be saturated or unsaturated. Major saturated algal fatty acids are Lauric acid (12:0) CH_3 $(CH_2)_{10}$

COOH, Myristic acid (14:0) $CH_3(CH_2)_{12}COOH$, Palmitic acid (16:0) $CH_3(CH_2)_{14}COOH$, Stearic acid (18:0) $CH_3(CH_2)_{16}COOH$, Arachidic acid (20:0), $CH_3(CH_2)_{18}COOH$.

Major unsaturated algal fatty acids are Palmitoleic acid (16:1), $CH_3(CH_2)_5CH{=}CH(CH_2)_7COOH$, Oleic acid (18:1) $CH_3(CH_2)_7CH{=}CH(CH_2)_7COOH$, Linoleic acid (18:2) $CH_3(CH_2)_3CH_2.CH{=}CH(CH_2)_7COOH$, α-Linolenic acid (18:3), $CH_3(CH_2CH{=}CH)_3(CH_2)_7COOH$. Marine algae contain C20 polynoic acids such as Arachidonic acid (AA) and Eicosapentaenoic acid (EPA) which are known to be either physiologically or nutritionally important (Pohl and Zurheide, 1979). Arachidonic acid has also been shown to be an important component of nutritional supplements and algae have been suggested as a source for these fatty acids.

Prawns, like most marine animals have a higher requirement for the n-3 series of fatty acids than terrestrial or freshwater aquatic species. The n-number in fatty acids indicates the number of carbon atoms between the terminal group and first double bond. Other series include n-6, n-7 and n-9 which are polyunsaturated fatty acids. Fatty acids with four or more double bonds are referred to as highly unsaturated fatty acids (HUFA).

Polyunsaturated fatty acids (PUFAs) are important for their role in preventing diseases and in overall health management. Ever since Dyerberg *et al.* (1975) found a characteristic composition in the plasma lipids in Greenland Eskimos, cis-518, 11, 14, 17–eicosapentaenoic acid (EPA) has been shown to be important in the prevention and treatment of human diseases. As EPA is commonly found in marine fish, marine fish oils are the current commercial source of EPA. Many species of micro algae have been reported to have a high content of EPA (Opute, 1974; Yongman Itchai and Ward, 1991b; Dunstan *et al.*, 1994) and are being considered as a promising source. In this respect the diatom *Navicula saprophila* is a promising candidate and EPA production was promoted under mixotrophic conditions in the presence of acetatae (Kitano *et al.*, 1997). Among the micro algae which are food source of PUFAs, *Phaeodactylum tricornutum* has been the object of study in recent times due to its simple nutrient requirements and fast growth rate (Regan, 1988). Diatoms are characterized by an unusual distribution of fatty acids compared to green algae and higher plants. The C_{14}, C_{16} and C_{20} acids comprise the bulk of diatom fatty acids while the unsaturated C_{18} acids, particularly linolenic acid, 18:3 are either absent or present at very low levels. Among other groups of algae, only the Chrysophyta, Rhodophyta and Pyrophyta are known to have low 18:3 content (Ackman *et al.*, 1964; Cheuecas and Riley, 1969a). The presence of a unique 21:6 hydrocarbon (hexeicosohexane) was demonstrated in diatoms and other, but not all groups of algae (Blumer *et al.*, 1971; Lee and Loeblich, 1971). This hydrocarbon has been traced up to the food chain to certain species of Copepods (Blumer *et al.*, 1970; Lee *et al.*, 1974). Several algae are being investigated as potential source of protein for both livestock and human consumptions. Seaweeds have potential in integrated culture systems by low technology/input techniques, yet still have high market potential.

The use of algal oils containing long-chain polyunsaturated fatty acids (PUFAs) as nutritional supplement has been given primary focus (Cohen *et al.*, 1988; Behrens and Kyle, 1996). The DHA is a dominant fatty acid in neurological and heart muscle tissues and sperm cells. The fact that humans are not capable of synthesizing DHA de novo implies the dependency on adequate supply of DHA through dietary sources (Emken *et al.*, 1994). The fatty acid is also a conditionally essential nutrient during infancy (Innis, 1994; Makridas *et al.*, 1995). A number of nutritional and professional organizations including the World Health Organisation have recommended the inclusive of supplementary DHA infant formula. Dinoflagellates are looked upon as a source of DHA for use in infant formulas (APT and Behrens, 1999). Changes in EPA levels are known to significance after the coronary vascular status of individuals (Salem, 1989) and a number of algae have been proposed for the production of EPA (Cohen, 1990; Sukenik, 1991; Boswell *et al.*, 1992; Molina *et al.*, 1995; Tan and Johns, 1996) and used in nutraceutical production. Arachidonic acid has also been shown to be an important component of nutritional supplements and algal here been suggested as a source for these fatty acids (Pohl and Zurheide, 1979). These acids function as precursors for some prostaglandins in animals. These fatty acids are successively transferred and accumulated in tissues of higher consumer through food-chain within the marine hydrosphere (Kayama *et al.*, 1963) and finally they can be used for nutritional supplement by human beings. Hence, the study of distribution and content of fatty acids among marine algae is important. The fatty acid composition of lipids from marine algae has been well documented (Pohl and Zurheide, 1979) but few systematic studies at the lipid class level (Peltit and Harwood, 1989; Arao *et al.*, 1987; Sato *et al.*, 1988; Smith and Harwood, 1984) have been made. In higher plants (Roughan and Slack, 1984; Joyard and Douce, 1987), Cyanobacteria (Murata and Nishida, 1987) and unicellular green algae (Harwood and Jones, 1989), it has been known that polyunsaturated molecular species of lipid are formed after the de novo synthesized fatty acids are incorporated into certain lipid classes which are NGDG and PC, in order to receive successive unsaturation on these lipids.

In a research project entitled "Biotechnological utilization of algae as fish feed" sponsored by DBT, Government of India, the juvenile of the freshwater fish *Labeo rohita* was investigated for survival and further growth using algal diets in the algal research laboratory, Berhampur University. Several reports are available on the nutritional value of marine algae to emphasize the suitability of those in mariculture (Johns *et al.*, 1979; Ben Amotz *et al.*, 1997, Volkman *et al.*, 1989 and Brown *et al.*, 1997). Except that of Ahlgren *et al.* (1992), the reports on the fatty acids and chemical composition of the freshwater microalgae are rarer and are limited to special algal growth (Kenyan *et al.*, 1972; Cramwell *et al.*, 1990).

The algal isolates from Cyanophyceae, Chlorophyceae, Bacillariophyceae and Rhodophyceae were investigated for fatty acid composition in order to know the suitability of these algae in aquaculture industries. The algal isolates maintained as bacterial free cultures are *Anabaena circinalis*, *Anabanea* sp., *Nostoc muscorum*, *Oscillatoria agardhii* and *Microcystis* sps. from Cyanophyceae, *Scenedesmus quadricauda*, *Chlamydomonas* sps., *Enteromorpha intestinalis*, *Chaetomorpha linum*, *Pithophora oedogonia* from Chlorophyceae, *Skeletonema costatum*, *Nitzschia pinnata*, *Nitzschia acuta*, *Navicula longa* from Bacillariophyceae, *Gracilaria verrucosa*, *Grateloupia filicina*, *Ceramium elegans*, *Polysiphonia sertularioides* and *Chondrus crispus* from Rhodophyceae.

For lipid, the algae were extracted with a chloroform–methanol mixture (2:1) containing 0.005 per cent butylated hydroxy toluene (BHT) (Boberg *et al.*, 1985).

The homogenate was added with 15 ml of 0.2 M sodium dihydrogen phosphate. After thorough mixing the non-aqueous chloroform layer was pipetted off, evaporated to dryness and weighed to determine the total lipid. The fatty acid in the lipid extract was transmethylated at 60°C overnight. After addition of 2 ml of 5 per cent H_2SO_4 in methanol (Boberg *et al.*, 1985) after addition of 1.5 ml of distilled water, the methyl esters were extracted in 3 ml of petroleum ether (BP 40-60°C) containing 0.005 per cent BHT. The individual fatty acids were identified by comparing the retention times with commercially available reference standard of FA mixture (Supeko 37 compound, Supeko, U.SA.) and individual FA standards (Sigma USA) as described by Gopal Kumar *et al.* (1972) and comparing the area of peaks with the peak of the international standard (Ahlgren *et al.*, 1992).

Figure 13.1 shows the SAFA, MUFA and PUFA of the algal organisms belonging to Cyanophyceae, Chlorophyceae, Bacillariophyceae and Rhodophyceae. All the five blue green algal species had higher levels of SAFA (40.2-55.7 per cent) of the total fatty acids and relatively higher levels of MUFA. This is in agreement with the earlier reports on freshwater algae (Ahlgen *et al.*, 1992) and Marine algae (Volkman *et al.*, 1989). The green algae *Scenedesmus quadricauda* had SAFA 51.8 per cent of total fatty acids, MUFA, 22.05 per cent and PUFA 25.1 per cent. All the five green algal species had higher level of SAFA (40.6-51.8 per cent) of the total fatty acids and relatively high levels of PUFA. The four diatoms contain high levels of PUFA ranging from 31.7–37.4 per cent PUFA, MUFA 22.6-44.3 per cent. All the red algal species had higher levels of PUFA ranging from 44.1 per cent to 61.1 per cent of total fatty acids.

From fatty acid analysis on blue-green algae, certain unicellular forms have very simple fatty acid patterns. The most favourable algal compositions from the human point of view is presence of high proportion of PUFA and a high n3/n6 ratio. Among the algal isolates examined comparatively the diatom *Navicula longa* and *Nitzschia acuta* contained large amounts of EPA and DHA. The blue greens *Oscillatoria and Microcystis* were characterized by relatively high n3/n6 ratios, whereas *Anabaena* and *Nostoc* had only traces of n3 acids. In comparison to Cyanophytes, the Chlorophytes had a higher content of PUFA. The red algal species also contained higher amounts of PUFA and the maximum amount was present in *Grateloupia filicina.*

PUFA of both n3 and n6 types are important components in biomembranes, particularly in the vascular and nervous systems (Crawford *et al.*, 1989; Vorgroesen, 1989). Fishes and cells derived from species higher up in the evolutionary hierarchy cannot introduce double bonds into the carbon chain of FA closer to the methyl end than carbon atom number 7. This means that Mammalian cells cannot transform n3 to n6 FA or the reverse (Drevon, 1990). Long chain polyunsaturates of fish that have nutritional value for human are derived from dietary sources and chain elongation and desaturation by fish (Stansby, 1986). If selected algae could be used as the dietary ingredients, those will not only serve as the source of protein but also the source EPA.

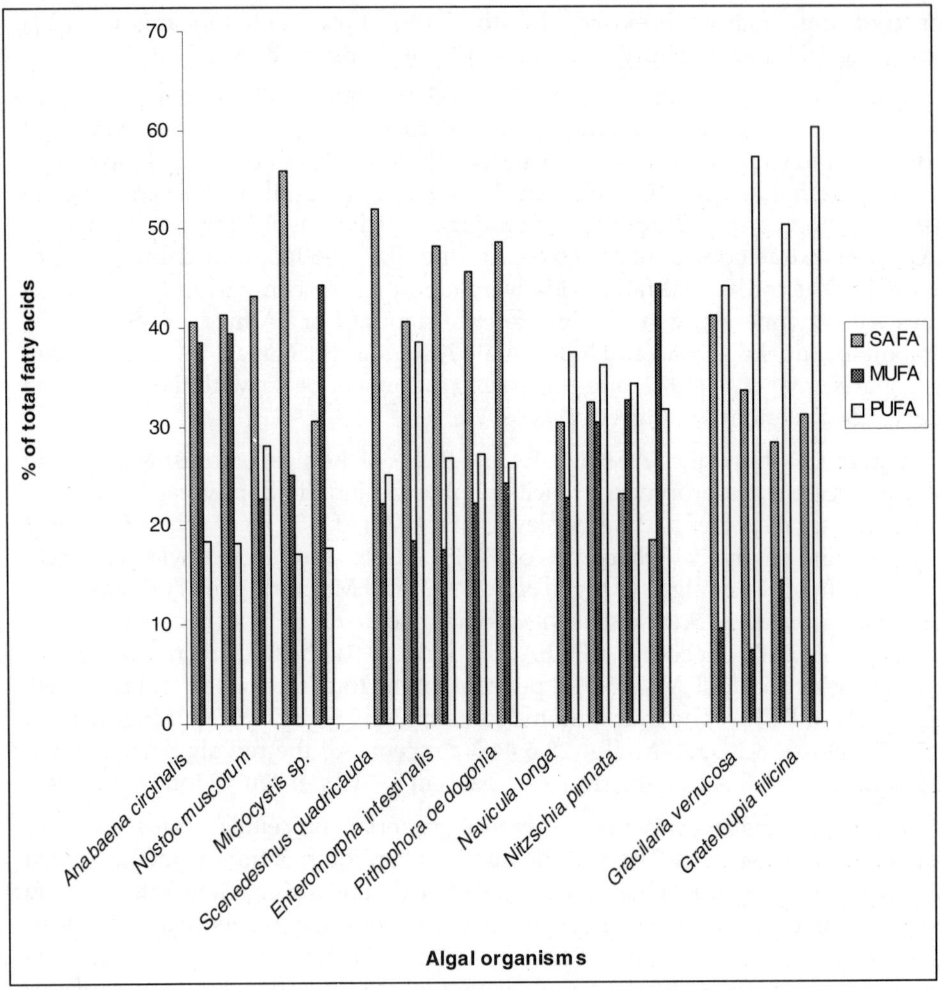

Figure 13.1: Relative Amounts of PUFA, MUFA and SAFA Content of Algae

2. Conclusion

Our study has suggested that the exact choice of the species is essential for preparing any fatty acid from the algal species on an industrial scale. Standardization of mass culture of some specific algae which have the ability to synthesize long chain n3 fatty acids is an important aspect.

Acknowledgement

The authors are thankful to DBT, New Delhi, Govt. of India for its financial assistance. The authors also extend their thanks to Coordinator and Head, Department of Botany, Berhampur University, for providing necessary research facilities.

References

Abbott IA (1996) Ethnoboatny of seaweeds: clues to use of seaweeds. Hydrobiologia 326-327: 15-20.

Ackman RG, Jangaard PM, Joyle RJ and Bruckerhoff H (1964) Origin of marine fatty acids. Analysis of the fatty acids produced as the Diatom *Skeletonema costatum.* J Fish Res Bid Can 21: 747-56.

Ahlgreen G, Gustafsson I and Boberg M (1992) Fatty acid content and chemical composition of freshwater microalgae.J Phycol 28: 37-50.

Apt KE and Behrens PW (1999) Commercial developments in microalgal biotechnology. J Phycol 35: 215-226.

Arao T, Kawaguchi A and Yamada M (1987) Positional distribution of fatty acids in lipids of the marine diatom *Phaeodactylum tricornatum*. Phytochemistry 26: 2573-2576.

Behrens PW and Kyle DJ (1996) Microalgae as a source of fattyacids. J Foodlipids 3: 259-272.

Belay A, Ota Y, Miyakawa K and Shimamatsy H (1994) Production of high quality spirulina at Earthrise forms. In: Algal Biotechnology in the Asia-pacific region. (Eds.) Phang SME, Lee YK, Borowitzka MA, Whitton BA Institute of advanced studies, University of Malaya, Kuala Lumpur: 5-15.

Ben Amotz A, Fishler R and Schneller A (1987) Chemical composition of dietry species of marine unicellular algae and rotifer with emphasis on fatty acids. Mar Biol (Berl) 95: 31-36.

Bennemann JR (1992) Microalgae Aquaculture feeds. J Appl Phycol 4: 233-245.

Bixler HJ (1996) Recent developments in manufacturing and marketing carrageenan. Hydrobiologia 326-327: 35-37.

Blumer M, Mullin MM and Guillard RRL (1970) A polyunsaturated hydrocarbon (3, 6, 9, 12, 15, 18 herico sahenane) in the marine freshwater. Mar Biol 6: 226-35.

Boberg M, Croon LB, Gustafsson IB and Vessby B (1985) Platelet fatty acid composition in relation to fatty acid composition in plasma and serum lipoprotein lipids in healthy subjects with special reference to the linoleic acid pathways. Clinical Science 68: 581-587.

Borowitzka MA (1997) Micro algae for aquaculture: opportunities and constraints. J Appl Phycol 9: 393-401.

Boswell KDB, Gladue R, Prima B and Kyle, DJ (1992) SCO production by fermentative microalgae.In: Industrial application of single cell oils. (Eds.) Kyle DJ and Ratledge C, American Oil Chemist Society, Champaign, Illinois, pp. 274-286.

Brown MR, Jeffery SW, Volkman JK and Dustan GA (1997) Nutritional properties of microalgae for mariculture. Aquaculture 151: 315-331.

Cheucas L and Riley JP (1969) The component combined aminoacids of some marine diatoms. J Mar Biol Assoc UK 49: pp 117-120.

Cohen Z (1990) The production potential of ecosapentanoic acid, arachidionic acid by red algae *Porphyridium cruentum*. J of Am Oil Soc 67: 916-920.

Cohen Z, Vonshak A, Boussiba S, Richmond A (1988) The effect of temperature and cell concentration on the fatty acid composition of outdoor cultures of *Porphyridium cruentum*. In: Algal Biotechnology (eds.) Stadler T, Mollion J, Verdus MC, Karamanos Y, Morvantt, Christiaen D, Elsevier Applied Science, London, pp.421-429.

Corner EDS and Coway CB (1968) Biochemical studies on the production of marine zooplankton, Biol Rev 43: 393-426.

Cranwell PA, Jawarski GHM and Bickley HM (1990) Hydrocarbon, sterols, esters and fatty acids in six freshwater chlorophytes. Phytochemistry 29: 145-151.

Crawford MA, Doyle W, Williams G and Drury PJ (1989) The role of fats and EFAs for the structures in the growth of *Fetus* sp. and neonate. In: Vergroesen AJ and Crawford M (eds.) "The role of fat in human nutrition". Academic press, London, pp. 81-115.

Day JG, Edwards AP and Rodgers GA (1991) Development of an industrial scale process for the heterotrophic production of micro-algal mollusk feed. Bioresource Technol 38: 245-249.

Dreveon CA (1990) n-6 and n-3 fatty acid–how much and which balance? Narings forskning 34: 56-61.

Dunstan GA, Volman JK, Barrette SM, Lerioi Jeffrey, SW (1994) Essential polyunsaturated fatty acids from 14 species of diatom. Phytochemistry 35: 155-161.

Dyerberg J, Bang HO, Hjorne N (1975) Fatty acid composition of the plasma lipids in Greenland Eskimos. Am J Clin Nutr 28: 958-966.

Emken FA, Adolf RO and Gulley RM (1994) Dietary linolenic acid influences desaturation and acylation of deuterium labeled linoleic and linolenic acids in young adult males. Biochem Biophys Acta 1213: 277-288.

Gopalkumar K Nair and Rajendranathan M (1972) Fatty acid composition of eight species of Indian marine fish. J Sc fd Agric 23: 493-496.

Harwood JL and Jones AL (1989) Lipid metabolism in algae. Advances in Botany Res 16: 1-53.

Heras H, Keanhowie J and Ackman RG (1994) The potential use of lipid microspheres as nutritional supplements of adult *Ostrea edulis*. Aquaculture 123: 309-322.

Holman RT (1981) In: Proj Lipid Res Pergamon Press, Oxford.

Horrobin DF (1983) J Reprod Med 28: 465-468.

Howell BR (1979) Experiments on the rearing of larval turbot, *Scophthalmus maximus* L. Aquaculture 18: 215-225.

Innis SM (1994) Fatty acids requirements of the new boon. J Physiol Pharmacol 72: 1483-1492.

Johnson EA and An GH (1991) Astaxathin from microbial sources. Crit Rev Biotechnol 11 (4): 297-326.

Johns RB, Nichols PD and Perry GJ (1979) Fatty acid composition of ten marine algae form Australian waters. Photochemistry 18: 799-802.

Jones DA, Kurmaly K, Arshard A (1987) Pareid shrimp hatchery trials using microencapsulate, diets. Aquaculture 64: 133-164.

Joyard J and Douce R (1987) Galatolipid synthesis in P.K. Stumpf (ed.) The biochemistry of plants, Vol. 9. Academic Press, New York, pp. 215-274.

Juario JV and Storch V (1984) Biological evaluation of Phytoplankton (*Chlorella* sp., *Tetraselmis* sp. and *Isochrysis galbana*) as food for milkfish (*Chanos chanos* fry). Aquaculture 40: 193-198.

Kayama M, Tsuchiya Y and Mead JF (1963) A model experiment of aquatic food chain with special significance in fatty acid conversion. Bull Jap Soc Sci Fish 29: 452.

Kenyon CN, Rippka R Stainer RY (1972) fatty acid composition and physiological properties of some filamentous blue-green algae. Arch Microbiol 83: 216-236.

Kitano M, Matsukawa R, Karube I (1997) Changes in eicosapentaenoic acid content of *Navicula sapprophila, Rhodomonas salina* and *Nitzschia* sp. under mixotrophic conditions. J Appl Phycol 9: 559-563.

Lee RF and Loeblich AR III (1971) Distribution of 21: 6 hydrocarbon and its relationship to 22: 6 fatty acid in algae. Phytochemisty 10: 593-602.

Lee RF, Nevenzel JC and Lewis AG (1974) Lipid changes during life cycle of Marine copepode, *Euchaeta taponica*, Marukawa, Lipids 9: 891-898.

Liao IC (1975) Experiments on induced breeding of the grey mullet: in Taiwan from 1963 to 1973. Aquaculture 6: 31-58.

Lynch R (1998) Aquasearch delaying Kan expansion deal. Honolulu star Bulletin (1990)–06-19: CA.

Makridas M, Neumann M, Simmer K, Pater J and Gibson R (1995) Are long chained polyunsaturated fatty acids essential nutrients in infancy? Lancet 345: 1463-1468.

Molina Geima F, Sanchez Perez JA, Garcia Camacho F, Fernandez Sevilla JM, Acien Fernandez FC and Urda Caedona J (1995) Biomass and icosapentaenoic acid productivities from an outdoor batch culture of *Phaeodactylum tricornutum* (UTEX 640 in an airlift tubular photobioreactor. Appl Microbiol Biotechnol 42: 658-663.

Murata N and Nishida I (1987) Lipids of blue-green algae (Cyanobacteria). In: Stumpf PK (ed) The biochemistry of plants, Vol.9, Academic press, New York, pp. 315-347.

Nash CE and Kuo CM (1975) Hypothesis for problem impeding the mass propagates of grey mullets and other finfish. Aquaculture 15: 119-133.

Nell JA (1993) The development of Oyster diets. Aust J agrie Res 44: 557-566.

Nell JA, Diemar JA, Heasman MP (1996) Food value of live yeasts and dry yeast-based diets fed to Sydney Rock Oyster *Saccostrea commercialis Spat*. Aquaculture 145: 235-243.

Opute FI (1974) Lipid and fatty acid composition of diatoms. J Exp Bot 25: 823-835.

Orus MI, Marco E and Martinez F (1991) Suitability of *Chlorella vulgaris*, UAM 101 for heterotrophic biomass production. Bio Resour Technol 38: 179-184.

Peltit TR, Jones AL and Harwood JL (1989) Lipid metabolism in the red marine algae *Chondrus crispus* and *Polysiphonia lanosa* as modified by temperature. Phytochemistry 28: 2053-2058.

Pohl P and Zurheide F (1979) Fatty acids and lipids of marine algae and control of their biosynthesis by environmental factors. In: Marine algae in pharmaceutical science (Eds) Hoppe HA, Lovering T and Tanaka J DC Gruyter Berlin. pp. 473-523.

Pullin RSV and Kuo CM (1981) Developments in the breeding of cultured fishes. Advance in food producing systems for arid and semi arid lands, Academic Press, London, pp. 899-978.

Regan DL (1988) In other micoralgae: In: Microalgal Biotechnology. (eds.) Borowitzka MA and Borowitzka LJ,Cambridge Univ. Press, New York, pp. 135-150.

Roughan PG and Slack CR (1984) Glycerolipid synthesis in leaves. Trends Biochem Sci 9: 383-386.

Salem (1989) Omega 3 fatty acids: Molecular and biochemical aspects. In: New protective roles for selected nutrient: Current topics in nutrition and disease. (Eds.) Spiller G and Scala J Vol. 22, Alan R. Leiss, New York, pp. 109-228.

Sato N, Nemoto Y and Furyuya M (1988) Lipids of *Chatonella antiqua* (Raphidophyceae). Plant Physiol Biochem 26: 93-98.

Smith KL and Harwood JL (1984): Lipids and lipid metabolism in the brown alga, *Fucus serratus*. Phytochemistry 23: 2469-2473.

Stansby ME (1986) Fatty acids in fish. In: Health effects of polyunsaturated fattyacids in seafoods. (Eds) Simopoulus AP, Kiefer RR and Martin RE. Academic press, London, pp. 389-401.

Sukenik A (1991) Ecophysiological consideration in the optimization of eciosapentaenoic acid production by *Nannochloropsis* sp. (Eustigmatophyceae). Bioresour Technol 36: 263-270.

Takeyama H, Iwamoto K, Hata S, Takano and Matsunaga T (1996) DHA enrichment of Rotifers: A simple two step culture using the unicellular algae *Chlorella reqularis* and *Isochrysis galbana* J Mar Biotechnol 3: 244-247.

Tan CK and Johns MR (1996) Screening of Diatoms for heterotrophic eicosapentaenoic acid production. J Appl Phycol 8: 59-64.

Venkataraman GS and Neelakandan S (1980) Save on N_2 fertilizer by the use of algae in rice fields. Indian Farming Oct: 27-30.

Vergroesen AJ (1989) Introduction. In: The role of fat in human nutrition. (eds.) Vergroesen AJ and Cranford M Academic press, London, pp. 1-44.

Volkman JK, Jeffrey SW, Nichols PD, Rogers GI and Garland CD (1989) Fatty acid and lipid composition of 10 species of micro-algae used in mariculture. J Exp Mar Biol Ecol 128: 219-40.

Young Manitchai W and Ward OP (1991b): Screening of algae for potential alternative sources of eicosapentaenoic acid. Phytochemistry 30: 2963-2967.

Algal Biotechnology: New Vistas (2010) *Pages* **206–222**
Editor: **Mihir Kumar Das**
Published by: **DAYA PUBLISHING HOUSE, NEW DELHI**

Chapter 14

Cyanobacterial Toxins

Anjana Pandey* and Archana Tiwari

Nanotechnology and Molecular Biology Lab, Department of Biotechnology, University of Allahabad, Allahabad – 211 002, U.P., India

ABSTRACT

Cyanotoxins are produced by some bloom forming cyanobacteria. The toxins are of three types- heptotoxins affect the liver; neurotoxins affect the nervous system, while the dermatotoxins target the skin. Significant progress has been made in characterization of cyanobacterial toxins in the last decade. Knowledge of the chemical structure and activity of the toxins should help scientists to devise more sensitive ways to measure the compounds in water and to develop antidotes to lethal doses. The toxins and their derivatives are being considered as potential medicines for Alzheimer's disease and other disorders. These days much attention is focused on revealing the molecular mechanism involved in toxin production and on showing the physiological, biochemical and genetical differences between toxic and non-toxic strains of cyanobacteria. In our laboratory bloom forming cyanobacterial strains were isolated from natural blooms of northern India. The toxin Microcystin was isolated from the ten purified strains belonging to genera *Synechococcus, Synechocystis, Merismopedia, Microcystis, Arthrospira, Nostoc, Anabaena,* and *Cylindrospermum.* The toxins were characterized by TLC, UV spectroscopy and mouse bioassay. Eight strains showed presence of toxin. The biochemical results were confirmed by Polymerase Chain Reaction (PCR) as molecular tool. The PCR results of peptide synthetase amplification show that eight among the ten strains studied showed presence of peptide synthetase genes. The amplification product of 760 bp was observed in *Synechocystis aquatalis, Merismopedia glauca, Microcystis aeruginosa, Microcystis aeruginosa* (O), *Arthrospira*

* E-mail: apandey70@rediffmail.com

(=Spirulina) platensis, Anabaena iiyengari, Anabaena variabilis and *Cylindrospermum muscicola*. Knowledge about the regulation of toxin production will prove very important in predicting and preventing toxic blooms. Much knowledge is indeed required in this area for better understanding of these toxic forms, their toxins and the ways to control these hazardous blooms.

Keywords: *Cyanotoxins, Blooms, Hepatotoxins, Neurotoxins, Dermatotoxins, Polymerase chain reaction.*

1. Introduction

Cyanobacterial toxins are the naturally produced poisons stored in the cells of certain species of bloom forming cyanobacteria. The first known reported incidence of cyanobacteria toxin poisoning was from an Australian lake in 1878 by George Francis. Since then investigators have confirmed that many genera of cyanobacteria include poisonous strains. The structure of toxic peptides of cyanobacteria was established by Jüttner (1987), Gleason and Wood (1987). The Cyanobacterial toxins or Cyanotoxins are very diverse group of toxins from both the chemical and toxicology standpoint. The toxins are classified on the basis of their effect on the living system.

2. Types of Toxins

Hepatotoxins mainly target the liver. Microcystins are a group of cyclic heptapeptides (7 amino acids). They are produced by *Microcystis aeruginosa, M. viridis, Anabaena flos aquae, Nostoc spp, Oscillatoria agardhii*. Nodularins are pentapeptide heptatoxins, produced by species of *Nodularia*. Structurally they are related to microcystins but differ in the sense that nodularin is composed of only five amino acids in the peptide ring. Both these toxins show the same hepatotoxic effect through the potent inhibition of protein phosphates. They are also reported to be carcinogenic (Ohta *et al.*1994). Microcystins and nodularins mainly known as tumour-promoting substances when combined with compounds that are able to initiate the cancer process usually by causing DNA-damage (Falconer and Buckley, 1989; Falconer *et al.*, 1981; Humpage *et al.*, 2000; Nishiwaki *et al.*, 1992). They are also inhibitor of serine/ threonine protein phosphatases 1 and 2A, which leads to hyperphosphorylation of proteins associated with the cytoskeleton in hepatocytes (Honkanan *et al.*, 1994, Nishiwaki *et al.*, 1992). Cylindrospermopsins are cyclic in alkaloids and like microcystins primarily affects the liver, although causes considerable damage to other major organs. They are produced by forms of *Cylindrospermum, Aphanizomenon, Umezakia, Raphidiopsis*. Four neurotoxins have been studied in detail. They affect the nervous system of the body. Anatoxin-a is group of neurotoxin produced by a number of cyanobacterial genera including *Anabaena, Oscillatoria*, and *Aphanizomenon*. These toxins cause death via a depolarizing blockage of neuromuscular transmission, and subsequent respiratory paralysis (Carmichael *et al.*, 1975). It is a highly potent and stereospecific agonist at nicotinic synapses. (Carmichael *et al.*, 1979; Spivak *et al.*, 1980; Swanson *et al.*, 1986). Anatoxin-a (s) is unique phosphate ester of cyclic N-hydroxyguanine produced by *Anabaena flos-aquae*. The toxin decomposes rapidly in basic solution but is relatively stable under neutral or acidic conditions. Anatoxin-a (s) induces salivation in mice by which it can be differentiated from other cyanobacterial neurotoxins (Mahmood and Carmicheal, 1987). It acts as an irreversible peripheral anticholinesterase (Mahmood *et al.*1988; Cook *et al.*, 1989). Saxitoxins are neurotoxic alkaloids known as PSP's (Paralytic Shelfish poisons) due to their occurrence and association with seafood. There are a number of STX variants generally divided into groups based on their structure or organism of origin. They are reported to be produced by *Aphanizomenon flos-aquae* and *Anabaena–circinalis.* The saxitoxins block the sodium channels in nerve cells, thus causing their neurotoxic effects. Dermatoxins affect the skin. Aplysiatoxins are produced by marine cyanobacteria such as *Lyngbia*. They are commonly known for their dermatotoxic activity causing inflammation of the skin. They are also potent tumour promoters. Debromoaplysiatoxins are reported to cause inflammatory activity (Mynderse *et al.*, 1977). These toxins are potent tumour promoters and protein kinase C activators (Fujiki *et al.*, 1990). Aplysiatoxins and debromoaplysiatoxins have been found associated with filamentous cyanobacterial species including *Schizothrix calicola* and *Oscillatoria nigroviridis*.

In the last decade significant progress has been made in characterization of cyanobacterial toxins. These days much attention is focused on revealing the molecular mechanism involved in toxin production and on showing the physiological, biochemical and genetical differences between toxic and non-toxic strains of cyanobacteria. There are five basic methods to analyze microcystins namely; those based on reactions with a fluorescent probe; enzyme–linked immunosorbent assays [ELISA]; inhibition of protein phosphatase and mass spectrometry (Dawson, 1998); and by polymerase chain reaction (Baker *et al.*, 2002, Pan *et al.*, 2002). Shimizu and colleagues (1995, as cited by Dawson, 1998) targeted conjugated dienes using a synthesized fluorogenic reagent called DMEQ–TAD. An and Carmichael (1994) has used a direct competitive ELISA to examine the specificity of the rabbit anti-microcystin-LR polyclonal antibodies. McDermott *et al.* (1995) described an ELISA potentially able to detect microcystins in water at a concentration as low as 100 pg/ml water.

Microcystins are inhibitors of protein phosphatase (Honkanen *et al.*, 1996). An and Carmichael (1994) had reported an IC50 of 6 ng/ml for microcystin–LR in their direct competitive ELISA. Whilst ELISA microcystin-LR/YR/RR detection limits of 0.10, 0.12, 0.14 and 0.20 ng/ml was reported by Yu *et al.* (2002). A screening method for microcystins in cyanobacteria has been developed based on the formation of 3-methoxy-2-methyl-4-phenylbutyric acid by ozonolysis (Harada *et al.*, 1996). The acid was detected by electron ionization–gas chromatography/mass spectrometry, using selected ion monitoring in a procedure that detected nanogram levels of microcystin in only 30 minutes.

Microcystins are the most widely distributed hepatotoxins (Gorham and Carmichael, 1988). The name 'Microcystin' is derived from the toxins that were first isolated from *Microcystis aeruginosa.* They are cyclic peptides with general structure described as cyclo- (D-Ala-L-X-erythro-b-methyl-Diso-Asp-L-Y-Adda-iso-Glu-N-methyl dehydro-Ala), where Adda refers to a b-amino acid with an apolar side chain (3-amino-9-methoxy 2, 6, 8- trimethyl-10-phenyl deca-4, 6-dienoic acid), and X and Y are two variable amino acids (Botes *et al.*, 1984, 1985). There are more than 60 microcystins reported till date (Codd *et al.*, 1999). The molecular weight of microcystins ranges from 800-1100 Daltons (Figure 14.1).

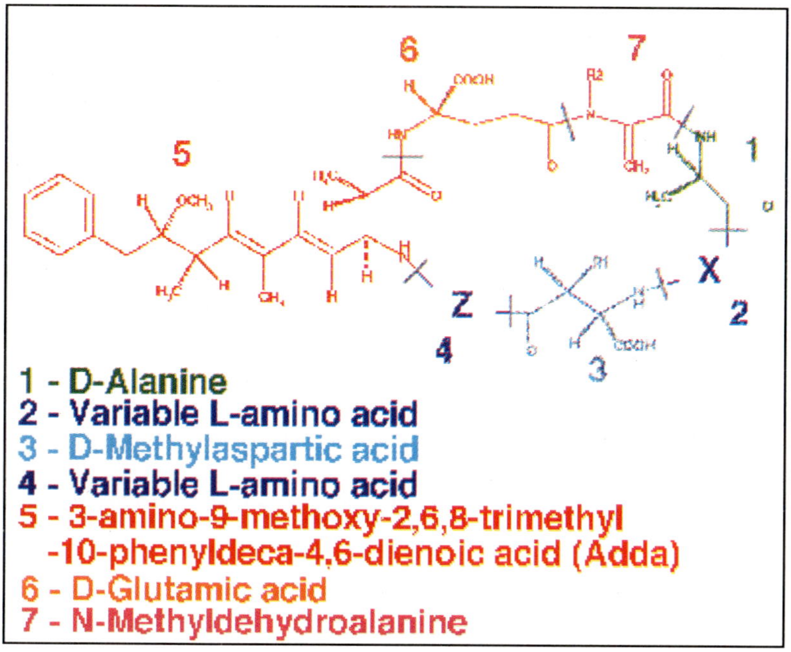

Figure 14.1: Structure of Microcystin

Microcystins are produced by some species of the genera *Microcystis* (Shirai *et al.*, 1991; Namikoshi *et al.*, 1992; Luukkainen *et al.*, 1994; Lee *et al.*, 1998), *Planktothrix* (Sivonen and Jones, 1999), *Anabaena* (Harada *et al.*, 1991a; Sivonen *et al.*, 1992; Namikoshi *et al.*, 1998), *Oscillatoria* (Meriluoto *et al.*, 1989; Luukkainen *et al.*, 1993) and *Nostoc* (Namikoshi *et al.*, 1990; Sivonen *et al.*, 1990; Beattie *et al.*, 1998).

Microcystins are inhibitors of serine/threonine protein phosphatase 1 and 2A (Honkanan *et al.*, 1994; MacKintosh *et al.*, 1990). Microcystins are reported as tumor-promoting substances when combined with compounds that are able to initiate the cancer process (usually by causing DNA-damage) (Falconer and Buckley, 1989; Nishiwaki-Matsushima *et al.*, 1992; Humpage *et al.*, 2000).

A number of environmental factors like light, temperature, nutrients and pH are reported to affect microcystin production (Watanabe and Oishi, 1985; Westhuizen and Eloff, 1983, 1985). The hepatotoxin, microcystin isolated from natural blooms and cultures of *M. aeruginosa* has been studied in detail (Ellenman *et al.*, 1978; Botes *et al.*, 1982, 1984; Krishnamurthy *et al.*, 1986; Runnegar *et al.*, 1987; Carmichael, 1988). The analysis and purification of microcystin involves organic solvent extraction and clean up followed by HPLC (Harada *et al.*, 1988, Krishnamurthy and Carmichael, 1986; Dierstein *et al.*, 1988; Birk *et al.*, 1988). Water-saturated 1- butanol containing 1 per cent acetic acid followed by lipid extraction by diethyl ether and purification of the toxin by HPLC was used by Birk *et al.* (1988). Consecutive freeze thaw technique was advocated by Ohtake *et al.* (1989). Further purification of the crude toxin was done by extraction with n- butanol-methanol-water from lyophilized cells followed by Bond Elut C18 cartridge and fractionated by Sephadex G-25 chromatography (Brooks and Codd, 1986). Toxins were extracted with 5 per cent aqueous acid only from lyophilized cells and purified by Baker 10 C18 Cartridge (Harada *et al.*, 1988). Methanol was used for the extraction of microcystins by Lawton *et al.* (1994). According to them methanolic extraction provides a rapid determination of the potent cyanobacterial microcystins and nodularin hepatotoxins. The toxin microcystin was extracted from frozen cells by using 5 per cent glacial acetic acid and purified by C18 cartridges by Utkilen and Gjolme, (1992).

Knowledge of the chemical structure and activity of the toxins should help scientists to devise more sensitive ways to measure the compounds in water and to develop antidotes to lethal doses. Improved understanding of how these chemicals function should also facilitate efforts to determine the long term effects of exposure to non-lethal doses (Carmichael, 1994). Research into the structure and activity of the toxins is sparking interest on other grounds as well. They and their derivatives are being considered as potential medicines for Alzheimer's disease and other disorders. The substances already serve as invaluable tools for exploration of questions in cell biology.

3. Molecular Studies on Bloom Forming Cyanobacteria

The bloom forming cyanobacteria grow luxuriantly in eutrophic water bodies, affecting the health of humans as well as live stocks. In the last forty years there has been a lot of initiation of research in this avenue. The biology and ecology of cyanobacteria have been extensively studied throughout the world during the last decades, but we still very little know about the factors and processes involved in regulating toxin production for many cyanobacterial species. There are basically very few molecular genetical studies on true planktonic cyanobacteria because of difficulties in the isolation and purification of strains. The DNA base composition of few cyanobacterial forms have been reported by Kaye *et al.* (1967), Edelman *et al.* (1967), Craig *et al.* (1969), Stanier *et al.* (1971), Rippka *et al.* (1974). Herdman *et al.* (1978) reported the DNA base composition of 176 strains of cyanobacteria determined by thermal denaturation or by CsCl density gradient centrifugation. Rouhianen *et al.* (1995) reported the characterization of toxin producing cyanobacteria by using an oligonucloetide probe containing a tandemly repeated heptamer. They used a 20 mer probe based on the heptamer repeat sequence to survey a large number of cyanobacterial strains, and all the filamentous *Anabaena, Nostoc* and *Calothrix* strains were found to have such tandem repeats at many chromosomal locations. The probes also helped in characterization of these strains. Neilan *et al.* (1995) studied the genetic diversity and phylogeny of toxic cyanobacteria determined by DNA polymorphisms within the Phycocyanin locus. They amplified the Phycocyanin-intergenic spacer (Figure 14.2) sequence via PCR, from extracted DNA and crude lysate of non-axenic environmental isolates of cyanobacteria. The Phycocyanin gene (Belknap and

Haselkorn 1987) usually occurs in a single copy but it may also be present in up to three copies (Apt *et al.*, 1995).

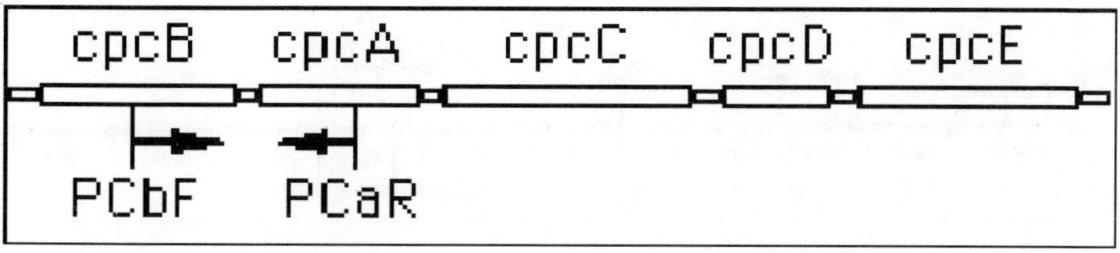

Figure 14.2: The Phycocyanin Operon (3.6 kb) consists of Five Open Reading Frames (Belknap and Haselkorn 1987). The Primers PCbF and PCaR binding on conserved regions of the beta and alpha subunit, respectively (Neilan *et al.*, 1995b) are indicated by arrows.

Meibner *et al.* (1996) investigated the occurrence of genes encoding peptide synthetase of the types known from other organisms in cyanobacteria, and differentiation of toxic and nontoxic strains. Three novel insertion sequences have been characterized from the unicellular cyanobacterium *Synechocystis* PCC 6803 by C. Cassier-Chauvat *et al.* (1997). Ulrich nübel (1997) developed a PCR procedure for the selective retrieval of cyanobacterial rRNA gene fragments from a variety of natural and artificial settings on the basis of published 16 S rRNA sequences. The Combination of this procedure with denaturing gradient gel electrophoresis (DGGE), a technique for the sequence-dependent separation of DNA molecules, proved useful to visualize the diversity of cyanobacterial 16S rRNA genes in environmental samples, to detect the uniqueness of isolated strains. Their approach can be a powerful tool for investigating the phylogenetic diversity in cyanobacteria and its ecological significance. Their approach is especially useful for unraveling the connections between cyanobacterial populations observed in nature, molecular sequence data, and physiology. Fergusson and Christopher (2000) phlylogenetically examined the cyanobacterial rpo C1 (DNA-dependent RNA polymerase) gene sequence focusing on the genus *Anabaena* and particularly the species *A. circinalis*. They also reported for the first time *A. circinalis* specific PCR assay targeting the rpo C1 gene to detect this organism directly in environmental water samples. Importantly, the rpo C1 gene provided sufficient sequence variation to differentiate the genus *Anabaena* and *A. circinalis* to species level. The gene cluster mcy ABCDEFGHIJ of *Microcystis aeruginosa* encodes the non-ribosomal peptide synthetase. This gene cluster has been recently identified and sequenced (Arment and Carmichael, 1996; Dittmann *et al.*, 1997; Moore *et al.*, 1991). Using RT-PCR and RACE, a complete transcriptional analysis of the mcy ABCDEFGHIJ cluster, has been performed (Melanie *et al.*, 2002). This study shows that the microcystin synthetase gene cluster consists of two polycistronic transcripts, mcy ABC and mcy DEFGHIJ, which are transcribed from a central promoter between mcy A and mcy D (Figure 14.3). Both polycistronic transcripts have alternate transcription start sites, which appear to be light dependent. Some putative intercistronic transcription start sites for individual mcy genes indicated complex gene regulation involving multiple and alternate messages. These findings may prove to be highly significant for understanding the regulation of microcystin synthetase gene cluster under particular light conditions and may increase our knowledge concerning a function of microcystin under these growth conditions.

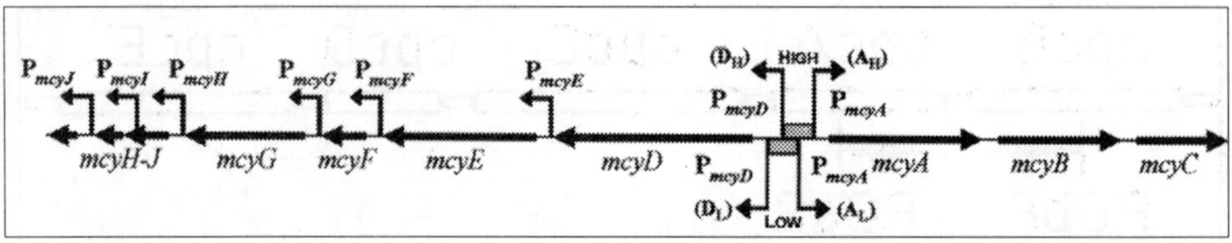

Figure 14.3: Organization of the Microcystin Synthetase Gene Cluster, *mcyABCDEFGHIJ*, Showing Putative Promoters for *mcyEFGHIJ* (P*mcyE* to P*mcyJ*) and the Alternate Promoters Identified for *mcyA* and *mcyD* (P*mcyA* and P*mcyD*) Under High Light Conditions (AH and DH) and Low Light Conditions (AL and DL)

A highly conserved sequence was detected in a region coding for the condensation domain of mcy A (referred to as mcy A-Cd for mcy A condensation domain) and used for the design of a primer pair that would allow the specific amplification of mcy A-Cd from microcystin synthesizing *Microcystis* and *Planktothrix* and other genera (Hisbergues *et al.*, 2003). Marahiel *et al.*, 1997 reported that this part of mcy A codes for the section of the condensation domain that contains the conserved motif C5 (Figure 14.4).

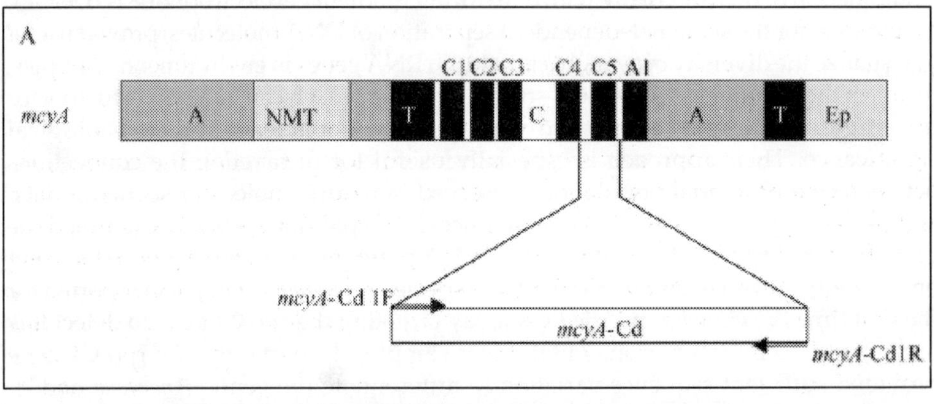

Figure 14.4: N-methyltransferase Domain Coded by mcy A. A: Adenylation domain; NMT: Methyl transferase domain; C: Condensation domain; T: Thiolation domain; Ep: Epimerisation domain; C1-5 and A1: Core motifs of the condensation and adenylation domains respectively (Marahiel *et al.*, 1997).

A single N-methyltransferase (NMT) domain (Marahiel *et al.*, 1997; Kleinkauf and Döhren, 1996) encoded by the microcystin synthetase gene mcy A (Tillet *et al.*, 1999) was used to design conserved molecular probes enabling identification of toxigenic *Microcystis* (Tillet *et al.*, 1999). In all 20 NMT–containing strains mcy C was adjacent to unknown *M. aeruginosa* ORF 1 called uma1 (Figure 14.5).

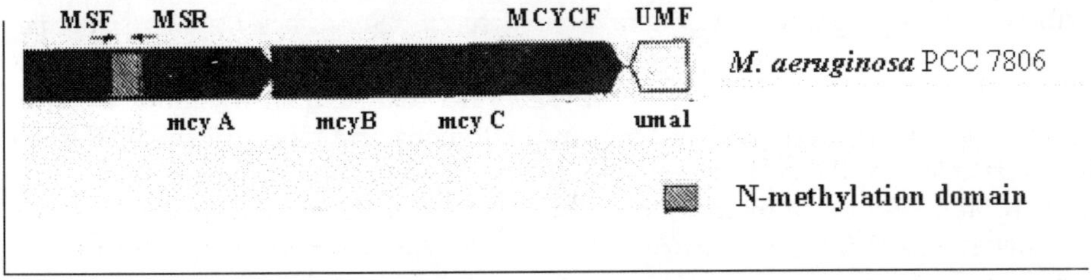

Figure 14.5: Chromosomal Arrangement of the mcyA, mcyB, mcyC and uma1 Genes in *M. aeruginosa* PCC 7806 (Tillet *et al.*, 1999)

The detection of strains with the genetic potential to produce microcystins, independent of their taxonomic category on environmental conditions in water bodies and investigation of genotypic composition of cyanobacterial populations, mainly of the genus *Microcystis*, through detection of mcy B was done by Dittmann *et al.*, 1999; Pan *et al.*, 2002; Oliveira, 2002. This PCR based gene detection procedures established a correlation between the existence of mcy B gene and microcystin production.

Vaitomaa *et al.* (2003) designed genus specific mcy E primers to identify and quantify the main microcystin producing genera in lakes. The mcy E gene primers are reported to be genus and mcy E gene specific. Single amplification product was observed when genomic DNA from a microcystin producing *Microcystis* and *Anabaena* strain were used as template in PCR with *Microcystis* or *Anabaena* specific primers.

Sielaff *et al.* (2003) amplified mcy F gene from *Microcystis* strain PCC 7806. The mcyF gene encodes aspartate recemase. Thus, *mcyF*, which is part of the mcy gene cluster, is involved in microcystin biosynthesis. The Taq Man PCR, or the Taq nuclease assay (TNA), was introduced to quantify specific genotypes of picocyanobacteria (Becker *et al.*, 2000) or microcystin-producing cyanobacteria in the field (Foulds *et al.*, 2002). This technique utilizes a sequence specific dually labeled fluorescent probe (Taqman probe) and primers to quantify the level of DNA template initially present in a sample. Recently a quantitative PCR approach has been used for quantifying the total population of *Microcystis spp.* as well as the subpopulation comprising all microcystin genotypes in terms of cell numbers and for monitoring the seasonal development of microcystin genotypes in lake Wannsee (Kurmayer *et al.*, 2003). Two independent TNAs, one to quantify the total population of *Microcystis spp* using the intergenic spacer region within the phycocyanin (PC) operon and the other assay to quantify microcystin genotypes using region of mcy B part of the microcystin synthetase gene cluster were developed. This study was the first attempt demonstrating the results of cell quantification obtained by TNA do correlate significantly with cell numbers determined by a more widespread cell counting method, both for laboratory and for field samples. This validation is necessary to estimate the reliability of quantitative real time PCR in field analysis. To date, no method for quantifying microcystin genotypes that is comparable in accuracy and feasibility to the technique developed in this study has been published.

In our laboratory bloom forming cyanobacterial strains were isolated from natural blooms of northern India. The strains were purified and identified taxonomically following recent monographs. The toxin Microcystin was isolated from the ten purified strains belonging to genera *Synechococcus, Synechocystis, Merismopedia, Microcystis, Arthrospira, Nostoc, Anabaena, and Cylindrospermum.* Toxins were isolated using protocol of Birk *et al.* (1989). Analysis of the toxic extract was done by Thin Layer Chromatography, Ultraviolet Spectrophotometry and mouse bioassay. Results showed that strain *Microcystis aeruginosa* (O) possesses maximum toxin (5.06 ± 0.06 ng/µg protein), followed by *Cylindrospermum* (4.23 ± 0.62), *Anabaena iyengarii* (3.65 ± 0.52), *Arthrospira (=Spirulina) platensis* (2.4 ± 0.35), *Anabaena variabilis* (1.80 ± 0.27), *Merismopedia glauca* (1.4 ± 0.19, *Synechocystis aquatalis* (1.09 ± 0.15) and *Microcystis aeruginosa* (0.8 ± 0.11). Two strains *Synechococcus elongates* and *Nostoc paludosum* did not show presence of the toxin.

The biochemical results were confirmed by Polymerase Chain Reaction (PCR) as molecular tool. The PCR results of peptide synthetase amplification show that eight among the ten strains studied showed presence of peptide synthetase genes. The amplification product of 760 bp was observed in *Synechocystis aquatalis, Merismopedia glauca, Microcystis aeruginosa, Microcystis aeruginosa* (O), *Arthrospira (=Spirulina) platensis Anabaena iiyengari, Anabaena variabilis and Cylindrospermum muscicola* (Figure 14.6).

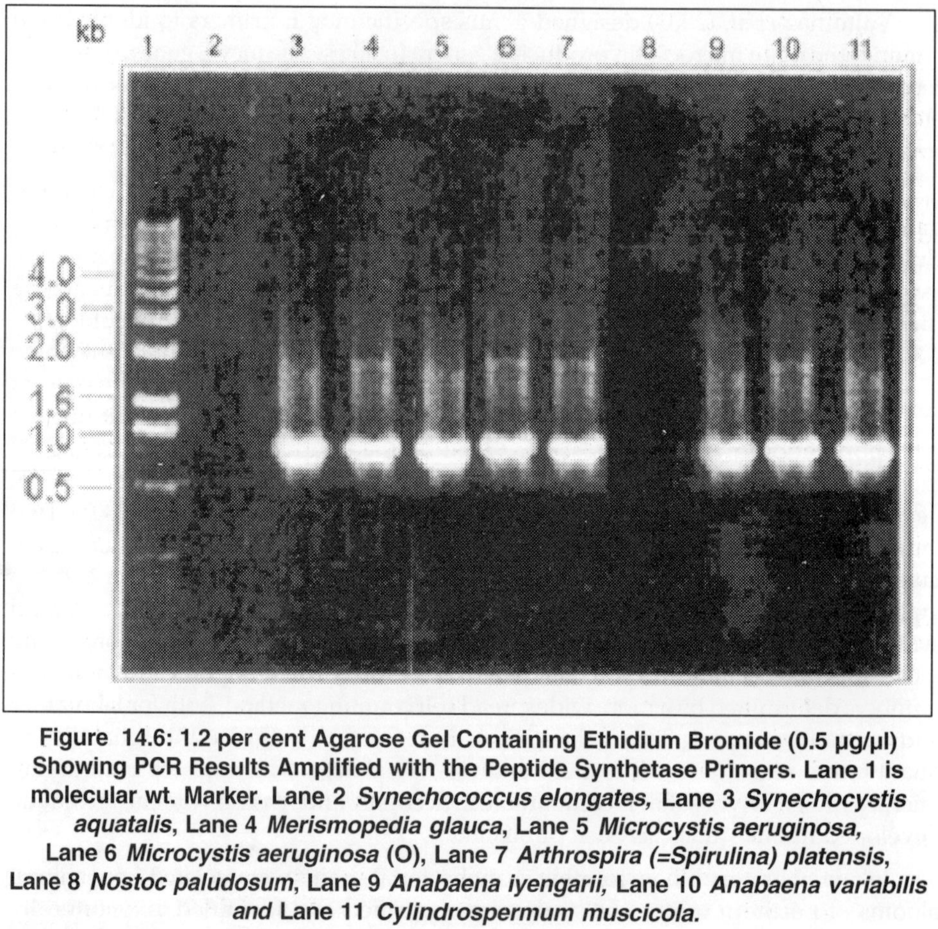

Figure 14.6: 1.2 per cent Agarose Gel Containing Ethidium Bromide (0.5 µg/µl) Showing PCR Results Amplified with the Peptide Synthetase Primers. Lane 1 is molecular wt. Marker. Lane 2 *Synechococcus elongates*, Lane 3 *Synechocystis aquatalis*, Lane 4 *Merismopedia glauca*, Lane 5 *Microcystis aeruginosa*, Lane 6 *Microcystis aeruginosa* (O), Lane 7 *Arthrospira (=Spirulina) platensis*, Lane 8 *Nostoc paludosum*, Lane 9 *Anabaena iyengarii*, Lane 10 *Anabaena variabilis* and Lane 11 *Cylindrospermum muscicola*.

The molecular results confirmed the biochemical testing of isolated toxins. Amplification was done from the isolated DNA based on conserved regions of peptide synthetase. The details of obtained results are enlisted in Table 14.1. However, *Synechococcus elongates* and *Nostoc paludosum* did not show any amplification product.

Table 14.1: PCR Results of Cyanobacterial Cultures Amplified by Peptide Synthetase

Strain	*Peptide Synthetase*	*Fragment Size*	*Toxigenicity*
Synechococcus elongates	–	–	Non-toxic
Synechocystis aquatalis	+	760 bp	Toxic
Merismopedia glauca	+	760 bp	Toxic
Microcystis aeruginosa	+	760 bp	Toxic
Microcystis aeruginosa (O)	+	760 bp	Toxic
Arthorspira platensis	+	760 bp	Toxic
Nostoc paludosum	–	–	Non-toxic
Anabaena iyengarii	+	760 bp	Toxic
Anabaena variabilis	+	760 bp	Toxic
Cylindrospermum muscicola	+	760 bp	Toxic

Much knowledge is indeed required in this area for better understanding of these toxic forms, their toxins and the ways to control these hazardous blooms. Molecular biological methods can detect the presence of genes and therefore the potential to produce the corresponding gene product (*e.g.* a peptide toxin), but for the estimation of actual toxin production, tests for corresponding mRNA, biochemical and physiological tests must be done.

References

An J and Carmichael WW (1994) Use of a colorimetric protein phosphatase inhibition assay and enzyme linked immunosorbent assay for the study of microcystins and nodularins. Toxicon 32: 1495-1507.

Apt KE, Collier JL and Grossman AR (1995) Evolution of the phycobiliproteins. Journal of Molecular Biology 248: 79-96.

Arment AR and Carmichael WW (1995) Evidence that microcystin is produced by a thio-template mechanism. Vth Cynobacterial Molecular Biology Workshop, 21-25 July 1995, pp 97. Asilomar, CA, USA.

Arment AR and Carmichael WW (1998) Evidence that microcystin is a thiotemplate product. J Phycol 32: 591-597, 1515-1518.

Baker JA, Entsch B, Neilan BA and McKay DB (2002) Monitoring changing toxigenicity bloom by molecular methods. Appl Environ Microbiol 68: 6070-6076.

Beattie KA, Kaya K, Sano T and Codd GA (1998) Three dehydrobutyrine-containing microcystins from *Nostoc*. Phytochemistry 47: 1289-1292.

Becker S, Böger P, Oehlmann R and Ernst (2000) PCR bias in ecological analysis: a case study for quantitative Taq nuclease assays in analysis of microbial communities. Appl Environ Microbial 66: 4945–4953.

Belknap WR and Haselkorn R (1987) Cloning and light regulation of expression of the phycocyanin operon of the cyanobacterium *Anabaena*. The EMBO Journal 6: 871-884.

Birk IM, Matern U, Kaiser I and Weckesser J (1988) The peptide toxin of the cyanobacterium *Microcystis aeruginosa* PCC 7941. Isolation and analysis by nuclear magnetic resonance and fast atom bombardment mass spectrometry. J Chromatography 449: 423-431.

Birk IM, Dierstein R, Kaiser I, Matern U, König WA, Krebber R and Weckesser J (1989) Nontoxic and toxic oligopeptides with D-amino acids and unusual residues in *Microcystis aeruginosa* PCC 7806. Arch. Microbiology 151: 411-415.

Botes DP, Kruger H and Viljoen CC (1982) Isolation and characterization of four toxins from the blue-green alga *Microcystis aeruginosa*. Toxicon 20: 945-954.

Botes DP, Tuinman AA, Wessels PL, Viljoen CC and Kruger H (1984) The structure of cyanoginosin-LA, a cyclic heptapeptide toxin from the cyanobacterium *Microcystis aeruginosa*. Journal of Chemical Society, Perkin Transactions 1: 2311-2318.

Botes DP, Wessels PL, Kruger H, Runnegar MTC, Santikarn S, Smith RJ, Barna JCJ and Williams DH (1985) Structural studies on cyanoginosins-LR, -YR, -YA, and -YM, peptide toxins from *Microcystis aeruginosa*. Journal of Chemical Society Perkin Transactions 1: 2747-2748.

Brooks WP and Codd GA (1986) Extraction and purification of toxic peptides from natural blooms and laboratory isolates of the cyanobacterium *Microcystis aeruginosa*. Lett Appl Microbiol 2: 1-3.

Carmichael WW, Biggs DF and Gorham PR (1975) Toxicology and pharmacological action of *Anabaena flos-aquae* toxin. Science 187: 542-544.

Carmichael WW, Biggs DF and Peterson MA (1979) Pharmacology of anatoxin–a produced by the cyanophyte *Anabaena flos-aquae* NCR 44-1.Toxicon 17: 229-236.

Carmichael WW, Beasley VR, Bunner DL, Eloff JN, Falconer IR, Gorham PR, Harada KI, Krishnamurthy T, Yu MJ, Moore RE, Rinehart KL, Runnegar MTC, Skulberg OM and Watanabe MF (1988) Naming of cyclic heptapeptide toxins of cyanobacteria (blue-green algae). Toxicon 26: 971-973.

Carmichael WW (1988) Toxins of freshwater algae. In: Tu AT (Ed), Marine Toxins and Venoms, *Handbook of Natural Toxins*. Marcel Dekker, New York, 121–147.

Carmichael WW (1994) The toxins of cyanobacteria. Sci Am 270: 64- 70.

Chauvat CC, Pontcelet M and Chauvat F (1997) Three insertion sequences from the cyanobacterium *Synecococcus* PCC 6803 support the occurrence of horizonal DNA transfer among bacteria. Gene 195: 257-266.

Codd GA, Bell SG, Kaya K, Ward CJ, Beattie KA and Metcalf JS (1999) Cyanobacterial toxins, exposure routes and human health. Eur J Phycol 34: 405–415.

Cook WO, Beasley VR, Lowell RA, Dahlem AM and Carmichael WW (1989) Consistent inhibition of peripherical cholinesterases by neurotoxins from the freshwater cyanobacterium *Anabaena flos-aquae*: studies of ducks, swine, mice and a steer. Environ Toxicol and Chem 8: 915-922.

Craig IW, Leach CK and Carr NG (1969) Studies with deoxyribonucleic acid from blue-green algae. Archiv für Mikrobiologie 65: 218-227.

Dawson RM (1998) The toxicology of microcystins. Toxicon 36: 953-962.

Dierstein R, Kaiser I and Weckesser J (1988) Rapid determination of *Microcystis aeruginosa* toxin by reversed phase liquid chromatography. FEMS Microbiol Lett 49: 143-147.

Dittmann E, Neilan BA, Erhard M, Doehren H von and Börner T (1997) Insertion mutagenesis of peptide synthetase gene that is responsible for hepatotoxin producation in the cyanobacterium *Microcystis aeruginosa* PCC 7806. Mol. Microbiol 26: 799-787.

Dittmann E Neilan BA and Börner T (1999) Peptide synthetase genes occur in various species of cyanobacteria. In: Peschek GA, Loeffelhardt W, Schemetterer G (Eds) The phototropic prokaryotes. Kluwer Academic/Plenum. New York, pp 615-621.

Edelman M, Swinton D, Schiff JA and Epstein HT and Zelden B (1967) Deoxyribonucleic acid of the blue-green algae (cyanophyta). Bacteriological Reviews 31: 315-331.

Ellenman TC, Falconer IR, Jackson ARB and Runnegar MTC (1978) Isolation, characterization and pathology of the toxin from a *Microcystis aeruginosa* (=*Anacystis cyanea*) bloom. Aust J Biol Sci 31: 209-218.

Falconer IR and Buckley TH (1989) Tumour promotion by *Microcystis* sp. a blue-green alga occurring in water supplies. Med J Aust 150: 351.

Falconer IR, Jackson ARB, Langley J and Runnegar MTC (1981) Liver pathology in mice in poisoning by the blue-green alga in *Microcystis aeruginosa*. Aust J Biol Sci 34: 179–187.

Fergusson MK and Saint PC (2000) Molecular Phylogeny of *Anabaena circinalis* and Its Identification in Environmental samples by PCR. Appl Environ Microbiol 66: 4145-4148.

Foulds IV, Granacki A, Xiao C, Krull UJ, Castle A and Horgen PA (2002) Quantification of microcystin-producing cyanobacteria and *E.coli* in water by 5'-nuclease PCR. J Appl Microbiol 93: 825-835.

Francis G (1878) Poisonous Australian lake. Nature 18: 11–12.

Fujiki H, Suganuma M, Suguri H, Yoshizawa S, Takagi K, Nakayasu M, Ojika M, Yamada K, Yasumoto T, Moore RE and Sugimura T (1990) New tumor promoters from marine natural products. In: *Marine Toxins: Origin, Structure and Molecular*

Pharmacology, Vol. 418 (Ed) Hall S and Strichartz G),. American Chemical Society, Washington, DC. pp 232-240

Gorham PR and Carmichael WW (1988) Hazards of freshwater blue-green algae (cyanobacteria). In: Lembi CA and Walland JR (Ed) Algae and human affairs. Cambridge University Press, Cambridge. pp 403-431.

Gleason FK and Wood JM (1987) Secondary metabolism in the cyanobacteria. In: Fay P, Baalen C van (Eds). The cyanobacteria. Elsevier Science Publishers B.V. (Biomedical Division), pp 437-469.

Harada KI, Matsuura K, Suzuki M Oka H, Watanabe MF, Oishi S, Dahlem AM, Beasley VR and Carmichael WW (1988) Chemical analysis of toxic peptides produced by cyanobacteria. Journal of Chromatography 448: 275-283.

Harada KI, Ogawa K, Kimura Y, Murata H, Suzuki M, Thorn PM, Evans WR and Carmichael WW (1991a) Microcystins from *Anabaena flos-aquae* NRC 525-17. Chemical Research in Toxicology 4: 535-540.

Harada KI, Murata H, Qiang Z, Suzuki M and Kondo F (1996) Mass spectrometric screening method for microcystins in cyanobacteria. Toxicon 34: 701-710.

Herdman M, Janvier M, Waterbury JB, Rippka R and Stanier RY (1978) Deoxyribonucleic Acid base composition of Cyanobacteria. Journal of Gen Microbiol 111: 63-71.

Hisbergues M, Guntram C, Leo R, Sivonen K and Börner T (2003) PCR based identification of Microcystin-producing genoptypes of different cyanobacterial genes, evolution, manipulation. J Bacteriol 185: 564-572.

Honkanan RE, Codispoti BA, Tse KT and Boynton A (1994) Characterization of natural toxins with inhibitory activity against serine/threonine protein phosphatases. Toxicon 32: 339-350.

Humpage AR, Hardy SJ, Moore EJ, Froscio SM and Falconer IR (2000) Microcystins in drinking water enhance the growth of aberrant crypt foci in the mouse colon.J Toxicol Environ Health 61: 155-165.

Jüttner F (1987) Volatile organic substances. In: Fay P, Baalen C van (Eds) *The cyanobacteria*. Elsevier Science Publishers BV (Biomedical Division), 437-469.

Kaye AM, Salomon R and Fridlender B (1967) Base composition and presence of methylated bases in DNA from blue-green alga *Plectonema boryanum*. J Mol Biol 24: 479-483.

Kleinkauf H and Döhren (1996) A non-ribosomal system of peptide biosynthsis. Eur J Biochem 236: 335-351.

Krishnamurthy T, Carmichael WW and Sarver EW (1986) Toxic peptides from freshwater cyanobacteria (blue green algae)-I. isolation, purification and characterization of peptides from *Microcystis aeruginosa* and *Anabaena flos-aquae*. Toxicon 24: 865-873.

Kurmayer R and Kutzenberger T (2003) Application of Real Time PCR for Quantification of Microcystis Genotypes in a polpulations of the toxic cyanobacterium *Microcystis* sp. Appl Environ Microbiol 69: 6723-6730.

Lawton LA, Edwards C and Codd GA (1994) Extraction and high-performance liquid chromatographic method for the determination of microcystins in raw and treated waters. Analyst 119: 1525–1530.

Lee TH, Chan YM and Chou HN (1998) First report of microcystins in Taiwan. Toxicon 36: 247-255.

Luukkainen R, Sivonen K., Namikoshi M, Fardig M, Rinehart KL and Niemela SI (1993) Isolation and Identification of eight microcystins from 13 *Oscillatoria agardhii* strains: structure of a new microcystin. Appl Environ Microbiol 59: 2204-2209.

Luukkainen R, Namikoshi M, Sivonen K, Rinehart KL and Niemelae SI (1994) Isolation and identification of 12 microcystins from four strains and two bloom samples of *Microcystis* spp.: structure of a new hepatotoxin. Toxicon 32: 133–139.

MacKintosh C, Beattie KA, Klumpp S, Cohen P and Codd GA (1990) Cyanobacterial microcystin-LR is a potent and specific inhibitor of protein phosphatases 1 and 2A from both mammals and higher plants. FEBS Lett 264: 187–192.

Mahmood NA and Carmicheal WW (1987) Anatoxin-a (S), an anticholinesterase from the cyanobacterium *Anabaena flos-aquae* NRC 525-17. Toxicon 25: 1221-1227.

Mahmood NA,Carmicheal WW and Pfahler D (1988) Anticholinesterase poisonings from a cyanobacterial (blue-green algae) bloom dominated by *Anabaena flos-aquae*. American Journal of Veterinary Research 49: 500-503.

Marahiel MA, Stachelhaus, and Mootz HD (1997) Modular peptide synthetases involved in non-ribosomal peptide synthesis. Chem Rev 97: 2651-2673.

McDermott CM, Feola R and Plude J (1995) Detection of cyanobacterial toxins (microcystins) in water of northeastern Wisconsin by a immunoassay technique. Toxicon 33: 1433-1442.

Melanie K, Dittmann E and Börner T (1996) Toxic and non-toxic strains of the cyanobacterium *Microcystis aeruginosa* contain sequences homologous to peptide synthetase genes. FEMS Microbiol Lett 15: 295–303.

Melanie K, Dittmann E, Börner T and Neilan BA (2002) Multiple Alternate Transcripts Direct the Biosynthesis of Microcystin, a cyanobacterial nonribosomal peptide. Appl Environ Microbiol 68: 449-455.

Meriluoto JAO, Sandstrom A, Eriksson JE, Remaud G, Grey Craig A, and Chattopadhyaya J (1989) Structure and toxicity of a peptide hepatotoxin from the cyanobacterium *Oscillatoria agardhii*. Toxicon 27: 1021-1034.

Moore RE, Chen JL, Moore BS, Patterson GML and Carmichael WW (1991). Biosynthesis of Microcystin-LR. Origin of the carbons in the Adda and Masp units. J Am Chem Soc 113: 5083-5084.

Mynderse JS, Moore RE, Kashiwagi M and Norton TR (1977) Antileukemia activity in the Oscillatoriaceae: isolation of debromoaplysiatoxin from *Lyngbya*. Science 196: 538-540.

Namikoshi M, Rinehart KL, Sakai R, Sivonen K and Carmichael WW (1990) Structures of three new hepatotoxins produced by the cyanobacterium (blue-green algae) *Nostoc* sp. strain 152. J Org Chem 55: 6135-6139.

Namikoshi M, Rinehart KL, Sakai R, Stotts RR, Dahlem AM, Beasley VR, Carmichael WW and Evans WR (1992) Identification of 12 hepatotoxins from a Homer Lake bloom of the cyanobacteria *Microcystis aeruginosa, Microcystis viridis,* and *Microcystis wesenbergii*: nine new microcystins. J Org Chem 57: 866–872.

Namikoshi M, Yuan M, Sivonen K, Carmichael WW., Rinehart K L, Rouhiainen L, Sun F, Brittain S and Otsuki A (1998) Seven microcystins possessing two L-glutamic acid units, isolated from *Anabaena* sp. Strain 186. Che Res Toxicol 11: 143-149.

Neilan BA (1995) Identification and phylogenetic analysis of toxigenic cyanobacteria by multiplex randomly amplified polymorphic DNA PCR. Appl Environ Microbiol 61: 2286-2291.

Neilan BA, Bass RA and Jacobs D (1995a) Putative genetic loci for cyclic peptide toxin biosynthesis found across several cyanobacterial genera. Poster: *1st International Congress on Toxic Cyanobacteria*, Rønne, Denmark.

Neilan BA, Jacobs D and Goodman AE (1995b) Genetic diversity and phylogeny of toxic cyanobacteria determined by DNA polymorphisms within the phycocyanin locus. Appl Environ Microbiol 61: 3875–3883.

Nishiwaki-Matsushima R, Ohta T, Nishiwaki S, Suganuma M, Kohyama K, Ishikawa T, Carmichael WW and Fujiki H (1992) Liver tumor promotion by the cyanobacterial cyclic peptide toxin microcystin-LR. J Cancer Res Clin Oncol 118: 420–424.

Ohta T, Sueoka E, Iida N, Komori A, Suganuma M, Nishiwaki R, Tatematsu M, Kim SJ., Carmichael WW and Fujiki H (1994) Nodularin, a potent inhibitor of protein phosphatases 1 and 2A, is a new environmental carcinogen in male F344 rat liver. Cancer Research 54: 6402-6406.

Ohtake A, Shrai M, Aida T, Mori N, Harada K, Kenji M, Suzuki M and Nakano M (1989) Toxicity of Microcystis species Isolated from natural blooms and purification of the toxin. Appl Environ Microbial 55: 3202-3207.

Oliveira MCB (2002) Detection of potential microcystin-producing cyanobacteria in Brazilian reservoirs with a mcy B molecular marker. Harmful Algae 2: 51-60.

Pan H, Song LR, Liu YD and Borner T (2002) Detection of hepatotoxic *Microcystis* strains by PCR with intact cells from both culture and environmental samples. Arch. Microbiol 178: 421-427.

Rippka R, Waterbury JB and Cohen-Bazire G (1974) A cyanobacterium that lacks thylakoids. Archievs of Microbiology 100: 419-436.

Rouhiainen L, Sivonen K, Buikema WJ and Haselkorn R (1995) Characterization of toxin-producing cyanobacteria by using an oligonucleotide probe containing a tandemly repeated heptamer. J. Bacteriol 177: 6021–6026.

Runnegar MTC, Andrews J, Gerdes RG and Falconer IR (1987) Injury to hepatocytes induced by a peptide toxin from the caynobacterium *Microcystis aeruginosa*. Toxicon 25: 1235-1239.

Shirai M, Ohtake A, Sano T, Matsumoto S, Sakamoto T, Sato A, Aida T, Harada KI, Shimada T, Suzuki M and Nakano M (1991) Toxicity and toxins of natural blooms and isolated strains of *Microcystis* spp. (cyanobacteria) and improved procedure for purification of cultures. Appl Environ Microbial 57: 1241–1245 .

Sielaff H, Dittmann E, Tandeau de Marsac N, Bouchier C, von Döhren H, Börner T and Schwecke T (2003) The *mcyF* gene of the microcystin biosynthetic gene cluster from *Microcystis aeruginosa* encodes an aspartate racemase. Biochemical Journal 373: 909–916.

Sivonnen K(1990) Effects of light, temperature, nitrate, orthophosphate, and bacteria on growth of and hepatotoxin production by *Oscillatoria agardhii* strains. Appl Environ Microbiol 56: 2658-2666.

Sivonen K, Namikoshi M, Evans WR, Carmichael WW, Sun F, Rouhiainen L, Luukkainen R and Rinehart KL (1992) Isolation and characterization of a variety of microcystins from seven strains of the cyanobacterial genus *Anabaena*. Appl Environ Microbiol 58: 2495–2500.

Sivonen K and Jones G (1999) Cyanobacterial toxins In: Chorus I, Bartram J (Eds) *Toxic cyanobacteria in water*, E and FN Spon, London and New York, pp 41–111.

Spivak CE, Witkop B and Albuquerque EX (1980) Anatoxin-a: a novel, potent agonist at the nicotinic receptor. Molecular Pharmacology 18: 384-394.

Stanier RY, Kunisawa R, Mandel M and Cohen-Bazire G (1971) Purification and properties of unicellular blue-green algae (order Chroococcales). Bacteriological Reviews 35: 171-205.

Swanson KL, Allen CN, Aronstam RS, Rapoport H and Albuquerque EX (1986) Molecular mechanisms of the potent and stereospecific nicotinic receptor agonist (+)-anatoxin-a. Molecular Pharmacology 29: 250-257.

Tillet D, Dittmann E, Erhard M, Döhren, Börner T and Neilan BA (1999) Structural organization of microcystin biosynthesis in *Microcystis aeruginosa* PCC 7806: and integrated peptide–polyketide synthetase system. Chem Bio 7: 753-764.

Utkilen H and Gjolme N (1992) Toxin production by *Microcystis aeruginosa* as a function of light in continuous cultures and its ecological significance. Appl Environ Microbiol 58: 1321-1325.

Vaitomaa JA, Rantala K, Halinen L, Rouhiainen P,Tallberg L, Mokelke, and Sivonen K (2003) Quantitative real-time PCR for determination of microcystin synthetase E copy numbers for *Microcystis* and *Anabaena* in lakes. Appl Environ Microbiol 69: 7289–7297.

van der Westhuizen AJ and Eloff JN (1985) Effect of temperature and light on the toxicity and growth of the blue-green alga *Microcystis aeruginosa* (UV-006). Planta 163: 55–59.

van der Westhuizen AJ and Eloff JN (1983) Effect of culture age and pH of culture medium on the growth and toxicity of the blue green alga *Microcystis aeruginosa*. Zeitschrift für Planzenphysiologie 110: 157-163.

Watanabe MF and Oishi S (1985) Effects of environmental factors on toxicity of a cyanobacterium (*Microcystis aeruginosa*) under culture conditions. Appl Environ Microbiol 49: 1342–1344.

Yu FY, Liu BH, Chou HN and Chu FS (2002) Development of a sensitive ELISA for the determination of microcystins in algae. J Agri Food Chem 50: 4176-4182.

Algal Biotechnology: New Vistas (2010) *Pages 223–240*
Editor: **Mihir Kumar Das**
Published by: **DAYA PUBLISHING HOUSE, NEW DELHI**

Chapter 15

Algae and the Human Affairs in the 21st Century

Surendra Singh*, Shipra Das and Anuradha Tiwari

Algal Biotechnology Laboratory,
Department of Biological Science, Rani Durgavati University,
Jabalpur – 482 001, India

ABSTRACT

Algae, one of the most primitive and diverse life form has no longer remained a mere subject of evolutionary interest but an influensive resource to rely on for better future. A worldwide attention is drawn towards the algal species for their high nutritional profile, therapeutic properties due to which they are being used as food, feed, fuel, fertilizers, colourant, antioxidants, a resource of toxins and bioactive compounds. Not also these properties but also some special properties that makes alga a much more reliable resource in its role in environment as bioindicators, bioremediation tool and global warming reducers. This review is intended to focus on biotechnological application of algae in human life.

Keywords: *Algae, Cyanobacteria, Bioactive compounds, Spirulina, Blue green algae.*

1. Introduction

Algae, which are regarded as important primary producers and early producer of oxygen by the oxygenic photosynthesis (Raymount, 1984) are being used as much efficient and promising biotechnological tool and hence symbolizing the sense of

* E-mail: dasshipra@gmail.com, anuashishpriya1091@gmail.com

pride and modernization for more than 75 per cent of the world. Due to their high demand, algal cultivation at commercial scale has developed as a modern technology and this technology has become an important part of economical strategy for several developed nations like U.S.A., U.K., Germany, Japan, Mexico, Israel and several developing countries like China, Malaysia, Taiwan and India.

Algae can be defined as a group of very simple chlorophyll containing photosynthetic organisms composed of one cell, or grouped together in colonies, or as organisms with many cells, sometimes collaborating together as simple tissues. The diverse group is composed of Xanthophyta (yellow green algae), Chrysophyta (golden brown algae), Dinophyta, Pyrrhophyta, Bacillariophyta, Cryptophyta, Haptophyta, Euglenophyta, Chlorophyta, Phaeophyta, Rhodophyta and Cyanophyta (Fristch, 1984).

1.1 Macroalgae

I. The larger algal species, visible to the naked eyes, are called macroalgae. The red, brown and green algae mainly represent microalgae, also known by the name of seaweed. The utilization of macroalgae dates back to 2,500 years in China (Tseng, 1981), whereas in Europe macroalgal utilization is 500 years old. Previously the macroalgae were mainly considered for their high protein content as well as carbohydrate, amino acids, minerals and vitamins (Table 15.1) (Walland 1981), but now microalgae are being utilized as a favorite feed for the rearing of commercially important animals (fish, prawn, poultry, cattle etc), as an important ingredient of medicine because of their stabilizing and emulsifying properties as well as in cosmetics and textile industries. Macroalgae are often used as an efficient green manure in order to increase the fertility of soil and as a bioindicator of pollution.

1.2 Microalgae

Microalgae are those algae whose cell size ranges from less than 1/1000th of 1mm to 2mm. The biotechnological potential of microalgae is a major focus for every nation because some genus of microalgae like *Dunaliella, Chlorella, Spirulina, Haematococcus, Porphyridium* etc are nature's wonderful gift to mankind as nutritious food supplement for human beings and animals, source of therapeutic agents, efficient nitrogen fixing agent, environmental cleaner, pollution indicators and lots more (Mitsui *et al.*, 1981; Subramanian and Uma, 1996). Accumulation of commercially important caroteinoids (β-carotene, astaxanthin, canthaxanthin and lutein) (Belay, 1997), which have a wide application as a natural colourants, antioxidants, production of phycobilin pigments (phycoerythrin and phycocyanin) having colouring and antioxidants properties due to which phycobilins are successfully used as a fluorescent dyes in flow cytometery and in immunological assays. Possession of high amount of lipids and a high proportion long chain unsaturated fatty acids like omega-3 polyunsaturated fatty acids (PUFAs), eicosapentaenoic acid, docosahexaenoic acid and arachidonic acids (Table 15.2) are of human interest. Production of bioactive compounds and toxins (Cohen, 1997) and the bioremedial properties of microalgae are some of the important properties of microalgae which make them a very efficient tool of modern biotechnology.

Table 15.1: Nutritional Components of Edible Algal Species

Components	*Chondrus cripsis*	*Lamin-ariales*	*Porphyra diocoa*	*Undoria pinnatifida*	*Spirulina*	*Chlorella*
Macronutrients						
Protein(g) and 1%	0.15g	0.17g	0.58(1%)	0.30g	72%	60.5%
Total fat (g) and%	0.02g	0.06g	0.03g	0.06g	0.77g	11.0%
Total carbohydrate (g) and%	1.23g	0.96g	0.51g	0.91g	2.39g	20.1%
Dietary fiber (g) and%	0.1	0.1	0	0.1	0.4	0.2%
Sugar (g) and%	0.06	0.06	0.05	0.07	0.31	Not known
Vitamins						
Vitamin c (mg)	0.03	0.03	3.9	0.3	1.0	15.6mg/100gms
Thiamin (mg)	0.002	0.005	0.010	0.006	0.238	1.5mg/100g
Riboflavin (mg)	0.047	0.015	0.045	0.023	0.367	4.8mg/100gms
Niacin (mg)	0.059	0.047	0.147	0.160		
Pantothenic acid (mg)	0.018	0.064	0.052	0.070	0.348	1.3mg/100g
Vitamin B6 (mg)	0.007	0	0.016	0.036	0.003	1.7 mg/100g
Folate (mcg)	18	12	15	20	9	26.9mcg/100g
Vitamin B12 (mcg)	0	0	0	0	0	125.9MCG/100g
Vitamin A (IU)	12	12	520	36	57	55.500.0 IU/100
Vitamin E (mg)	0.09	0.09	0.10	0.10	0.50	1.0 IU/100 g
Vitamin K (mg)	6.7	6.6	0.4	7.2	2.6 mg	Not known
Minerals						
Calcium (mg)	7	17	7	15	12	203mg/100 g
Iron (mg)	0.89	0.28	0.18	0.22	2.85	167mg/100g
Magnesium (mg)	14	12	0	11	20	203mg/100
Phosphorus (mg)	16	4	6	8	12	989.0mg/100
Potassium (mg)	6	9	36	5	136	Not known
Sodium (mg)	7	23	5	87	105	Not known
Zinc (mg)	0.20	0.12	0.10	0.04	0.20	71.0 mg/100g
Copper (mg)	0.015	0.013	0.026	0.028	0.610	0.08 mg/100g
Mangnese (mg)	0.037	0.020	0.099	0.140	0.190	Not known
Selenium (mcg)	0.1	0.1	0.1	0.1	0.7	Not known
Fatty Acids						
Saturated Fat (g	0.003	0.025	0.006	0.013	0.265	18.20%
Monounsaturated fat (g)	0.002	0.01	0.003	0.006	0.068	81.80%
Polyunsaturated Fat (g)	0.006	0.005	0.011	0.022	0.208	81.80%

Table 15.2: Algal Derived Products and their Commercial Importance

Algal Products	*Algal Species*	*Commercial Uses*
Agar agar	*Euchema, Geladium, Gracilaria, Hypnea, Gigartina* and *Macrocystis*	Base of culture medium, stabilizers,emulsifier, Bakery icing, dessert gel, ingredient in fruit juices, laxatives, anticoagulants, paper coating, adhesive, textile printing.
Carrageenan	*Chondurs, Sigartina* and *Gigartina*	Used in frozen food stuff, pasteries, syrup, jams and jellies, lotions and creams and in air freshing gel.
Aliginate	*Gelidium, Gracilaria, Fucus,Laminaria* and *Macrocystis*	Soups,sauces, sausages coating,surgical purpose as internal gauzes and in explosives.
Carotenoids (β-carotene) (Astaxanthin) (Canthaxanthin)	*Dunaliella salina, Haematococcous, Spirulina* and *Chlorella*	Used in food industries and cosmetic as a natural colourant and as antioxidant.
Phycobillins	*Spirulina, Porphyridium* and *Rhodella*	Used as a natural colourant, Flourescent,dye and as an antioxidant.
Fatty acids (Omega 3 PUFA, ecicoasapentaenoic, docosahjexaenioc and arachionic acid	*Porphyridium cruentum, Phaeodactylum tricomutum* etc.	Used as a neutraceutical and as therapeutic ingredient.
Hydrocarbon oils and liquid fuels	*Botryococcus braunii, chlamydomonas*	Will be used as an resource of H_2 production in future as fuel.

2. Algae as Food Supplement

In few countries like Japan and China algae are often regarded as "Ancient food" due to their nutritional profile (Table 15.1). According to a recent survey, seaweed and microalgae account for about 10 per cent of the Japanese diet and their consumption has reached up to an average of 3.5 kg per house hold with approx increase of 10 per cent every year (Coy, 1987). The most commonly used algal names are "Nori" (*Porphyra*), Kombu (*Laminaria*) and Wakane (*Undaria pinnatifida*). The tradition of consumption of algae as food is very ancient (1500 years) in China. While in west, seaweeds are regarded as a health remedy (Waaland, 1981). In these countries seaweeds are harvested from the sea and then dried in the form of powder which is then mixed up with fish meat dishes, soups and eaten as a vegetable and boiled rice. Laverbread mainly *Porphyra dioloca* and *Porphyra purpurea* as harvested and is used as an additive to soup. (Wakane soup), toasted (Yaki Wakane) and even coated in sugars. *Spirulina* and *Chlorella* are well known name in the hit list of food and feed industries. Their nutrients are very easily digested: Protein (biliprotein), Carbohydrate (Mucopolysaccharides, Rhamnose and Glycogen), 50 different minerals and trace minerals, β-carotene, chlorophyll, gamma-linolenic acid (GLA) and omega 3 fatty acid. According to standard nutritional test, the digestive absorption of *Spirulina* and *Chlorella* protein is four times greater than any other protein. They are low calories diet hence its consumption has now become an effective latest trend (Jeraci and Vensoest, 1986; Sanchez *et al.*, 2002; Thajuddin and Subramanian, 2005).

3. Algae as Feed

Ascophyllum laminaria and *Fucusare* used as a nutritional feed for stocks like sheep and other domestic cattle (Kumar and Singh, 1982). Whereas in Ireland and Scotland several feed processing algal based industries have been established for meeting the demand for commercial feed. The regular utilization of seaweed is mainly recommended for cattle as it has been reported that milk produced by cows that feeds on seaweeds is found to be richer in its fat content than that produced by cows feeding on conventional fodders. Seaweeds are also given to hens in the form of "Dalia" as it enhances the iodine content of eggs. Microalgae are known as an essential feed additive in the commercial rearing of many aquatic animals, mainly the larvae, spat of bivalve molluscks, penoid larvae, rotifers, cattles and poultry.

The regular diet of *Spirulina, Chlorella, Dunaliella, Haematococcus* and *Phaeodactylum* is recommended for the enhanced pigments of Prawn, Salmonid fish and ornamental fishes (Ciferri, 1983; Saxena *et al.*, 1983; Henrickson, 1994; Toyomizu *et al.*, 2001). Regular utilization of microalgae is also been in aquaculture and the poultry farms which increases the natural pigments, skin texture, feathers, yolk content of egg, fertility and healing ability (Nandeesha *et al.*, 2001).

4. Therapeutic Importance of Algae

Algae are well known for their great medical values. Brown algae are used as an important cure of goiter due to its high iodine contents (20,000 times the concentration in sea water). Agar-agar which is commonly produced by various algal species is an efficient prescription for prolapsed stomach (Bensky and Barlot, 1990). *Cladophora* and *Lyngbya* are known for their antibiotic activity against the strain of *Psuedomonas* and *Mycobacterium,* hence their antibiotic extracts are often prescribed as pills and ointments to the patient suffering from fungal and bacterial diseases. Carrageenan, another algal product is a boon to the surgical science as a best coagulant and used as internal gauzes to control severe bleeding. *Chondrus,* an algal species is always recommended for the treatment of kidney, bladder and fungal diseases and as a gut cleanser. Seaweeds are considered as a nature's gift to the digestive system: soothing, disinfecting and nourishing distressed surfaces, as well as maintaining a healthy balance between digestive yeast and bacteria in intestine is also a reported benefit of seaweed consumption Seaweed is a wise prescription for patients suffering from gastric ulcers, duodenal ulcers, ulcerated colon, colitis, constipation, watery stools and other intestinal problems (Bensky and Barolet, 1990). *Spirulina, Chlorella, Haematococcus* and *Dunaliella,* these each single names are a complete range of disease curing remedies, from simple cold and cough to AIDS, cancer etc. Polysaccharide, a bioactive compound which is produced by the Spirulina cells shows activity against HIV, Herpes Simplex virus, Human Cytomegalovirus, Influenza A virus, Mumps virus and measles virus (Henrikson, 1994; Hayashi, 1996b). Phycocyanin shows inhibitory effect on the cancer cell formation. Spirulina is often sold as a supplement (phytochemical) for preventing the fruits from presaging and cellular free radicals.

Spirulina and *Chlorella* in regular diet discourage the building up of the cholesterol in arteries and hence reduces the chances of hypertension. Regular consumption of these algae has reported an increased strength of immune system, purification of blood hence also regarded as Green Blood Tonic. Microalgae are good purifier of kidney toxins. *Spirulina* plays an important role in reducing anemia, sugar level, cholesterol level and hyperlipidemia. Chlorella has been reported to delay the naturel process of ageing. One species of marine cyanobacterium *Oscillatoria willei* BDU 1305114 has been reported to have anti male fertility factor thus can play an important role as oral male contraceptive (Deth, 1999).

5. Algae in Cosmetics and Toiletries

In cosmetics algae are used as an thickening agents, water-binding agents, coloring and antioxidants. Commonly known algal members used as cosmetic ingredients are *Ulva lactuca, Ascophyllum, Laminaria, Longicrusis, Laminaria saccharina, Laminaria digiata, Alaria esculenta, Porphyra sp., Spirulina, Chlorella, Dunaliella, Condrus cripus, Mastrocarpus stellatus* and many other species. Other than this polysaccharide derivatives are broadly appreciated for their skin friendly properties like antiwrinkle and minimizing the hyperpigmentation. Algal biotechnology has produced a complete series of toiletries, including shampoos, bath foam in new spectrum of colors. Oil contents obtained from the algal rich sea water is now been in a new beauty technique called " Spa Bath Treatment" as the seawater is reported to be rich in essential beauty oil.

6. Algae in Fuel Production

Recent potential of algal biomass was introduced by a conceptual engineering analysis presented in 1950. *Botryococcus braunii, Chlamydomonas reinhardhii, Emilianea huxleyi* and *Chlorella* are some of the algal species that has given a hope of ecofriendly fuel production (Ghirardi *et al.*, 2000), although the fuel production has only been witnessed at a very small scale under laboratory conditions. The metabolic reaction that triggers the production of hydrogen is photosynthesis. In fuel producing algae photoevolution of H_2 and O_2 occurs in the same cell and production of either of these two components inhibits the production of the other (Greenbaum and Lee, 1998). An oxygen sensitive enzyme hydrogenase catalyses hydrogen production process (David *et al.*, 2004). In order to inhibit the production of oxygen in the algal cell the growth medium used for culturing is made sulphur deficient due to which algal cell do not perform usual respiration and results in the production of hydrogen (Ghirardi *et al.*, 2000).

7. Algae as a Resource of Bioactive Compound

Marine environment is known to accumulate immense number of organism including marine algae that produce a variety of chemical compounds (Paul and Puglisi, 2004; Kubanek *et al.*, 2003; Steinberg, 2002; Engel *et al.*, 2002; Ren *et al.*, 2001; Nagale and Paul, 1999 and Hay, 1996). These bioactive compounds include toxins as well as compounds of medical importances (Table 15.3) (Bhadury and Wright, 2004). These biogenic compounds have proven their potential as a therapeutic agent, as antibiotic, antiviral, anticancer, anti-inflammatory, hypocholesterolemic enzymes inhibitory activities (Takamatsu *et al.*, 2003; Umemura *et al.*, 2003; Mayer and Gustafson, 2003). 1,500 natural products have been produced by these bioactive compounds (Table 15.3), which can be used as antifouling paints (Kubanek *et al.*, 2003). Other impressive properties are antibacterial, algicidal, antimacrofouling, antifungal and antiprotozoan properties.

Table 15.3: The Bioactive Compounds of Marine Algae and their Activity

Activity	*Bioactive Compounds*	*Algal Species*
Antibacterial	Malyngolide	*Lyngbya majuscule*
	Methanolic extract	*Nostoc spongiaeforme vartenue*
		Nostoc muscorum
	Aqueous extracts	*Synechocystis leopoliensis*
	Methanolic extracts	*Pleurastrum terrestre*
	Methanolic extracts	*Dictyosphaerium pulchellum*
	Methanolic extract	*Klebsormidium crenulatum*
	Aqueous extracts	*Chlorococcum sp.*
	α-Linolenic acid	*Chlorococcum HS-101*
	Acetone extracts	*Chlorokybus atmophyticus*
	Halogenated metabolites	*Laurencia nidifica*
	Brominated diterpene	*Laurencia yonaguniensis*
	Halogenated metabolites	*Laurencia mariannensis*
	Dichloromethane fraction	*Laurencia pinnatifida*
	Laurencienyne	*Laurencia obtuse*
	Elatol	*Laurencia rigida*
	Bromophenols	*Rhodomela confervoides*
	Ethanolic fractions	*Chondrus crispus*
	Chloroform extract	*Gracilaria corticata*
	Lipid extract	*Cytoseira balearica*
	Lipid extract	*Zanardnia prototypes*
	Steroidal glycosides	*Codium iyengarii*
	Benzene, diethyl ether and Petroleum ether extract	*Ulva reticulata*
Antifungal	Hapalindole	*Hapalosiphon fantinalis*
	Tolytoxin (Scytophycin)	*Syctonema ocellatum*
	Scytophycins	*Syctonema pseudohofmanni*
	Toyocamycin	*Tolypothrix tenius*
	Carazostatin	*Hyella caespitose*
	Tjipanazoles	*Tolypothrix tjipanasensis*
	Goniodomin A	*Goniodoma pseudogoniaulax*

Contd...

Table 15.3–Contd...

Activity	Bioactive Compounds	Algal Species
	Polyether compounds (gambieric acid A and B)	*Gambierdiscus toxicus*
	Polyether compounds	*Prorocentrum lima*
	Polyether compounds	*Dinophysis fortii*
	Lobophorolide	*Lobophora variegate*
Anti algal	Nostocine A	*Nostoc spongiaeforme*
	Cyanobacterin LU -1	*Nostoc linckia*
	Fischerellin	*Fischerella muscicola*
	Galactosyldiacyl glycerols	*Phormidium tenue*
	Ether extracts	*Oscillatoria sp*
	Aponin	*Gomphosphaeria aponina*
	Water-soluble extracts	*Peridinium bipes*
	Methylene chloride extracts	*Ceramium rubrum*
	Polyhalogenated monoterpenes	*Plocamium hamatum*
	Ethanolic extracts	*Laminaria pinnatifida*
	Methylene chloride extract	*Laminaria ochroleuca*
	Mucilage extract	*Sargassum hornei*
Antifouling	Cyanobacterin	*Scytonema hofmanni*
	Crude extract	*Calothrix brevissima*
	5β-Hydroxyaplysistatin	*Laurencia obtuse*
	Palisol	*Laurencia obtuse*
	Palisadin A	*Laurencia obtuse*
	Elatol, deschloroelatol	*Laurencia rigida*
	Halogenated monoterpenes	*Plocamium costatum*
	Phlorotannins	*Saragassum vestitum*
	Phlorotannins	*Saragassum natans*
	Phlorotannins	*Ecklonia radiate*
	Diterpene alcohols	*Dictyota menstrualis*
	Galactosyl and sulfoquinovosyl-diacylglycerols and sulfoquniovosyldiacylglycerols	*Costaria costata*
	Phlorotannins	*Fucus spiralis*
	Phlorotannins	*Ralfsia spongiocarpa*
	Diterpene	*Dilophus okamurai*

8. Algae in Agriculture

The usage of algae in agriculture is very old. They are mainly applied directly in the form of compost. The regular utilization of algae shows positive results by increasing the quality and quantity of crops as they are rich source of nitrogen, potash, gelatinous substances as well as valuable source of amino acids, auxins, gibberlin and vitamins. The most economic method of seaweed use is in the form of liquid extract sprays. Seaweed utilization not only increases the fertility of soil and heavy crop yield but also provide a degree of frost resistance to tomato plants. Comparatively greater resistance has also been reported in crop treated with seaweed against insects, pests and other diseases than that of crop treated with conventional methods.

8.1 Blue Green as Biofertilizers

Regular utilization of chemical fertilizers has resulted into the reduction of the natural fertility of the soil and also became a great reason of environmental pollution. Harsh chemical fertilizers cause death and danger to the agriculture friendly microbes and insects that favors the enhancement in the fertility of soil. Cyanobacteria or blue green algae (gram-negative, oxygenic, photosynthetic prokaryote) as biofertilizer is a much more economically and ecologically sound alternative to harsh chemical fertilizer with a true promise of increased productivity of paddy, barley, oats, tomato, raddish, cotton, sugarcane, maize, chilli and lettuce (De, 1939). Nitrogen is one of the most required element in the plants and few cyanobacterial species fixes nitrogen from the environment by the uptake of ammonia through passive diffusion method (Mishra and Pabbi, 2004) and then provide it to the needed crop Anabaena-Azollae (a cyanobacterial symbiont) fixes atmospheric nitrogen estimated between 120-312 kgN_2/hectare (Thajuddin and Subramanian, 2004). The metabolic process by which nitrogen sources are converted into useful nitrogenis called "Nitrogen Fixation" (Venkataraman, 1969). This metabolic process is carried out in the heterocyst of nitrogen fixing cyanobacterial species and is catalyzed by a high molecular weight, oxygen sensitive, metalloprotein enzyme known as "Nitrogenase" (Venkataraman, 1972). *Anabaena, Nostoc, Calothrix, Cylindrospermum, Gloeocapsa, Scytonema* and *Tolypothrix* are known as nitrogen fixers not only produce useful nitrogen but also a huge supply of green manure (Thajuddin and Subramanian, 2005). This process of enriching the fertility of soil is known as algalization. Intensive researches have proved tat the nitrogen fixing cyanobacteria can be established permanently if inoculation is done continuously for 3-4 cropping season (Watanabe, 1951).

9. Algae as Ecological Friends

Earth's atmosphere has gone through several changes from the time of earth's evolution. Some of these changes can be justified by the production of fresh oxygen and ozone layer to protect us from harmful radiations. But now the change that are occurring in our atmosphere due to human activity cannot be justified, as these changes has resulted in several pollution born diseases, global warming, ozone depletion and acid rain. Algae serves the nature as a bioindicator of pollution, as a reducing agent of global warming, heavy metal binder by which algae not only indicates the ecological imbalance but also reduce it.

9.1 Algae as Global Warming Reducers

The increasing temperature has become a major concern issue for every eminent ecologist and common man. After a lot of researches it has been proved that increasing global temperature can be regulated by the increased evolution of oxygen, absorption of carbondioxide and formation of clouds. Clouds are composed of water vapors that absorb plenty of planetary radiation near infra-red part of solar system and carbondioxide. Regular photosynthetic ability of algae minimizes the earth temperature by the evolution oxygen and reduces the concentration of carbondioxide. Algae has also been reported to secrete dimethyl sulphoniopropionate (DMSP) (Phillips, 1992), this DMSP breaks down into three compounds: sulphur dioxide, sulphates and methane sulphonic acid, then the last two componds (sulphates and methane sulphonic acid) combines with water vapors forms the clouds which acts as an excellent agent to reduce global warming. Algal blooms regulate the global warming by the production of sulfite that is about 20-50 million tons from the oceans to the atmosphere each year (Fell and Liss, 1993). This sulphite not only has great potential not only has great potential to reduce global warming but also regulates the acid rain (Wells, 1997). By the production of DMSP and sulfite, algae are future hope to minimize global warming without any age sensation (Bernard, 1993).

9.2 Algae in Bioremediation

Heavy metals are one of the important pollutants that come directly or indirectly from various sources like sewage effluents, thermal pollution, eutrophication, petroleum, faecal pollution, pesticide runoff, acid mine drainage, radioactive substances as well as heavy metal such as cadmium, copper, lead and zinc and when enters from water sources by the food chain enters our body and causes diseases. Some algal species *Spirulina platensis, Nostoc linckia, Nostoc rivularis* and *Aulosira fertilisima* shows metal binding tendency against above mentioned heavy metals. In several experiments it has been proved that concentration of heavy metal from the polluted sites have been reduced by biomineralization and by a process that is often termed as a response mechanism to toxicity. In this specific physiological process when the concentration of any heavy metal increase above the desired level then the cell wall of these metal binding cyanobacterial species accumulates that excess concentration in cells through surface binding process by chemical functional groups present on the cell wall and the absorbed metal is converted into inactive immobilized pellets form and stored in the cell. Hence the concentration of heavy metal is reduced.

10. Negative Aspects of Algae

10.1 Production of Toxins

Some of the blue green algal species (*Microcystis, Anabaena* and *Aphanizomenon*) have been reported to produce several neurotoxins such as anatoxin-a and oranatoxin-a (Chamorrow *et al.*, 1996). Animals poisoned with anatoxin-a show several symptoms like rapid occurance of rigidity and muscle tremors followed by paralysis, cyanosis and cholinesterase inhibition. Some species of blue green algae produce hepatotoxins such as microcystin or nodularin. Microcystin or nodularin poisoning has not been reported to be as fatal as that of anatoxin. This poisoning can be diagnosed by gastrointestinal disturbances, vomiting (depends on species), abdominal pain and diarrhoea (often bloody). This is followed by weakness, destruction of hepatocytes, lack of responsiveness and death. Rapid necrosis of the liver cells often results in severe hemorrhage into the liver, which can cause shock and death. In some cases, there is significant edema around the gal bladder.

10.2 Biodeteriorating Effect of Algae

Algae represent their biodeteriotating effect by the production of Extracellular Polymeric Substances (EPS). This substance is reported to be comprised of uronic acids, sulphated residues or carboxylic groups, this compound facilitates fast hydration as well as delays the moisture release which results in the formation of biofilms. These biofilms not only results in the ugly appearance of the building but also traps dust soot in it, due to which the natural process of biodeterioration is enhanced. These microbial biofilm are also known to produce favorable condition for the growth of decay causing bacterial species, which also cause adverse effect historically important building (Tomaselli *et al.*, 2000b).

11. Mass Production of Commercially Important Microalgae

Today culturing of important algal species has become one of the modern technologies. As the world has become aware of the benefits of the algal utilization, it is impossible to depend on the natural resources for the fulfillment of the market in the case of microalgae. Whereas, for the microalgae or the seaweed production, world market only depends on the oceans as no other artificial production techniques and strategies have been developed. In case of microalgae this condition changes as the cell size of microalgae is very small, so several new production techniques and strategies have been developed for the optimization of factors regulating the growth of desired microalgae (Vonshak, 1997).

In order to enhance the production rate of desired microalgae the optimization of following physical parameters are important.

11.1.1 Temperature

Temperature shows a longer impact on the process of respiration and photorespiration than photosynthesis (Pulz, 2001). Increase in temperature results in increase in the rate of respiration whereas Calvin cycle increases marginally which reduces the net efficiency of photosynthesis. An ideal condition of temperature is said to be that temperature at which the desired organisms are collected from the nature and hence temperature be that particular temperature, the ideal temperature varies from region to region, polar region (<10°C), temperate (10-25°C), tropical (>25°C). In temperate region ambient temperature is generally considered as optimum.

11.1.2 Light

In the life cycle of any photoautotroph light plays a significant role because rate of photosynthesis is directly proportional to light intensity. For most of the microalgal species the saturation point of photosynthesis is at the light intensity of 1,700-2,000$\mu Em^{-2}s$ (that is 30 per cent of the total terrestrial solar radiation) (Pulz, 2001). Hence in microalgal cultivation the optimization of this factor is very essential. For the laboratory cultures, natural light is sufficient, but direct exposure to sunlight is always prohibited. Light intensity of 1660 µEm^{-2}s is suggested as optimum light range (Ayala).

11.1.3 CO_2/O_2 Balance

An appropriate balance between CO_2/O_2 is very important for the high photosynthetic rate. The balance of CO_2/O_2 should be adjusted in the way that the Rubisco (prime carboxylating enzyme) furnishes CO_2 for the Calvin cycle without using oxygen for photorespiration. For the optimum algal growth sufficient CO_2 must be available while the oxygen evolved from photosynthesis must be removed before reaching to inhibitory concentration. The concentration of evolved oxygen can be minimized from algal suspension by increasing the turbidity of culture or by a simple injection of small amount of CO_2 into high cell density microalgal cultures (Straka *et al.*, 2000).

11.2.1 Nutrient Media

The culture medium of a desired algal species is composed of several macro and micronutrient elements, trace elements, chelators, vitamins as well as seawater base, soil extract and buffer (in some selected species). The concentration of above mentioned contents varies from culture to culture.

11.2.2 Seawater Base

Seawater base is usually prepared by the quality improvement of seawater by aging, by autoclaving or by filtration through Whatmann filter paper and acid washed charcoal. Sea water based medium may also be prepared artificially by mixing various salts with deionized water.

11.2.3 Nutrient

This term is mainly used for chemicals which are essential in large quantities for algal growth. The nutrient medium used for culturing should be rich in nitrogen source, inorganic (orthophosphate) and sulphur source (Borowitzka, 1992).

11.2.4 Trace Metal

Trace metals are also known as micronutrients, which are required in very small quantity for algal growth. Fe, Mn, Zn, Cu, Co, Mo and Se are some of the important trace element.

11.2.5 Chelators

Several researches have revealed that chelators acts as trace metal buffer and hence maintains a constant concentration of free ionic metals. The most widely used chelator in algal culturing is E.D.T.A.

11.2.6 Vitamins

After several researches on the growth cycle of several microalgal species it was found that vitamin B_{12} is required for the removal of methyl group and methylating toxic elements such as arsenic, mercury, thallium, platinum, gold and tellurium (Brand, 1986). Whereas the other important vitamins required in the algal culture is thiamine and biotin.

11.2.7 Soil Extract

Soil extract is an important component of culture media as it provides macronutrients, vitamins and trace metals, chelators in undefined quantities. Soil extract is prepared by heating, boiling or autoclaving 5-30 per cent slurry in freshwater or sea water.

11.2.8 Buffer

Buffer is essentially required in the culture media to attain constant pH. While optimizing the appropriate culture system various factors that effects the culture growth like biological features of alga, water requirement, nutrient supply, temperature, light intensity, pH as well as other external features like land cost, charge of labour and energy input rate should be kept in mind. The chosen culture system should also be capable of maintaining the desired temperatures, the hydrodynamic stress on the algae and most important the axenic condition should always be maintained.

The advancement in the production technology has resulted into two culture systems.

11.3.1 Open Culture System

Open system is very much similar to natural milieu of microalgae. Although there are certain variations in shapes but most widely accepted technical design for open system is raceway cultivators aerated by paddle wheels (Pulz, 2001). This system is being successfully used all over Asia and Ukraine. Due to several disadvantages like significant evaporation lost, diffusion of CO_2 to the atmosphere, threat of contamination and pollution, lack of maintenance, open culture is less appreciated.

11.3.2 Closed Culture System

Closed culture system is characterized by the regulation and control of nearly all the biotechnologically important parameters, bioreactors and for the following basic benefits (Pultz, 1992).

(*a*) Reduced contamination risk

(*b*) No CO_2 loss

(*c*) Reproducible cultivation system

(*d*) Regulated hydrodynamics, temperature and pH

(*e*) Technical design

Closed culture systems are more widely used throughout the world. In this system algal cultivation is done in tubular photobioreactor (glass, plastic bags), flattened, plate-type systems and ultrathin immobilized configurations for the optimum biomass production (Tredici and Materassi, 1992; Gabel and Tsoglin, 2000).

12. Future of Algal Based Industries and Algal Products in World Market

In India, the government-run Council for Scientific and Industries Research (CSIR) with the association of India's leading pharmaceutical companies namely

Ranbaxy Laboratories Limited, Delhi and Nicholas Piramal India Limited, Delhi multinationals such as Heinz and Herbal medicine manufacturers such as Dabur India Limited, Gurgaon has started a five year programme and a grand project worth Rs 14- crore, "Positioning Indian Nutraceuticals and Nutragenomics" for the development and promotion of health- enhancing substances and to discover foods specific for people with specific genetic makeup in 2003. This leading initiative decision has made a large group of people, school students, youngsters starting out on their first jobs and even pregnant women aware and buyers of a range of products such as Spirulina, vitamin supplements, protein drinks. After manufacturing microalgal based products the annual turnover of Parry Nutraceuticals Limited increased from 7 crore in 2000 to Rs 17 crore in 2001, whereas Sami Labs annual turnover is being expected to touch Rs 200 crore in 2009 which was 66 crore in 2003. Merina Benny of Live Long Nutraceuticals, Alwaye, Kerala estimates, companies are expected to make an approx profit 30-40 per cent from one product, NatuRose which is a brand name of natural astaxanthin used in aquaculture to impart a pink to red color to pen raised fish and shrimp. This natural astaxanthin is a strong competent to synthetic astaxanthin whose world wide annual are estimated to be more than $ 150 million. Natu-Rose sales accounted for 13 per cent, 16 per cent, 22 per cent,30 per cent of net sales in year 04, 05,06 and 07 respectively. Cyanotech introduced another algal astaxanthin product for the human nutrition market by the name of bioAstin sales accounts for 27 per cent, 25 per cent, and 21 per cent of net sales in years 06, 05 and 04 respectively. According to Mr Ram Bajekal (CEO, Parry's Nutaceuticals), the global nutraceuticals market is around $500 billion. Within this vitamin and nutraceuticals market was estimated to be $500 million and was estimated to double to per cent 1 billion by 2008-2009, while Indian nutraceuticals market is valued at Rs2,300 crore and is estimated to double to 4, 500 crore by 2009. For the Murgappa group nutraceuticals is one of the focused business areas apart from financial services and bio-pesticides. Hawaii based Cyanotech Corporation in intended to come out with new algae-based nutraceuticals by every three years. *Spirulina* sales for the fourth quarter of fiscal 2007 were $ 3, 252,000, up 27 per cent from sale of $2,560,000 for the fiscal 2006. Today world's leading companies are truly agreeing with fact that microalgal resources are capable of providing maximum output profit than any other resource.

13. Importance of Algal Resource Development with Present Scenario

The unique range of environment and biodiversity as well as number of natural resources to rely on has blessed our country. Our land is one of the highest fertile lands of the world, which produces a good yield thrice a year is one of the facts to feel proud of. But today conditions are changing; increasing load of population and shortage of water resources has made ourself to think again on the success of the green revolution. In order to strengthen our economy its very essential that new non-conventional resources must be explored. According to a recent survey report by 2020 there will be an immense increase in the population within an unpredictable need for food which will be impossible to fulfill as by that time the fertile land area will shrink to their limits.Hence a sincere efforts is required to promote the awareness and market for the development of unconventional resources like algae.

Acknowledgement

Thanks are due to the Head, Department of Biological Sciences, R.D University, Jabalpur (M.P).

References

Arasaki S and Arasaki T (1983) Vegetables from the Sea. Japan Publ. Inc., Tokyo.

Ayala F (1998) Guía sobre el cultivo de *Spirulina.* In Biotecnología de Microorganismos Fotoautótrofos. Motril, Granada, España. pp 3-20.

Ayala F (2000) Solarium Biotechnology, La HuaycaRegión,Chile.Communication Personal

Baardseth E (1970) Synopsis of biological data on knobbed wrack *Ascophyllum nodosum* (Linnaeus) Le Jolis. FAO Fisheries Synopsis 38: 1-38.

Becker EW (1994) Microalgae. (Ed) Cambridge University Press, Cambridge.

Belay A (1997) Mass cultivation of *Spirulina platensis* outdoors- The earthrise farm experience. In: Vonshak A (Ed) *(Arthrospira).* Physiology, Cell Biology and Biotechnology.Taylor and Francis, London pp 131-158.

Benemann JR (1980). Enzyme Microb Technol 2: 103-111

Bensky D and Barolet R (1990) Chinese Herbal Medicine: Formulas and Strategies rev. (ed)., Eastland Press, Seattle, WA.

Bernard Harold WJr (1993) Global warming unchecked, Indiana University Press, Bloomington and Indianapolis.

Borowitzka M (1992) Algal growth media and sources of algal cultures. In: Borowitzka M, Borowitzka L (Eds). Microalgal Biotechnology. Cambridge University Press, Great Britain. pp 456-465

Caswell Margaret. (1988) Algoculture, Economic research services, USDA

Ciferri O (1983) *Spirulina,* the edible microorganism. Microbiol Rev 47: 551-578.

Cohen Z (1997) The chemicals of *Spirulina.* In: Vonshak A (Ed) *Spirulina platensis (Arthrospira)*: Physiology, cell-biology and biotechnology. Taylor and Francis. London. pp 175.

Coy.MC (1987) The commercial algae. Prospects for one of the oldest industries. Aqu-Mag July/Aug pp.46-47.

De PK (1939)The role of blue green algae in nitrogen fixation in rice fields.Proc Roy Soc London B, Vol 127: 129-139.

Engel S,Jensen PR, Fenical W (2002) Chemical ecology of marine microbial defense. J Chem Ecol 28: 1971–1985.

Fristch FE (1984) The present day classification of algae. Bot Rev 10: 2333-77

Fell Nolan and Peter Liss (1993) Can algae cool the planet.New Science pp14.

Ghirardi ML, Togasaki RK and Seibert M (1997) Appl Biochem Biotechnol 63: 141–151.

Greenbaum E. and Lee J.W. (1998). BioHydrogen, Zaborsky OR(Ed) Plenum Press, New York, NY, pp 235–241.

Hay ME (1996) Marine chemical ecology: what's known and what's next? J Exp Mar Biol Ecol 200: 103-134.

Henrikson R (1994) Microalga Spirulina, superalimento del futuro. Ronore Enterprises. 2ª ed. Ediciones Urano, Barcelona, España. pp 222.

Hills C and Nakamura H (1978) Food From Sunlight. World Hunger Research Publ., Boulder Creek, CA.

Kubanek J, Jensen PR, Keifer PA,Sullards MC,Collins DO,William Fenical (2003) Seaweed resistance to microbial attack: a targeted chemical defense against marine fungi.Proc Natl Acad Sci USA 100: 6916–6921.

Kumar HD, Singh HN (1982) The economic importance of algae. A text book on algae.pp 40-54.

Materassi R, Balloni W, Pushparaj B, Pelosi E and Silli C (1980) Coltura masiva di Spirulina in sistemi colturali aperti: In: Cantarelli, C., Ciferri, O., Florenzano, G., Kapsiotis, G., Materassi, R., Treccani, U., Eds. Consiglio Nazionale delle Richerche, Progetto finalizzato "Ricerca di nuove fonti proteiche e di nuove formulazioni alimentari". Atti del Convegno: Prospettive della coltura di *Spirulina* in Italia. Firenze-Academia dei Georgofili, CNR Tipografia Coppini; pp. 49-82.

Mayer AM,Gustafson KR (2003) Marine pharmacology in 2000: antitumor and cytotoxic compounds.Int J Cancer 105: 291–299.

Mishra Upasana and Pabbi Sunil (2004) Cyanobacteria: A Potential Biofertilizer for Rice.Resonance.

Nagle DG, Paul VJ (1999) Production of secondary metabolites by filamentous tropical marine cyanobacteria: ecological functions of the compounds.J Phycol 35: 1412–1421.

Oswald W.J. and Golueke C.G (1960) Adv Appl Microbiol 11: 223-242

Palmer CM (1980) Algae and water pollution. Castle House Publication,Ltd England. pp 123.

Paul VJ and Puglisi MP (2004) Chemical mediation of interactions among marine organisms. Nat Proc Rep 21: 189–209.

Phillips Victor D,Magaard, Lorenz Mc Kinley, Kelton R and Takahashi, Patrick.K (1992) Oceans-A global thermostat sea technology. pp 29.

Pulz O (1992) Cultivation techniques for microalgae in open and closed ponds. Proceedings of the 1st European workshop on microalgal biotechnology. Potsdam, pp 61.

Pulz O, Scheibenbogen (1998) Photobioreactors: design and performance with respect to light energy input. In: Scheper Ts S.and Pire R 2001, Fatty acid composition of *Chlorella* and *Spirulina*. Microalgae species J.aOAc Int 84: 1708-1714.

Pulz O and Scheibenbogen K (1998) Photobioreactors: design and performance with respect to light energy input. In: Scheper T(Ed) Bioprocess and algae reactor technology, apoptosis.(Advances in biochemical engineering/biotechnology, vol Springer, Berlin Heidelberg New York, pp 123.

Pulz O (2001) Photobioreactors: production systems for phototrophic microorganisms. IGV Institute for Cereal Processing, Arthur-Scheunert-Allee 40/41.4558 Bergholz Rehbrücke, Germany.

Rai LC and Kumar (1976) Algal growth as a mean of evaluation of nutrient status of the effluent of a fertilizers factory near Shahpuri, Varanasi.Phykos 27: 38-43.

Raymont JEG (1984) Plankton and Productivity in the Oceans. Vol 1. Phytoplankton. Pergamnon Press, Oxford.

Ren D,Sims JJ, Wood TK (2001) Inhibition of bio film formation and swarming of *Escherichia coli* by (5Z)-4-bromo-5-(bromomethyl-ene)-3-butyl-2(5H)-furanone.Environ Microbiol 3: 731–736.

Sánchez M,Jaime BC, Camilo R, Ignacio R (1986) *Spirulina (Arthrospira)*AN edible microorganism. A review university de la Salle.Columbia

Saxena PN., Ahmad MR, Shyan R and Amla DV (1983) Cultivation of *Spirulina* in sewage for poultry feed. Experientia 39: 1077-1083.

Singh PK, Dhar DW, Pabbi S, Prasann R and Arora A (2000).Biofertilizers: (Eds) Blue Green Algae and Azolla. IARI, New Delhi. Venus Printers and Publishers, New Delhi.

Stanier RY and Van Neil (1962) The concept of bacterium. J Arch Microbial 42: 437-66.

Steinberg PD and de Nys R (2002) Chemical mediation of colonization of seaweed surfaces.J Phycol 38: 621–629.

Steve Morton (1998) Modern uses of cultivated algae. Ethnobotanical leaflet. Southern Illinois University Carbondale.

Takamatsu S, Hodges TW, Rajbhandari I, Gerwick WH, Hamann MT, Nagle DG (2003) Marine natural products as novel oxidant prototypes.J Nat Prod 66: 605–608.

Thajuddin N and Subramanian G (2005) Cyanobacteria biodiversity and potential applications in biotechnology.Jr Current Sciences 89(1): 47-57.

Torzillo G and Carlozzi P (1996) Productivity of *Spirulina* in a strongly curved outdoors tubular photo bioreactor. Appl Microbiol Biotechnol 45: 18-23.

Toyomizu M, Sato K Taroda H, KatoT and Akiba Y (2001) Effects of dietary of *Spirulina* on meat color muscle of broiler chickens. Br Poul Sci 42: 197-202.

Tseng CK (1981) Commercial cultivation. In: The Biology of Seaweeds. CS Lobban and Wynne MJ (Eds.) Blackwell Scientific Publication, Oxford. Pp 680-725.

Tziveleka LA,Vagias C,Roussis V (2003)Natural products with anti-HIV activity from marine organisms.Curr Top Med Chem 3: 1512–1535.

Umemura K,Yanase K,Suzuki M,Okutani K,Yamori T,Andoh T (2003)Inhibition of DNA topoisomerases I and II and growth inhibition of human cancer cell lines by a marine microalgal polysaccharide.Biochem Pharmacol 66: 481–487.

Venkatraman GS, Goyal SK, Kaushik BD and Raychoudhary P (1974) Algal forms and functions. Delhi. pp 49-83.

Venkataraman GS (1972) Algal biofertilizers and rice cultivation.Today and Tommorrow Printers and Publishers, Faridabad, India,

Waaland JR (1981) Commercial Utilization. In: The Biology of Seaweeds. CS Lobban and Wynne MJ (Eds) Blackwell Scientific Publication, Oxford. pp. 726-741.

Watanabe AIto R and Konishi C (1951) Effect of nitrogen fixing blue green algae on the growth of rice plants. *Nature* 168: 748-749.

Watanabe Y and Hall D (1996) Photosynthetic production of the filamentous cyanobacterium *Spirulina platensis* in a cone-shaped helical tubular photobioreactor. Appl Microbiol Biotechnol 44: 693-698.

Wells.N (1997) The atmosphere and Oceans. John Wiley and Sons, NY.

Yang YF (2002) Chinese Herbal Medicines: Comparisons and Characteristics, Churchill-Livingstone, London.

Zarrouk C.(1966).Contribution à l´étude d´une cyanophycée influence de divers facteurs physiques et chimiques sur la croissance et la photosynthèse de *Spirulina maxima* (Setch. et Gardner) Geitler (Ph. D. thèse). Université de Paris.

Algal Biotechnology: New Vistas (2010) *Pages 241–252*
Editor: **Mihir Kumar Das**
Published by: **DAYA PUBLISHING HOUSE, NEW DELHI**

Chapter 16

Responses of Rice Field Cyanobacteria to Insecticides

Mihir Kumar Das

P.G. Department of Botany, G.M. College (Autonomous), Sambalpur – 768 004, Orissa, India

ABSTRACT

Cyanobacteria form a living constituent of soil biotypes. They grow luxuriantly in submerged conditions of rice fields, which provide congenial habitat for their growth. They form most efficient system providing biologically fixed nitrogen to the crop. However, cyanobacteria in the rice fields are subjected to a variety of biotic and abiotic stresses. Use of insecticides has become imperative for providing protection to plants and achieving higher production. Indiscriminate uses of the insecticides are known to affect the cyanobacterial population in the rice fields. Under laboratory conditions, they affect cyanobacteria differently, either by changing the growth rate and metabolic rates or alter the population equilibrium. Their effect on growth of these organisms may be stimulatory, inhibitory or neutral. In this review article, the range of insecticide-induced biological responses, induction of biochemical changes in cyanobacteria, the tolerance of cyanobacteria to insecticides and interaction with other stress conditions are highlighted.

Keywords: Cyanobacteria, Insecticides, Toxicity.

* E-mail: mihir_gmc@rediffmail.com

1. Introduction

Cyanobacteria (cyanophytes/blue-green algae) constitute the largest, most diverse and widely distributed group of photosynthetic prokaryotes. They have the ability to invade other-wise uninhabitable sections of the environment, namely, hot springs, ice lands, volcanic soils and bare rocks where the combined nitrogen content is either absent or the temperature is either of the extremes or a variety of natural or artificial chemicals are present (Whitton, 1992). In addition, these organisms multiply rapidly in the flood water in neutral to alkaline soils and release nitrogen slowly for meeting the nutrient requirement by rice (Tirtol *et al.*, 1982; Whitton *et al.*, 1988). Many genera are aerobic nitrogen-fixers and some others are known to grow on molecular nitrogen in the anaerobic and symbiotic state (Fogg *et al.*, 1973). Agronomic potential of cyanobacteria in rice cultivation was first recognized by De (1939), who attributed the natural fertility of tropical rice fields to biological nitrogen fixation by these organisms. Since then, long term fertility experiments and nitrogen fixation measurements have confirmed the importance of cyanobacteria in maintaining a moderate, but constant rice production in fields receiving no fertilizer (Roger, 1988) and many trials have been conducted to increase rice yields by inoculation of cyanobacteria (Venkataraman, 1972; 1975; 1981).

2. Pesticides

Modern high-yielding rice varieties which are highly responsible to fertilizers have become an integral part of the cultivation practices. With higher fertilizer inputs, disease and insect/pest problems have become prominent which endangered the yield necessitating the large scale use of pesticides. Until 1850s, the use of pesticides against pests was not in practice. Before 1940, they were used in relatively small amount and considered as "Safe Pesticides" which were derivatives of various plants. But from 1940 onward, the development of numerous synthetic organic compounds or so called "second generation pesticides" completely changed the basis and strategies of pest control. Thus, during 1950s and 1960s, huge amount of organic pesticides were widely used in agriculture and horticulture to facilitate the production and protection of food, feed and fibre in greater quantity. During 1960s, it was realized that any one single pesticide did not completely eliminate all the species of a target pest and also the concept of resistance was recognized. This has produced a number of trends in use and development of new pesticides for effective control. As a result, the number of chemical pesticides has increased during the past 40 years. In India, the use of pesticides steadily increased from a meagre 4000 tonnes of technical grade agrochemical in 1960 to more than 82,000 tonnes in 1991, out of which 68,000 tonnes are insecticides (SPPL, Bombay, 1992).

Though pesticides were the miracle drugs for the plants during 1950s, doubts were raised concerning the hazardous effects of these chemicals on the environment from the early 1960s. Richael Carson was one of the first to become aware of the danger of pesticide pollution and in her famous book entitled 'Silent Spring", she produced evidences of toxic effects of pesticides upon wild life and potential hazards to man (Misra and Mani 1994). Despite increasing research efforts towards crop improvement by methods obviating the use of pesticides (Day, 1987), agriculture remains heavily dependent on these agrochemicals (Gadkari, 1988). The soil, being the ultimate sink in the rice fields, harbors many beneficial cyanobacteria, which are potentially susceptible to pesticides. Cyanobacteria are known to contribute to the maintenance of soil fertility (Singh, 1961; Venkataraman, 1972). It is therefore, imperative that regular monitoring be undertaken to evaluate the influence of pesticides on these organisms.

In general, pesticide toxicity is regarded only as the lethal effects of the chemicals upon a particular organism. The direct effects concern changes in growth rate and changes on specific metabolic rates *i.e.* photosynthesis and respiration. Secondly, pesticides have a toxicity which is selective for certain groups of microorganisms (Padhy, 1985). The effects of this kind may alter population equilibrium in an indirect manner by changing the competitive efficiency of one or other group. Thirdly, pesticides may promote either directly or indirectly, the growth of one or more types of organisms. The type favored, could be either beneficial or undesirable in paddy field ecosystem. Finally, toxic action of pesticides causing death of certain groups of organisms would result in the release of nutrients by the decay of those organisms, thus inducing a corresponding increase in the number of pesticide tolerant forms.

Many works have been carried out in various laboratories of the world on the responses of pesticides/herbicides to cyanobacteria through *in vitro* determinations and/or under *in vivo* conditions. Most of the works pertain to the determinations of lethal doses of pesticides on different species/strains of cyanobacteria. The more recent studies have been more biochemical and physiological in nature than the earlier ones, departing from examination of pesticide-induced qualitative and quantitative fluctuations in soil microbial populations and concentrating more on factors such as the impairment of nitrogen-fixation (Kapoor and Sharma, 1980; Kaushik and Venkataraman, 1983). Although studies on the interaction of cyanobacteria with agrochemicals continue to be widely conducted, there has been an increase in the amount of attention paid to pesticides, particularly insecticides. Since 1979, emphasis has also been laid on the examination of the response of cyanobacteria to mixtures of pesticides (Stratton, 1983; Tandon *et al.*, 1988). This is encouraging, since in rare instances, would any pesticide be presenting isolation in the soil.

3. Insecticides

Insecticides which constitute 83 per cent of total pesticides produced have been studied quite extensively with respect to their effect on cyanobacteria.

Carbamate group of insecticides have also been used for testing their influence on cyanobacteria in paddy field ecosystem. Kar and Singh (1979a), working with carbofuran (Furadan, 3 per cent active ingredient) observed that its 30 ppm was algistatic to *Nostoc muscorum* and *Wollea bharadwajae*. So far the influence of various environmental factors are concerned, Kar and Singh (1978) found that the insecticide at 100-500 ppm was the most toxic at p^H 7.5 to 10; the toxicity increased with decreasing light intensity and at critical population levels. The toxicity of carbofuran was reduced by increasing the concentration of dibasic potassium phosphate (10 to 40ppm) and calcium chloride (55 to 330 ppm). These findings further illustrate the complexity of pesticide interactions with other soil chemicals and in general, the involvement of environmental factors in determining their efficacy. Adhikary *et al.* (1984) found that sevin (50 per cent w/v) at 10 mg/ml increased the survival, growth and nitrogen fixation while higher concentrations (50 mg/ml) showed an inhibitory effect on *Anabaena* sp. and *Westiellopsis prolifica*. Since the recommended doses of sevin in the rice field was less than the lethal doses, Adhikary *et al.* (1984) suggested that excessive use of the insecticides may affect the overall nitrogen economy of rice field soils. Rath and Adhikary (1994) screened the relative tolerance of twenty two species of cyanobacteria belonging to seven different genera, *viz. Anabaena, Nostoc, Cylindrospermum, Calothrix, Scytonema, Aulosira* and *Westiellopsis* to commercial grade Furadan containing 3 per cent active ingredient of carbofuran and observed varied responses among the different species, tolerance values ranged from 15 to 70 mg/ml of the pesticide. All the test species of *Calothrix* showed maximum tolerance up to the higher doses of carbofuran (70 mg/ml), while *Anabaena fertilissima* and *Aulosira* sp. UU 1437 were proved to be quite susceptible to the pesticides, indicating the ensheathed forms to be the ideal strains for using as biofertilizer. *Nostoc muscorum* was susceptible to carbaryl at 120 ppm (Vaishampayan, 1985). Carbaryl acted on *N. muscorum* by causing cell wall repture and cell lysis with the induction of various mutation. The major concern was that of paddy field pesticides such as carbaryl can elicit toxicity and/or mutagenicity in paddy field cyanobacteria, this may also be the case with the main crop, leading to a possible carcinogenic hazard to mankind (Vaishampayan, 1985). Begum and Islam (1993) studied the effect of basudin and furadan on the growth and nitrogen fixing capacity of *Nostoc commune* and *Calothrix marchica*, which are very common in rice fields of Bangladesh. Recommended dose (for field) of furadan showed inhibitory effect in the growth and nitrogen fixation of both the algae. On the contrary, basudin did not generally affect these activities in both the organisms.

Sahu *et al.*, 1992, while studying the effect of pesticide on 17 cyanobacterial species, observed that in the presence of furadan and sevin upto 100 ppm in the test soil, the occurrence of microalgae remained almost unaffected. With increase in pesticide level, a proportionate decrease in the occurrence of cyanobacterial forms was noted. At the 500 ppm level of the carbamate insecticides tested, most of the unicellular and filamentous heterocystous forms could not grow in the soils. Only the algal species *Gloeocapsa atrata, Lyngbya major, Calothrix parietina* and *Scytonema pascheri*, which possess well-defined sheaths around the cells/trichomes could tolerate 1000 ppm of sevin and furadan. To the contrary, Das and Adhikary (1996), in another set of experiment, found that *Nostoc muscorum* which does not posess a thick slime layer, tolerated higher concentration of sevin which implies that the presence of only additional surface features around the trichome of the trichome can not be the only factor for tolerance to higher levels of pesticide.

Extensive research on the interaction of rice field cyanobacteria with organophosphate insecticides have been undertaken in various laboratories of India because of their wide application in the tropical rice fields. These chemicals, being biodegradable and non-persistent in the soil, are preferred by farmers for pest management practices. From among the organophosphate insectides, effect of malathion and phorate have been studied by Gangawane (1979), Sardeshpande and Goyal (1982), Kaushik and Venkataraman (1983), Tandon *et al.* (1988). Malathion was found to have a differential toxicity to *Anabaena* sp. and *Aulosira fertilissima*; former survived up to 500 ppm, where as latter recovered to control levels at 10 ppm of the pesticide (Tandon *et al.*, 1988). Sardeshpande and Goyal (1982) reported the inhibition of growth and nitrogen fixation in *Hapalosiphon* sp. by 1.0 and 0.5 ppm phorate respectively. However, 1 ppm phorate stimulated the growth of *Anabaena iyengarii* and *Calothrix membranacea*. This level of phorate was not tolerated by *Nostoc* sp. in laboratory culture (Gangawane, 1979). Subramanian *et al.* (1994) while studying the effect of 50 ppm of monocrotophos and malathion on ten strains of filamentous heterocystous cyanobacteria found that *Aulosira fertilissima* ARM 68 and *Nostoc muscorum* ARM 221 grew maximally with the pesticide even in the absence of inorganic phosphate in the medium. This suggested that the pesticides were used by these organisms as a sole source of phosphorous in the absence of inorganic phosphate in the medium and as an additional source of phosphorous when inorganic phosphorous was available in the medium. Another interesting observation was that phosphorous starvation increased alkaline phosphatase activity while the pesticides induced acid phosphatase activity in the test cyanobacteria (Subramanian *et al.*, 1994). Das (2008) studied the effect of Rogor(dimethoate 30 EC) on four different cyanobacterial strains *viz. Anabaena variabilis, Nostoc muscorum, Calothrix parietina* and *Westiellopsis prolifica*. They showed their relative tolerance to the test pesticides. Of these organisms, *Calothrix parietina* was comparatively more tolerant, while *Anabaena variabilis* was more sensitive to Rogor.

As a group, organochlorine insecticides are considered hazardous to these organisms due to their toxicity and persistence (Lal and Saxena, 1980). It appears that exposure of microorganisms, altered call membrane permeability and integrity and interfered with the synthesis of DNA, RNA and proteins (Lal and Saxena, 1980). One of the organochlorines examined by a number of workers for its effect on soil cyanobacteria is BHC and its isomer, lindane or the commercial preparation of Gammexane (5 per cent active ingredient). Kar and Singh (1979a) in a study on the detoxification of HCH (as gammexane) by *Nostoc muscorum* and *Wollea bharadwajae* found that 4 ppm HCH was algistatic to these two organisms. The toxicity of HCH to *N. muscorum* was reduced by increasing the concentration of individual nutrient (dibasic potassium phosphate from 10 to 40 ppm and calcium chloride from 55 to 330 ppm) (Kar and Singh, 1979b). When higher concentrations of these two nutrients were used in combination in the growth medium, HCH became more toxic to cyanobacteria. The authors suggested that individual nutrients at suitable concentrations may in some way have obstructed the movement of HCH in to the cells or interact with it to form a less toxic product and these effects were lessened by antagonism when the nutrients were used in combinations (Kar and Singh, 1979b). Sharma and Gour (1981), using *Anabaena doliolum, Aulosira fertilissima* and *Nostoc* species isolated from paddy fields found that maximum levels of tolerance of these organisms to Lindane was 9, 15 and 10 ppm respectively. Sardeshpande and Goyal (1982) noted significant endosulfan-induced reduction in the growth of *Anabaena iyengarii, Hapalosiphon intricatus* and *Calothrix bharadwajae*. Tandon *et al.* (1988) also observed an inhibitory effect of 1 ppm endosulfan on an *Anabaena* sp. and also on *Aulosira fertilissima*. Photosynthesis and nitrogenase activity were impaired by endosulfan in *A. fertilissima,* but not in *Anabaena* sp. They also pointed out the seriousness of such observations considering that organochlorine pesticide, endosulfan is widely used in countries such as India following the ban of endrin and restricted use of DDT and BHC.

Sahu *et al.* (1992) observed that endotaf at 10 ppm was toxic to almost all test cyanobacteria, except *Lyngbya major, Scytonema pascheri and Calothrix parietina,* which could tolerate upto 250 ppm. The mechanism of resistance of the ensheathed cyanobacteria to pesticide is not known; the presence of a sheath outside the cell wall of these organisms might be protecting them from adverse effects of the toxic chemicals. Das and Adhikary (1996) also observed that EC_{50} level of Hildan decreased the growth and chlorophyll-a content of ten test cyanobacteria more intensely than Roger and Sevin. Reddy *et al.* (1999) conducted short term experiments to study the effect of HCH on *Synechococcus aeruginosus* and found that HCH inhibited photosynthesis and respiration, which enhanced with concentration and exposure duration. Besides, HCH also affected the quantities of chlorophyll, proteins, carbohydrates, nucleic acids and enzymes like a-amylase and acid phosphatase. Significant reduction in the cellular metabolites of this alga reflects the impact of HCH on their metabolic pathways. Three cyanobacterial strains *viz. Aulosira fertilissima, Anabaena variabilis* and *Nostoc muscorum* were studied for their stress responses to an organochlorine pesticide 'endosulfan' with special reference to oxidative stress, role of proline and antioxidant enzymes in endosulfan induced free radical detoxification.(Kumar *et al.*, 2008). Reduction in growth, photosynthetic pigments and carbohydrate of the test microorganisms were accompanied with increase in their total protein, proline, malondialdehye (MDA), superoxide dismutase (SOD), ascorbate peroxidase (APX) and catalase (CAT) in higher endosulfan doses. Increased amount of MDA is indicative of formation of free radicals, while increased level of CAT, APX, SOD and proline indicated their involvement in free radical scavenging mechanism. In lower concentrations, test pesticide showed increase in photosynthetic pigments. Order of tolerance was *Nostoc muscorum* > *Anabaena variabilis* > *Aulosira fertilissima.*

Gupta and Saxena (1974) studied the effect of Panacide (Dichlorophen) on 5 green and blue-green algae *viz. Scenedesmus obliquus, Chlorella pyrenoidosa, Myxosarcina spectabilis, Aulosira prolifica* and *Nostoc* sp, and these were exposed to concentrations of Panacide ranging from 1-80 ppm. A conc. of only 10 ppm killed 100 per cent of *Nostoc* cells, and a conc. of 20 ppm or more strongly inhibited the growth of the other algae, degeneration setting in within 7 days of treatment. Panacide may, therefore, be useful for the control of algae in garden tanks, ponds, lakes, etc. The response of blue-green algae, *Anabaena* (Arm 310) and *Aulosira fertilissima* to DDT, fenitrothion and chlorpyrifos, revealed that algae are quite sensitive to insecticides and the effects depend on the type and nature of the insecticide, the organisms and the experimental conditions (Lal *et al.*, 1987). DDT inhibited the growth of *Anabaena* whereas it was stimulatory to *Aulosira*. Fenitrothion and chlorpyrifos were, however, quite toxic even at concentrations of 100 times less than DDT. Organisms recovered from the toxic effect if the treatment continued for 35 days.

DDT at all concentrations inhibited photosynthesis in *Anabaena* and *Aulosira*. Fenitrothion was extremely toxic to both these organisms as it inhibited photosynthesis by more than 75 per cent at the highest concentration. Chlorpyrifos was comparatively less toxic to *Aulosira* than *Anabaena* as it inhibited $^{14}CO_2$-uptake at 10 ppm by 76.0 and 69.4 per cent, respectively. Nitrogenase activity was stimulated by DDT in *Anabaena* but inhibited by it in *Aulosira*, whereas fenitrothion and chlorpyrifos inhibited nitrogenase activity in both the organisms.

Since soil is a repository for the pesticides thus applied, it is highly likely that the non-target microorganisms, implicated essentially in all major biospheric nutrient cycles, react with these agrochemicals. Thus, it has been necessary to determine the effects of agriculturally needed combinations of pesticides, applied at recommended concentrations, in order to establish the ecological significance of the pesticidal effects. However, most research concerning the effects of pesticides, in general, on non-target microorganisms has dealt only with individual pesticide and, in fact little data are available for pesticide combinations, particularly on microalgae (Stratton, 1983). An attempt was made therefore to determine the impact of insecticide combinations involving a synthetic pyrethroid (cypermethrin or fenvalerate) and an organophosphate (monocrotophos or quinalphos) on two phototrophic microorganisms, *Scenedesmus bijugatus* (Turpin) Kuetzing (unicellular green alga) and *Synechococcus elongates* Nageli (unicellular cyanobacterium), isolated from a black soil used to cultivate cotton. This has been to further the studies, reported recently, demonstrating the differential effect of these selected insecticides when tested alone on the above organisms. Megharaj *et al.* (1989) studied the interaction effects of insecticide combinations involving an organophosphate (monocrotophos or quinalphos) and a pyrethroid (cypermethrin or fenvalerate) on the growth of *Scenedesmus bij'ugatus* (unicellular green alga) and *Synechococcus elongatus* (unicellular cyanobacterium). In general, an organophosphate and a pyrethroid in combinations interacted additively, antagonistically, and synergistically on the growth of the two phototrophic microorganisms. However, combinations of quinalphos and cypermethrin or fenvalerate yielded additive interaction response in only *S. bijugatus*.

Ma *et al.* (2006) studied the effect of five carbamate insecticides on three cyanobacteria, *Anabaena flos-aquae*, *Microcystis flos-aquae*, and *Mirocystis aeruginosa*, and five green algae, *Selenastrum capricornutun*, *Scenedesmus quadricauda*, *Scenedesmus obliquus*, *Chlorella vulgaris*, and *Chlorella pyrenoidosa* which were subjected to 96 h acute toxicity The average acute toxicity of the carbamate insecticides to the cyanobacteria and the green algae was in descending order carbaryl>carbofuran, propoxur, metolcarb>carbosulfan. Wide variations in response to the tested carbamate insecticides occurred among the eight individual species of cyanobacteria and green algae. The sensitivity of various species of algae exposed to carbofuran, propoxur, metolcarb, and carbaryl varied over one order of magnitude, and that of algae exposed to carbosulfan varied over two orders of magnitude. With regard to the diffrential sensitivity of cyanobacteria and green algae, the cyanobacteria were less sensitive than green algae to carbosulfan and propoxur. The pollutants may initiate a shift of algal group structure; especially, a shift from dominance by green algae to dominance by cyanobacteria and may sustain cyanobacterial blooms at particular times. Therefore, the descending order of the ecosystem risk was carbosulfan > propoxur > carbofuran > carbaryl, metolcarb. There was a strong variance between toxicity and ecosystem risk; *i.e.*, "low toxicity" does not imply "low ecosystem risk."

Effects of butachlor, bensulfuron-methyl and dimethoate on the growth, photosynthesis, and photoinhibition of the edible cyanobacterium Ge-Xian-Mi (*Nostoc*) were examined by Chen *et al.* (2007) in order to gain insight into the relationship between its productivity reduction and the abusive use of pesticides in the field. Severe inhibition of growth was found in the presence of four- to six-fold field concentration of butachlor and very high concentrations of bensulfuron-methyl and dimethoate. Mild stimulation of photosynthesis was observed over a limited range of low concentrations of these three pesticides. They found that PSII and PSI were, respectively, the inhibitory sites of 150 μM butachlor and 150 μM bensulfuron-methyl. However, the inhibitory site of 2000 μM dimethoate seems to be situated at the terminal of the whole chain or dark reaction. The colonies exposed to 150 μM butachlor were more sensitive to high light than control cells and those exposed to bensulfuron-methyl, dimethoate, or low butachlor concentration. Dim light-induced rapid recovery of photoinhibited colonies was observed for the control, 10 μM butachlor, bensulfuron-methyl, and dimethoate treatments. However, the maximal PSII photochemical efficiency of photoinhibited colonies treated with 150 μM butachlor was maintained at a relatively stable value in low light. Their findings suggested that the abusive utilization of butachlor might be an important factor limiting the productivity of Ge-Xian-Mi in the field.

Dhanaraj *et al.* (1989) studied the bioconcentration and metabolism of aldrin and phorate by the blue-green algae *Anabaena* (ARM 310) and *Aulosira ferti lissima* (ARM 68). *Anabaena* sp. and *Aulosira fertilissima* showed marked ability to bioconcentrate aldrin and phorate from culture medium. Uptake was directly related to the insecticide concentration in the medium and was inversely related to the water solubility of the insecticides. Bioconcentration of aldrin in *Anabaena* sp. and *A. fertilissima* ranged from 3.9 to 247.5 μg g^{-1}, respectively. Maximum concentration of aldrin in the blue-green algae was reached after 8–16 h and that of phorate after 16–32 h. Aldrin was metabolised to dieldrin by both blue-green algae but no metabolism was noticed in the case of phorate. Lakshmi and Annmalai (2007) made a comparative study on the cyanobacterial isolates, *Anabaena ambigua* Rao (A100) and *Oscillatoria foreaui* Fremy (A1340) on their responses to an organophosphorus insecticide Divap 100 at different concentrations of 1, 5 and 10 ppm on the growth and release of extracellular products such as protein, nitrogen, ammonia, carbohydrates, amino acids and phenols. The experiment was conducted up to 28 days and observed at intervals of 4 days, which showed to be deleteriously affecting the release of these extracellular products in both the organisms. However, in *A.ambigua*, the release of extracellular products was affected at higher concentrations of 5 and 10ppm of Divap 100. It also revealed that these products were released in higher amounts only at late exponential phases even in untreated control conditions in both the isolates.

4. Conclusion

Pesticide contamination in aquatic ecosystem including paddy fields is a serious global environmental concern. Cyanobacteria are also affected by pesticides as non-target organism. For better exploitation of cyanobacteria as biofertiliser, it is indispensable to select tolerant strains along with understanding of their tolerance. The above reports on the interaction of pesticides (Insecticides) with cyanobacteria have shown that:

1. Agrochemicals affect differently the growth and nitrogen fixation of different cyanobacterial species,
2. There exists an inter-and intra specific strain variation for pesticide toxicity.
3. Recommended field dose of many pesticides did not generally affect the biological activity of these organisms; however, excessive application had an adverse effect on nitrogen fixing cyanobacteria, thereby decreasing crop productivity.

Acknowledgement

The author expresses his deepest sense of gratitude to Prof. S.P. Adhikary, P.G. Department of Botany, Utkal University, Vani Vihar, Bhubaneswar for his constant encouragement and valuable suggestions and the Principal, G.M. College (Autonomous), Sambalpur for his support during the preparation of this manuscript.

References

Adhikary SP, Dash P and Pattnaik H (1984) Effect of the carbamate insecticide Sevin on *Anabaena* sp. and *Westiellopsis prolifica*. Acta Microb Hung 31: 335-338.

Begum NT and Islam MMS (1993) Study on the effect of basudin and furadan on *Nostoc commune* Vaucher and *Calothrix marchica* Lemm. In: Hypertrophic and polluted freshwater Ecosystems: Ecological Bases for Water Resource Management. Ed. Tilzer MM and Khondker M in Proc Int Symp Limnol, pp 103-109.

Chen Z, Juneau P and Qiu B (2007) Effects of three pesticides on the growth, photosynthesis and photoinhibition of the edible cyanobacterium Ge-Xian-Mi (*Nostoc*). Aquatic Toxicol 81: 256-265.

Das MK and Adhikary SP (1996) Toxicity of three pesticides to several rice field cyanobacteria. Trop Agric 73: 155-157.

Das MK (2008) Differential response of cyanobacteria to an organophosphate pesticide, Rogor(Dimethoate 30 EC). Nature Environ and Pollution Tech 7(1): 139-143.

Day PR(1987) Crop improvement constraints and challenges. In: British Crop Protection Conference on weeds. BCPC Publ Surrey, UK, pp 3-12.

De PK (1939) The role of blue-green algae in nitrogen fixation in rice fields. Proc Roy Soc, London 127: 121-134.

Dhanaraj PS, Kumar S and Lal R (1989) Bioconcentration and metabolism of aldrin and phorate by the blue-green algae *Anabaena*(ARM 310) and *Aulosira fertilissima*(ARM 68). Agriculture, Ecosystem and Environment 25(2-3): 187-193.

Fogg GE, Steward WDP, Fay P and Walsby AE (1973) The Blue-Green Algae. Academic Press, New York and London.

Gadkari D (1988) Assessment of the effects of photosynthesis–inhibiting herbicides Diuron, DCMU, Metamitron and Metribuzin on growth and nitrogenase activity of *Nostoc muscorum* and a new cyanobacterial isolate strain G4. Biol Fertil Soils 6: 50-54.

Gangawane LV(1979) Tolerance of Thimet by nitrogen-fixing blue-green algae. Pesticides 13: 33-34.

Gangawane LV (1980) Tolerance of nitrogen-fixing blue-green algae to Brassicol, Bavistin and Fytolan. J Ind Bot Sci 59: 157-160.

Gangawane LV and Saler RS (1979) Tolerance of certain pesticides by nitrogen-fixing blue-green algae. Curr Sci 48: 306-308.

Gupta GS and Saxena PN (1974) Effect of Panacide on some green and blue-green algae. Curr Sci 43(15): 492-493.

Kapoor K and Sharma VK(1980) Effect of certain pesticides on survival, growth and nitrogen fixation of blue-green algae *Anabaena doliolum* Bharadwaja. Z Allg Mikrobiol 20: 465-469.

Kar S and Singh PK (1978) Toxicity of carbofuran to blue-green alga *Nostoc muscorum*. Bull Environ Contam Toxicol 22: 707-714.

Kar S and Singh PK (1979a) Detoxification of pesticides carbofuran and hexachlorocyclohexane by blue-green algae *Nostoc muscorum* and *Wollea bharadwajee*. Microb Lett 10: 111-114.

Kar S and Singh PK (1979b) Effect of nutrients on the toxicity of pesticides carbofuran and hexachlorocyclohexane to blue-green alga *Nostoc muscorum*. Z Allg Mikrobiol 19: 467-472.

Kaushik BD and Venkataraman GS (1983) Response of cyanobacterial nitrogen fixation to insecticides. Curr Sci 52: 321-323.

Kumar S, Habib K and Fatma T (2008) Endosulfan induced biochemical changes in nitrogen-fixing cyanobacteria. Sci Total Environ 403(1-3): 130-138.

Lakshmi PTV and Annamalai A (2007) Biochemical studies on the response of organo-phosphorus insecticide and release of extra-cellular products by cyanobacteria. Res J Fisheries and Hydrobiol 2 (1): 13-17.

Lal R and Saxena DM (1980) Cytological and biochemical effects of pesticides on micro-organisms. Residue Review 73: 49-86.

Lal S, Saxena DM and Lal R (1987) Effects of DDT, Fenitrothion and Chlor pyrifos on growth, photosynthesis and nitrogen fixation in *Anabaena* (Arm 310) and *Aulosira fertilissima*. Agric Ecosyst and Environ 19(3): 197-209.

Ma J, Lu N, Qin W, Xu R, Wang Y and Chen X(2006) Differential responses of eight cyanobacterial and green algal species to carbamate insecticides. Ecotoxicol and Environ Safety 63(2): 268-274.

Megharaj M, Venkateswarlu K and Rao AS (1989) Interaction effects of insecticide combinations on the growth of *Scenedesmus bijugatus* and *Synechococcus elongatus*. Plant and Soil 114: 159-163.

Misra SG and Mani D (1994) Agricultural Pollution Vol.II Ashish Publ House, New Delhi, pp 11.

Moore RB (1967) Algae as biological indicators of pesticides. J Phycol 3: 4.

Padhy RN (1985) Cyanobacteria and Pesticides. Residue Reviews 95: 1-44.

Panigrahi KC (1984) *Physiological and genetical effect of pesticides on blue-green algae: effect of carbamate pesticides.* Ph.D. Thesis, Berhampur University, Berhampur (Orissa) India.

Rath B and Adhikary SP (1994) Relative tolerance of several nitrogen-fixing cyanobacteria to commercial grade furadan (carbofuran, 3 per cent). Ind J Expt Biol 32: 213-215.

Reddy TRK, Jyothi BLB and Reddy GRK (1999) Influence of hexachlorocyclohexane on metabolic activity and cell contents of microalgae. Phykos 38 (1&2): 43-54.

Roger PA (1988) Reconsidering the utilization of blue-green algae in wetland rice cultivation. In: Biological N_2-fixation associated with rice production. Eds. Dutt SK and Sloger C, Oxford and IBH Publ Co Pvt Ltd, New Delhi, pp 119-137.

Roger PA and Kulasooriya SA (1980) Blue-green algae and Rice; IRRI, Manila. Los Banos.

Sahu J, Das MK and Adhikary SP (1992) Reaction of blue-green algae of rice field soils to pesticide application. Trop Agric 69: 362-364.

Sardeshpande JS and Goyal SK (1982) Effect of insecticides on the growth and nitrogen fixation by blue-green algae. In: Proc Natl Sym Biol N_2 fixation. Ed. Kaushik BD, Today and tomorrow's Printers and Publishers, New Delhi. pp 588-605.

Sharma VK and Gaur YS (1981) Nitrogen fixation by pesticide-adapted strains of paddy field cyanophytes. Int J Ecol Environ Sci 7: 117-122.

Singh RN (1961) Role of blue-green algae in nitrogen economy of Indian agriculture, ICAR, New Delhi.

SPPL, Bombay (1992) Indian Pesticide Directory (Ed.3) Scientia Publ Pvt Ltd, Bombay, pp 225-367.

Stratton GW (1983) Interaction effects of permethrin and atrazine combinations towards several non-target micro-organisms. Bull Environ Contam Toxicol 31: 297-303.

Subramanian G, Sekar S and Sampoornam S (1994) Biodegradation and Utilization of organophosphorous pesticides by Cyanobacteria. Int Biodet Biodeg 33: 129-143.

Tandon RS, Lal R and Rao VVSN (1988) Interaction of endosulfan and malathion with blue-green algae *Anabaena* and *Aulosira fertilissima*. Environ Poll 52: 1-9.

Tirol AC, Roger, PA and Watanabe, I (1982) Fate of nitrogen from a blue–green alga in a flooded rice soil. Soil Sci Plant Nutr 28: 559-569.

Vaishampayan A(1985) Mutagenic activity of alachlor, butachlor and carbaryl to a nitrogen-fixing cyanobacterium, *Nostoc muscorum*. J Agric Sci 104: 571-576.

Venkataraman GS(1972) Algal biofertilizer and rice cultivation. Today and tomorrow's Printers and Publishers, New Delhi.

Venkataraman GS (1975) The role of blue-green algae in tropical rice cultivation. In: Nitrogen fixation by free living micro-organisms. Ed. Stewart WDP, Cambridge Univ Press, pp 207-218.

Venkataraman GS (1981) Blue-green algae: A possible remedy to nitrogen scarcity. Curr Sci 50: 253-256.

Venkataraman GS and Rajyalakshmi B (1971) Tolerance of blue-green algae to pesticides. Curr Sci 40: 143-144.

Venkataraman GS and Rajyalakshmi B (1972) Relative tolerance of nitrogen-fixing blue-green algae to pesticides. Ind J Agric Sci 42: 119-121.

Whitton BA (1992) Diversity, Ecology and Taxonomy of the cyanobacteria. In: Photosynthetic prokaryotes. Ed. Mann NH and Carr NG, Plenum Press, New York, pp 1-51.

Whitton BA, Aziz A, Francis P, Rother JA, Simon JW and Tahmida ZN (1988) Ecology of deep water rice fields in Bangladesh. I. Physical and Chemical environment. Hydrobiol 169: 3-67.

Algal Biotechnology: New Vistas (2010) *Pages 253–259*
Editor: **Mihir Kumar Das**
Published by: **DAYA PUBLISHING HOUSE, NEW DELHI**

Chapter 17

Cyanobacterial Toxins and Public Health

Mukesh Kumar

Department of Botany, Sahu Jain College, Najibabad – 246 763 U.P., India

ABSTRACT

Cyanobacteria are also known as blue-green algae, so named because these organisms have characteristics of both *i.e.* algae (ability to photosynthesize like plants) and bacteria (prokaryotic cell organization). They have the ability to grow in most diverse ecological conditions and are very beneficial to mankind in many aspects, besides the fact that they produce some toxins, which cause deleterious effects on human and animal health. Cyanobacterial blooms can produce health and environmental hazards in water and affect potability of water and recreational purposes. How, why, and when? The blooms are produced and how to deal with them are questions whose answers are necessary for the safety of human and animal health in regions where these organisms occur in abundant. The blooms are linked to eutrophication of water *i.e.* the biological response of water to over enrichment by plant nutrients. The present chapter deals with the discussions regarding the growth of cyanobacterial blooms, their nature, distribution and impact on human health.

Problems with cyanobacteria occur only under specific environmental conditions and are dose related, seen only when animals ingest water containing large quantities of organisms. Hot, dry, calm day stimulate reproduction of the organism, and under normal conditions, the algae are homogenously suspended in the water. When large number of algae starts to die, gas is produced inside the cells and the colonies tend to float to the pond surface. After this, even a gentle

* E-mail: mukesh.najibabad@gmail.com

wind will concentrate the organisms down to form a "scum" on or just below the pond surface. It makes the water poisonous and dangerous for man and his pet.

Keywords: Cyanobacteria, Cyanotoxins, Health hazards.

1. Introduction

Cyanobacteria are truly ubiquitous being distributed throughout the globe under widely varying environmental conditions. Cyanobacteria have a worldwide distribution, living in sea, in brackish and freshwater areas, in damp soil, glaciers, deserts and hot water springs. They belong to the most archaic organisms on the earth. The Proteozoic era is some times called the 'age of cyanobacteria' because at that time, between 2.5 and 1.5 billion years ago, the blue-green algae dominated the biota (van den Hoek *et al.*, 1993). The great antiquity of cyanobacteria is well documented. The earlier unicellular and filamentous forms of cyanobacteria were reported in sedimentary rocks formed 3500 million years ago. Heterocystous forms and, forms with true branching were observed more recently after the Precambrian era.

Cyanobacteria are prokaryotes with oxygen evolving photosynthetic system and are found in wide range of environments. They are the dwellers of diverse ecological habitats, constituting larger and fascinating group of photosynthetic nitrogen fixing prokaryotes. Their ability of adaptations such as UV absorbing sheath pigments, increase their fitness in the relatively exposed land environment. Many species are capable of living in the soil and other terrestrial habitats. Most organisms show structural, physiological, biochemical and genetic adaptations to their habitat.

The most important factor, which favors their distribution widely through out the world is their ability to survive in the extreme conditions of the environment. Along with this some environmental factors also play an important role in the growth of cyanobacteria as light, temperature, pH etc.

2. Cyanobacterial Toxins

Canobacteria occur naturally in many ponds. There are a number of species, some are harmless while other produce toxins affecting nervous system and the liver. Cyanobacteria are particularly troublesome organisms because apart from producing large amount of biomass, they secrete very toxic substances to water that are known as cyanobacterial toxins. Any release of toxin into surrounding water can present a significant hazard to humans drinking the same water.

Cyanobacterial toxins are the group of organic compounds with very diverse chemical structure. They are divided into two groups *i.e.* cyanotoxins and biotoxins. Cyanotoxins are not lethal for people and animals. They are enzymes, antibiotics and anticarcinogenic factors with very complicated chemical structure. Biotoxins are very toxic for people and can even cause lethal effect. They are divided into neurotoxins (affecting nervous system), hepatotoxins (affecting liver) and dermotoxins (affecting skin).

High pH of water, scarcity of CO_2 dissolved in water and high amount of phosphorus is conductive to the production of the biomass of cyanobacteria (Carmichael 1994, Gajdek 2000). Cyanobacteria survive in the periods of unfavorable conditions in the form of spores and then, in favorable conditions quickly achieve quantitative mass development. At this time they give water distinct colour, referred to as bloom.

Blooms can discolour water, form surface scums, produce unpleasant taste and odor to water, and create problems to aquatic life. They may vary in colour from green to blue, red, brown, dark green or black. Species of cyanobacteria or blue-green algae may dominate and increase excessively in water when:

1. Nutrient levels, particularly phosphorus and nitrogen are sufficient to support growth.
2. Water is still or turbulence is minimal.
3. Weather pattern is stable for a long time.
4. Weather is warm.

The longer the period of calm weather conditions, the greater is the 'bloom forming potential'. Such potential exists most in flowing rivers or lakes, dams, weirs and reservoirs.

Several hypotheses have been put forward to explain algal toxin production. Turner and Tester (1997), Krik and Gilbert (1992), Karlson *et al.* (1995) have suggested that Grazer repellence is probably the most studied hypothesis. Other hypotheses suggest that toxins are responsible for allelopathic activity (Windust *et al.*, 1996, Pushapraj *et al.*, 1999) or functioning as a nitrogen source (Dale and Yentsch 1978) or those toxins are the results of different metabolic processes (Turner and Tester 1997).

3. Distribution of Cyanotoxins

CyanobacteriaL toxins can grow rapidly in favorable conditions such as calm nutrient rich freshwaters in warm climates or during the late summer months in cooler parts of the world. Cyanobacteria tend to occur repeatedly in the same water, posing repeated risk to human populations. Cyanobacterial toxins in lakes, ponds and dugouts in various parts have long been known to cause poisoning in animals and human beings. One of the earliest incidents of toxic affect of cyanobacteria was reported from China, at about 1000 years ago (Chorus and Bartarm 1999).

4. Influence of Cyanotoxins on Various Body Organs

Cyanobacterial toxins are classified by how they affect the human body. Hepatotoxins are produced by some strains of cyanobacteria as *Microcystis, Anabaena, Oscillatoria, Nostoc* and *Cylendrospermopsis*. Neurotoxins are produced by some strains of *Aphanizomenon* and *Oscillatoria*. Cyanobacteria *Cylindrospermum raciborsku* may also produce toxic alkaloids, causing gastrointestinal symptoms or kidney diseases in human. Not all cyanobacteria of these species form toxin and it is likely that there are unrecognized toxins.

People are mainly exposed to cyanobacterial toxins by drinking or bathing in contaminated sources including algal food tablets. Some species form a scum on the water, but high concentration may also be present throughout the affected water. Surface scums, where they occur, represent hazards to human health and it is because of their particularly higher toxin contacts.

5. World Senario of the Problem

Cyanobacteria have been linked to illness in various regions through the world, including South America, Africa, Australia, Europe, Scandivania and China. There is no reliable fixed number of people affected world wide. The only documented and scientifically substantial deaths due to cyanobacterial toxins have been due to exposure during dialysis. An extensive work has been carried out to identify the poisonous strains of cyanobacteria, to isolate the toxins they produce and to study their effects upon living organism.

6. Cyanotoxins and Human Health

Many cyanobacterial species produce toxins that affect animals and humans. Consumption of contaminated water by people in any form may cause cyanobacterial toxicity. The most frequently adverse health affects are caused by drinking water containing the toxins or by recreational water contact.

7. Diseases Caused by Cyanotoxins in Human Beings

Ailments due to cyanobacterial toxins, varies according to the type of toxin and the type of water related exposure *i.e.* drinking, skin contacts etc. Humans are affected with a range of diseases including skin irritation, stomach cramps, vomiting, nausea, diarrhoea, fever, sore throat, muscle and joint pains, blisters of mouth and liver damage. Swimmers in water containing cyanobacterial toxins may suffer from allergic reactions, such as asthma, eye irritation, rashes around the mouth and nose. Water contaminated by cyanobacteria constitutes a health risk to water users in the following ways:

(*i*) Drinking of Water

This causes a number of diseases as nausea, headache, vomiting, abdominal pain, diarrhoea, gastroenteritis, muscle weakness, pneumonia and paralysis. It is said that boiling the water destroys toxins but it is not true. Boiling will kill algae but in doing so will release toxins into the water. In fact, some toxins become more dangerous after boiling the water.

(*ii*) Skin Contact

Contact with cyanobacteria (toxic and non-toxic forms) through water based activities can cause problems such as skin rashes, swollen lips, eye irritation and redness, ear ache and itchiness, sore throat, hay fever symptoms, asthma and possibly promotion of skin tumors. The risk increases as contact time increases (Bell and Cidd 1994). Wearing wet suits may result in greater risk of skin irritation because algal material trapped inside the wet suit may be in close contact with the skin for a long period. The risk of some water activities have been presented in Table 17.1.

Table 17.1: Frequency of Risks Caused by Different Activities of Human Beings

Sl.No.	*Level of Risk*	*Human Activities*
1.	High	Swimming, Diving, Sail boating, Water skiing
2.	Medium	Canoeing, Sailing, Rowing
3.	Low	Fishing, Pleasure cruising, Passive shoreline recreation

(*iii*) Some Other Affects

In addition to the above mentioned cases acute poisoning there is evidence of more long-term chronic human health hazards produced by cyanobacterial toxins. It is found that cyanobacterial toxins produce more chromosome damage than done by benzene. All the results suggest that toxins produced by cyanobacteria may be a more serious environmental hazard than generally recognized (Repavic *et al.*, 1990). The impact of toxins on public health has also been discussed by Piotois *et al.* (2000) and Falconer (1999).

8. Precautions

1. Reducing nutrient built-up on lakes and reservoirs, especially by management of wastewater disposal systems and control of pollution by fertilizer from agriculture.
2. Educating the staff in the health and water supply sectors, as well as the public aware of drinking, bathing or water sports in water likely to contain high densities of cyanobacteria.
3. Water treatment to remove the organisms and their toxins from drinking water should be appropriate.

9. Composite Remedies for Control of Cyanobacteria

The objectives of control and treatment of cyanobacteria are to:

- ☆ Minimize nutrient concentrations in streams and storages.
- ☆ Encourage water conservation measures.
- ☆ Develop or improve techniques to treat or detoxify contaminated water.
- ☆ Removal of cyanotoxins in the process of coagulation, adsorption, biodegradation and microfilteration (Nawrocki *et al.*, 2000).
- ☆ Degradation of hepatotoxins under the influence of chlorination with chlorine and chloramines (Nicholson *et al.*, 1994).
- ☆ The photocatalytic degradation of toxins.
- ☆ Integrated catchments management that is aimed at reducing a managing nutrient flows into waterways from natural and artificial sources.
- ☆ Promoting natural algal grazers through selective fish stocking, aquatic plant restoration and reduction in the contamination of waterways with pesticides.

- ✰ Supporting the maintenance and restoration of riparian vegetation *i.e.* the plants near or at the edges of water bodies.
- ✰ Community involvement in tackling catchments management issues.
- ✰ Research programs focused on toxin identification, rapid testing procedures for field use, toxin stability and degradation under different environmental conditions.

10. Conclusion

It has been clearly explained in the chapter that along with the vast beneficial aspects of cyanobacteria, they may also be harmful to human health by producing different kinds of toxins. Applying certain measures, their growth can be stopped. To conclude, prevention of blooms is the primary aim but as they form part of natural ecosystem, this may not always be possible. It is therefore important to be aware of the potential dangers associated with blooms. By adhering to the procedures as outlined in the chapter, dangers and risks can be minimized and we, along with our pets can live a healthy life.

References

Bell SG and Codd GA (1994) Cyanobacterial toxins and health hazards. Reviews in Medical Microbiology 5: 256-264.

Carmichael WW (1994) The toxins of Cyanobacteria. Scientific Americana 270: 78-86.

Carlsson P, Graneli E, Finenko G and Maestrine SY (1995) Copepod grazing on a phytoplankton community containing the toxin *Dinophysis acuminate*. J Plankton Res 17: 1925-1938.

Chorus I and Bartram (1999) Toxic cyanobacteia in water: A guide to their public health consequences, monitoring all management. World Health Organization, Geneva.

Dale B and Yentsch M (1978) Red tide and paralytic shellfish poisoning. Oceanus 21: 41-49.

Falconer IR (1999) An overview of problems caused by toxic blue-green algae (Cyanobacteria) in drinking and recreational water. Environmental Toxicology 14: 5-12.

Gajdek P (2000) Mikrocystyny sinic W zbiornikach wodnych. Wiadomollci Chemiczne 54: 637-650.

Van den Hoek C, Mann DG and Jahnm HM (1993) Algae- An Introduction of Phycology, 2nd Ed. Cambridge University Press, Cambridge, pp. 327.

Kirk KL and Gilbert JJ (1992)Variation in herbivore response to chemical defences. Zooplankton Foraging on Toxic Cyanobacteria. Ecology 73: 2208-2217.

Nawrocki J and Bilozor S (2000) Uzdanianie wdy Procesy Chemiczne I Biologiczne. Warszawa: PNW.

Nicholson BC, Rositano J and Burch MD (1994) Destruction of cyanobacterial peptide hepatotoxins by chlorine and chloramines. Wat Res 28: 1247-1303.

Piotois S, Jackson MH and Wood BJB (2000) Problems associated with the presence of cyanobacteria in recreational and drinking waters. International Journal of Environmental Health Research 10: 203-218.

Pushparaj B, Pelosi E and Juttner F (1999)Toxicological analysis of the marine cyanobacterium *Nodularia harveyana*. Journal of Applied Phycology 10: 527-530.

Repavich WW, Sanzogni WC, Standridge JH, Wedepohl RE and Meisner LF (1990) Cyanobacteria in Wisconsin waters. Acute and chronic toxicity. Water Research 24(2): 225-231.

Turner TJ and Tester PA (1997) Toxic marine phytoplankton grazers and elagic food webs. Limnological Oceanography 42: 1203- 1214.

Windust AJ, Wright JLC and Mc Lachan JL (1996) The effect of the diarrheic shellfish toxins, Okadaic acid and dinophysistoxin-1, on the growth of microalgae. Marine Biology 126: 19-25.

Algal Biotechnology: New Vistas (2010) *Pages* **260–276**
Editor: **Mihir Kumar Das**
Published by: **DAYA PUBLISHING HOUSE, NEW DELHI**

Chapter 18

Cyanobacteria for Biofertilizer, Bioremediation and Bioactive Compounds

Kaushal Kishore Choudhary*

Department of Botany, B.R. Ambedkar Bihar University, Muzaffarpur – 842 001, Bihar, India

ABSTRACT

Cyanobacteria are contributing unanimously since their origin on the earth surface about 3.5 billion years ago in proterozoic era. Cyanobacteria originated in oxygen and nitrogen lacking environments and converted earth atmosphere into oxygenic and rich in nitrogenous compounds. They enhance the fertility of soil by adding nitrogenous compounds fixed in the 'Heterocyst' with the help of enzyme 'nitrogenase complex . Cyanobacteria also mobilize organic phosphate present in the soil and water into inorganic orthphosphates and add growth-promoting substances in the agriculture fields. The progress in physiological and biochemical study of cyanobacteria disclosed their significant potential for colourants, foods, feeds, fuels, vitamins, polysaccharides, pharmaceuticals and pharmacological probes. Recent advances in cyanobacteria for bioactive compounds have revealed the potency of cyanobacteria to produce a large variety of cytotoxic compounds active against a number of disease causing bacteria, viruses and fungi. These include antiAIDS, antibacterial and antiviral compounds. Interest in cyanobacteria is also mounting because cyanobacteria constitute a

* E-mail: kkc1970@gmail.com, k_k_c1970@yahoo.co.in

source of valuable products such as phycobiliproteins, polysaccharides and protein for feed and food. The present article aims at significant potential of cyanobacteria towards biofertilizers, bioremediation, pigments and production of diverse primary and secondary metabolites of human interest.

Keywords: *Biofertilizer, Bioremediation, Colourants, Cyanobacteria; metabolites.*

1. Introduction

Rapidly growing human population and economic development have led to the alteration in physico-chemical properties of the air, soil and water leading to the environmental degradation. Indiscriminate application of natural resources, uncontrolled urbanization and industrialization, expansion and massive intensification of agriculture and the destruction of forests have adversely affected the sustainability of diverse and productive ecosystems. The addition of undesired substances into air, water and soil are limiting the growth and development of living systems ultimately leading to threat or disappearance of biodiversity (Pimentel *et al.*, 2005). The reduction in biodiversity and contamination of environment is the increasing concern of present day environmental biotechnology programme change.

Another important aspect of present day biotechnological programme is the development and formulation of bioactive compounds to cope with the diseases caused due to deterioration in environments. Regular deterioration in environments and their harmful effect on human health, soil fertility and water productivity and maintenance of rich bioreserves for future generation has necessitated the need for formulation and development of biodegradable biological resources to cope with hazardous degradation (Lombi *et al.*, 2001). Current advances in understanding of metabolism of different group of plants including microbes (bacteria, cyanobacteria and fungi) have attracted worldwide scientists to utilize living system to fulfill the requirements of human beings and for sustainable agriculture practices. Among living system, cyanobacteria hold a good position in biological world due to their wide occurrence and ability to fix atmospheric carbon and nitrogen.

2. Cyanobacteria

Cyanobacteria are geographically widespread photosynthetic oxygenic prokaryotes that share the metabolic property of the eukaryotes and cellular structure of prokaryotes. Cyanobacteria originated about 3.5 billion years ago as freshwater unicells in "proterozoic era" termed as 'Age of cyanobacteria' (Schopf, 2000). Cyanobacteria appeared on the earth surface in extreme condition of nitrogen and oxygen lacking environments and build up the platform for the proliferation of oxygenic plants and higher forms by generating oxygen. During evolutionary consequences, cyanobacteria diverged and colonized all possible illuminated habitats on the earth surface.

2.1 Forms and Habitats Diversity

Originating as freshwater unicells (Sanchez-Baracaldo, 2005), cyanobacteria modified their forms from unicellular to multicellular while maintaining their unicellular forms in natural habitats. Their morphology ranges from unicellular to colonial, parenchymatous to filamentous, simple uniseriate filamentous to multiseriate branched filamentous forms (Thajuddin and Subrmanian, 2005). *Chroococcales, Chamaesiphonales* and *Pleurocapsales* represent unicellular and colonial forms whereas *Nostocales* and *Stigonematales* represent filamentous forms. Filamentous forms represent non-heterocystous and heterocystous species. Non-heterocystous filamentous forms are representative members of *Oscilatoriaceae* whereas heterocystous forms belong to *Nostocales* (except *Oscillatoriaceae*) and *Stigonematales*. Further morphological differentiation of filamentous forms is categorized into unbranched (*Nostocaceae*), pseudobranched (*Scytonemataceae* and *Rivulariaceae*) or truly branched (members of Stigonematales).

Cyanobacteria are cosmopolitan in distribution and occur in nature as free-living, endosymbionts or epiphytic forms. They are commonly found in aquatic and terrestrial environments. Besides, they inhabit extreme environments of hot (Castenholtz, 1996) and cold (Laamanen, 1996), acid and alkali (Gerasimenko *et al.*, 1996), marine (Hoffman, 1999; Thajuddin and Subramanian, 1995) to brine (Paerl *et al.*, 2003). They have also been reported from Antarctic and Arctic regions (Taton *et al.*, 2006). They are reported to survive in long periods of darkness and certain cyanobacteria show a distinct ability for heterotrophic nutrition (Fay, 1965). They inhabit the environments where no other organisms including microbes can survive. Cyanobacteria also occur in symbiotic associations with plants of different groups such as algae, fungi, bryophytes, pteridophytes, gymnosperms and angiosperms (Rai, 1990).

2.2 Cellular Structure of Cyanobacteria

Cyanobacteria are pigmented bacteria capable of oxygenic photosynthesis. They differ from eukaryotes in lacking double layer membrane-bound organelles. Their cell wall is composed of peptidoglycan as found in the eubacteria. They possess a discoid nucleus present in the cytoplasm. Nucleic acid *i.e.* DNA lack histone protein as observed in eukaryotes. The protein synthesis machinery *i.e.* ribosomes are 70S (Allen, 1984) rather than 80S as in eukaryotes. The photosynthetic machinery remains localized in thylakoids distributed throughout the periphery of the cell membrane in the cytoplasm (Douglas, 1994). Cyanobacterial cell contain various prominent cytoplasmic inclusions within their cytoplasm. The important among them are glycogen granules, lipid globules, cyanophycin granules, carboxysomes and polyphosphate bodies (Fay and Baalen, 1987). The accumulation of glycogen and lipids takes part in absence of supply of nitrogen source. Some of the aquatic species of cyanobacteria possess gas vesicles that regulate buoyancy in them. The multiplication in cyanobacteria is asexual. In filamentous forms, the multiplication occurs by trichome fragmentation or by formation of special hormogonia. Hormogonia differ from fragmented filaments and show gliding movements upon their liberation.

3. Potential Applications of Cyanobacteria

Cyanobacteria are supposed to be the first oxygen evolving organisms that dominated the earth surface long ago and transformed anoxygenic environments into oxygenic. Later study demonstrated their significant role in primary productivity in marine and freshwater ecosystems. With the establishment of agronomic potential of cyanobacteria (De, 1939), distribution (Singh *et al.*, 2001), diversity (Choudhary, 1999) and their role as biofertilizer (Mishra *et al.*, 2005) were studied by different workers. In addition to contributing available sources of nitrogenous compounds to the soil and water, they mobilize organic phosphates into inorganic orthophosphates thereby providing energy rich compounds required for the proliferation of microbes and higher plants (Mishra *et al.*, 2005). Cyanobacteria were also reported to contribute growth-promoting auxin-like substances in the soil (Gupta and Lata, 1964).

Further study on distribution of cyanobacteria in polluted environments demonstrated that cyanobacteria could be utilized as bioremediating agents (Micheletti *et al.*, 2008) for cleaning the polluted sites or degrading the pollutants at contaminated sites. Pigments of cyanobacteria particularly carotenoids and phycocyanin has been elevated at industrial scale to use them as colourants (Emodi, 1978), dietary supplements, sources of vitamins, pharmacological probes and in immunoassays (Choudhary *et al.*, 2007). Detailed study on polysaccharidic mucilaginous envelope present around the cell revealed their significant contribution in soil fertility by enhancing soil aggregation property of the soil and subsequently increasing the water holding capacity (Saxena and Kaushik, 1992). Additionally, cyanobacterial polysaccharides are used for encapsulation and formulation of drugs (Trabelsi *et al.*, 2009). Recent advances in bloom forming cyanobacteria have attracted worldwide attention for their ability to produce bioactive compounds (Yoshida *et al.*, 2008). Present section of this article aimed towards the discussion of cyanobacterial potential in agriculture, environment and biotechnology.

3.1 Cyanobacteria and Biofertilizer

Fast and ever-growing population of human beings is posing tremendous pressure on limited agriculturally cultivated area to produce enough food grains to meet human demands for food and fodder. This led to the application of excessive chemical fertilizer that certainly increased the production of food grains and make the country self dependent on food grains. However, chemical fertilizers started slowly displaying their ill effects on human health and productivity of the agricultural fields (Pimentel, 1996). Application of excessive chemical fertilizers led to the alteration in air, soil and water. Plants do not take up all the nutrient ions applied with chemical fertilizers rather some of them (harmful substances) accumulate in soil, leached into water bodies (Hallberg, 1987) or may be lost to the atmosphere by volatilization. Accumulation of ions in the soil adversely effect the soil porosity, water holding capacity and fertility of the soil whereas their accumulation in the water bodies cause eutrophication, that is, an increasing environmental concern worldwide.

Accumulation in soil and water badly effects the microbial population that play significant role in nutrient cycling. In addition, many chemical fertilizers contain acids such as sulfuric acid and hydrochloric acid that tends to increase the acidity of the soil thereby decreasing the microbial population and interferes with plant growth. Importantly, the accumulation of nitrate or nitrite in the soil may leach into the ground water or might move into the seed that may cause harmful effects on health of animals and humans on consumption. Considering harmful effects of chemical fertilizers, approach towards application of biological systems to ameliorate the polluted sites and fertilize the agriculture fields for production of enough foods was proposed. Recent advances in microbial potential to supply nitrogen and phosphorus along with some growth promoting substances have received much more attention among scientists worldwide to design, develop and upgrade the biological system as alternate or supplementary sources of fertilizers.

Nitrogen is most important essential nutrients for all living beings as it is required for nucleotide and amino acid synthesis. Atmospheric molecular dinitrogen gas is the major component of the Earth's atmosphere but these inert gases are not readily available sources of nitrogenous compounds that can be used by plants or animals for their growth and survival. Hence, conversion of dinitrogen into complex nitrogenous compounds is the ultimate requirement for the sustainable practices of ecosystem. Conversion of atmospheric molecular nitrogen occurs through biological and non-biological (lightning, Haber-Bosch process and combustion) processes. However, biological nitrogen fixation is safe, reliable, eco-friendly and biodegradable. Biological nitrogen fixation is catalyzed by enzyme-complex termed 'nitrogenase' (Stacey *et al.*, 1992). A number of different prokaryotes including bacteria, actinobacteria and certain types of anaerobic bacteria carry out biological nitrogen fixation naturally. Microbial systems fix atmospheric nitrogen as free-living or get associated with higher plants to fix and supply nitrogenous compounds to host plants. Among nitrogen fixers, cyanobacteria are simplest of all living system in terms of nutritional requirements and considered as most promising group of organisms for the biofertilizers (Adams and Duggan, 1999). Agronomic potential of cyanobacteria was first established by De (1939). Afterwards, cyanobacteria received attention of worldwide scientist to study the morphological and biochemical characteristics of these tiny organisms for biofertilizers.

The establishment of significant contribution by cyanobacteria in nitrogen economy of the soil witnessed rapid advancement in cyanobacterial diversity of agriculture fields and quantitative estimation of supplementation of nitrogen to the soil under field and laboratory conditions. Earlier developments in cyanobacteria witnessed the distribution pattern of heterocystous forms in agriculture fields particularly rice fields. Diversity studies on cyanobacteria was recorded both in fertilized and unfertilized agriculture fields. It is now well-established that unfertilized fields showed greater diversity of heterocystous forms than fertilized fields. Singh *et al.* (2001) recorded the dominance of heterocystous forms of cyanobacteria in paddy fields of 24 Parganas of West Bengal. Cyanobacterial diversity and their application as biofertilizer in agriculture fields have been studied worldwide (Singh, 1961; Singh *et al.*, 2001; Roger and Kulasooriya, 1980; Watanable and Yamamoto, 1971; Kannaiyan,

1983 and Kaushik, 1986). Most common heterocystous forms inhabiting the rice fields are *Anabaena, Nostoc, Cylindrospermum, Calothrix, Rivularia, Gloeotrichia, Scytonema* and *Tolypothrix*. Discoveries in recent years have disclosed that nitrogen fixation is not the monopoly of heterocystous forms rather it is also fixed by some unicellular and non-heterocystous forms (Bergman *et al.*, 1997; Taniuchi *et al.*, 2008). Some of them are *Aphanothece, Gloeocapsa, Plectonema, Trichodesmium* and others (Singh, 1973, El-Shehawy *et al.*, 2003).

Nitrogen fixed by cyanobacteria is released in the soil environment by autolysis, extracellular transport or through microbial decomposition after the cell death (Choudhary, 2009). Cyanobacteria contribute about 15-30 Kg nitrogen hectare^{-1} per season in linear fashion. However, the amount of nitrogen fixed and released into immediate environments varies differently from fields to fields and from one species to others. Kolte and Goyal (1989) showed that nitrogen fixing capacity (mg N 50 ml^{-1} 28 days^{-1}) ranged from 0.84-2.45 in *Anabaena*, 0.77-5.83 in *Calothrix*, 0.44-4.04 in *Nostoc*, 0.33-3.41 in *Scytonema* and 2.55-5.67 in *Tolypothrix*. Further advancement in cyanobacterial study showed their significant contribution in mobilizing organic phosphates deposited to the soil after death of living system. Living system requires inorganic phosphates for their growth and survival as a source of energy. However, most of the sources of phosphates remain in complex organic forms that cannot be utilized by living systems. Cyanobacteria along with algae and bryophytes are capable of converting complex organic phosphates into simple inorganic forms. The mobilization of organic phosphates is of increasing environmental concern because of large-scale accumulation in terrestrial and aquatic ecosystems as agricultural, municipal and industrial effluents.

Phosphates accumulation in agriculture fields is badly affecting the fertility of soil by decreasing the porosity whereas in aquatic system, accumulation leads to the eutrophication that favors the massive growth of particular living system. Establishment of cyanobacteria in terrestrial system particularly agriculture fields might maintain the diversity of native flora by generating the inorganic phosphates from organic one with the help of enzyme 'Phosphatases'. The stimulation of 'phosphatases' in the soil after supplementation of cyanobacteria as biofertilizer was studied under fields (Mishra *et al.*, 2005) and laboratory condition (Whitton *et al.*, 1991). Cyanobacteria can work under both acidic and alkaline condition as they contain both acid and alkaline phosphatases. In addition, cyanobacteria contribute growth-promoting substances into the soil. Gupta and Lata (1964) showed enhanced germination of rice seedlings presoaked in extracts of non-heterocystous cyanobacterium, *Phormidium sp*. Kartikeyen *et al.* (2008) recently reported presence of different potential growth promoting amino acids in culture filtrates of cyanobacterial strains isolated from wheat rhizosphere. In this way, cyanobacteria play a significant role in development of sustainable agriculture practices to support human life on the earth.

3.2 Cyanobacteria and Bioremediation

Rapid expansion of agriculturally cultivated land and increasing sophistication of industries in the past century has remarkably increased the amount of toxic substances into the air, water and soil beyond the tolerance of living system. This has been termed as 'Environmental pollution and contamination' in context of modern environmental biotechnology. As discussed earlier, human activities have led to the decline in fertility and productivity of the air, soil and water, which are to be maintained sustainable for better perpetuation of living system on the earth. Removal or degradation of toxic substances from polluted or contaminated sites (remediation) is the increasing concern of present day environmental biotechnology programme. Several chemical approaches applied to remove or degrade the toxic substances but none found suitable for effective remediation.

Recent advances in physiological, biochemical and molecular understanding of microbes and their diverse habitats attracted the attention of worldwide scientists to exploit and utilize these naturally occurring or tailor-made microbes for removal or degradation of pollutants or contaminants from the polluted sites (Bioremediation). Bioremediation might be processed in *in-situ* or *ex-situ* conditions with fungi, bacteria or cyanobacteria. Among them, cyanobacteria (also known as blue-green algae) seem to be better option for bioremediation because some of them are capable of fixing carbon and nitrogen together. Cyanobacteria are geographically widespread and are oxygen-evolving organisms that respond to stress conditions such as light deprivation (Borbely *et al.*, 1990.). These cells have developed natural methods of responding to metals such as copper, lead, and cadmium through passive accumulation in cells and through surface binding to various functional groups. Field and laboratory studies on remediation of polluted sites by cyanobacteria have disclosed their significant contribution in removal or detoxification of pollutants or contaminants.

Bioremediation is an eco- friendly, cost-effective and natural technology targeted to remove or detoxify heavy metals, radionuclides, xenobiotic compounds, organic waste, pesticides etc. from contaminated sites or industrial discharges by using microbes or plants. Since this technology is used in *in-situ* conditions, it does not physically disturb the site unlike conventional methods *i.e.* chemical or mechanical methods. Bioremediation involves higher plants or microbes alone or in combination. Biological system used to remove the hazardous material are grown in the polluted sites and metal-loaded plants are harvested and incernated to reduce the volume of waste. The accumulated wastes are either disposed off as hazardous materials or used for recovery of precious metals. Some of the cyanobacterial potential in removal of toxic substances from polluted sites has been discussed below.

Cyanobacteria are considered as bioremediating agents not only for their ability to accumulate or detoxify or degrade the toxic and hazardous substances, rather it enhances fertility of polluted sites by contributing nitrogen and phosphorus along with some growth-promoting substances. Removal of Cu(II) has been reported by two exopolysaccharide producing cyanobacteria *Cyanospora capsulata* and *Nostoc* PCC 7936 (De Philippis *et al.*, 2003) and *Nostoc calcicola* (Verma and Singh, 1990). Micheletti *et al.* (2008) showed sorption of Cu by exopolysachharide producing cyanobacteria *Cyanothece* and *Nostoc* sp. El-Enany and Issa (2000) reported the enhanced accumulation of 30 fold and 10-fold of Zn and 10-fold and 2-fold of Cd by *N. linckia* and *N. rivularis* respectively grown in growth medium with 75 per cent sewage water over growth medium with 25 and 50 per cent sewage water.

Another contaminating toxic metal is lead which has been also shown to be absorbed by a cyanobacterium, *Gloeocapsa* (Raungsomboon *et al.*, 2008). Cyanobacteria are capable of removing Cr, Zn, Cd, Co, Mn, Hg and degrading PAHs (Polycyclic Aromatic Hydrocarbons) (Cerniglia, 1984) and crude oils (Reghukumar *et al.*, 2001). *Spirulina platensis*, a cyanobacterium, was shown to contain detectable levels of mercury and lead when grown under contaminated conditions (Slotton *et al.*, 1989), implying that this cyanobacterium was taking up the toxic metal ions from its environment.

Cyanobacteria have developed multiple strategies to cope with metals and toxic substances induced stress conditions. Cyanobacteria can both adsorb and take up metal ions (Bender *et al.*, 1994) in the field conditions. Detailed cellular structure and metabolic activities proposed that carboxyl groups present on cyanobacterial cell envelope are responsible for binding to various ions (Gardea-Torresdey *et al.*, 1990). Further studies revealed that possession of intracellular polyphosphates support cyanobacteria in metal sequestration whereas extracellular polysaccharides serve to chelate or bind metal ions (Zhang and Majidi, 1994; Kaplan *et al.*, 1987). Gardea-Torresdey *et al.* (1996) showed that strains of *Synechocystis spp.* develop a thickened calyx whereas *Synechococcus sp.* PCC 7942 develop a copper-transporting P-type ATPase in the thylakoid membrane (Bonilla *et al.*, 1995) when exposed to copper-stressed growth conditions. In some cases, cyanobacteria produce low-molecular weight proteins or polypeptides (6,000-8,000 amu), metallothioneins, which bind metal ions in metal-thiolate clusters (Turner and Robinson, 1995). Some of the cyanobacteria produce extracellular macromolecules, bioflocculants that contains numerous negatively charged binding sites ultimately empowering cyanobacteria in removal of positively charged heavy metals from polluted sites (Bender *et al.*, 1989). Such findings show the possibility of manipulating or overexpressing existing resistance mechanisms and the use of such organisms to remove harmful metals from the environment.

3.3 Cyanobacterial and Colourant

Photosynthetic pigments are present in chloroplasts that capture light energy necessary for photosynthesis. Pigments are chemical compounds which reflect only certain wavelengths of visible light as the result of selective colour absorption. Pigments differ from fluorescence, phosphorescence and other forms of luminescence in which the material itself emits light. Cyanobacteria carry out oxygenic photosynthesis and carbon fixation with the help of chlorophyll a in presence of light. The pigments localized in the thylakoids, remain free in the cytoplasm or attached to the plasma membrane. Cyanobacteria consists of three kinds of pigments such as Chlorophyll a, carotenoids and phycobiliproteins (Phycocyanin, allophycocyanin and phycoerythrin). Phycobiliproteins are embodied in phycobilisomes, which are located on the outer surface of the thylakoids (Douglas, 1994). On the basis of participation in photosynthesis, pigments have been divided into two categories. Chlorophyll a is necessary pigment whereas carotenoids and phycobiliproteins are accessory pigments. Chlorophyll a and carotenoids are soluble in organic solvents whereas phycobilliproteins are water-soluble. Pigments in cyanobacteria not only assist in carbon fixation but make them colourful. Carotenoids are usually red, orange or yellow pigments and contain a compound carotene. Among phycobiliproteins, phycocyanin is commonly blue in colour whereas phycoerythrin is brown.

Colourants become most important part of any commodity not only for its appeal but it also enhances consumer acceptability (Clydesdale, 1993). Colourants are used not only for aesthetic beauty to the foodstuff but also to prevent them from microbial spoilage, maintenance of original food appearance during storage, protection of flavour and vitamins from light and increase acceptability of food as appetizing item (FNB, 1971; Chattopadhyay *et al.*, 2008). Use of colourants began in the late 18th century with the mechanization of the textile industries. Rapid spread of cotton mills and increased productivity in textile manufacturing encouraged chemists to investigate the compositon of natural dyes and synthesize synthetic dyes to fulfill increasing demand. Use of colourants slowly became part of many more industries like food, cosmetics, beverages, textiles, plastics, paints and medical devices. But, advancement in knowledge regarding toxic effect of synthetic colourants on health and increasing impact of energy and high raw material cost have shown decline in production and new introduction of dyestuff for industries and paved way for scientists to develop other sources particularly bioresources for colourants production. Also, limiting natural resources of charcoal–raw material for synthetic dye is the reason behind formulation of biological colourants for different industries.

Biocolourants are colouring agent obtained from biological sources. These are necessary and accessory photosynthetic pigments that directly or indiretly participate in photosynthesis. Biocolourants are commonly of two types *viz.* natural and artificial biocoulorants. Natural biocoulorants are produced from bioresources and are natural or nature identical (biotechnological) whereas artificial biocoulorants are synthesized biologically but can also be synthesized chemically. Further differences between natural and artificial biocoulorants lie in their physiological and biochemical activities. Natural biocoulorants are more active than artificial biocoulorants. Besides colouring agent, natural biocoulorant have a multidimensional function in the food industries. They serves as sources of vitamins, aminoacids, antioxidants etc. Natural biocoulorants have been reported to show antagonistic activity to certain bacteria, fungi and viruses thereby protecting food from microbial spolage (Bridle and Timberlake, 1997).

Cyanobacteria are considered as a rich and unexhaustible source of biocoulorants. In spite of wide range of diversity and forms, only a few cyanobateria, specifically *Spirulina* has been studied extensively. Cyanobacteria are rich source of carotenoids and phycocyanin pigments that have wide commercial and biotechnological implications. Carotenoids are used as natural food colouring, as feed adhesives, to enhance flesh colour of fish and egg yolk colour (Emodi, 1978). They also serve as second line of defence against cancer, enhance immunity, as photoprotective agents and growth promoters. Carotenoids are also used as constituents in vitamins and dietary supplements. Phycocyanin find application as colouring agents, as fluoroscent dye (phycoflour probe) in immunoassays and are important components of diagnostic kits. Recent developments in carotenoids study has revaeled their significant contribution in reducing risk of heart disease, cancer, eye disease and other diseases. The other pigments produced by cyanobacteria are scytonemin and chlorophyll d. Advances in biotechnological application of pigment as colourants and other health and hygiene effects, cyanobacteria are receiving world wide attention among scientists to develop a cost effective method for mass production of pigments.

3.4 Cyanobacteria and Bioactive Compounds

Besides enhancement of fertility and amelioration of polluted ecosystem, cyanobacteria produce a variety of novel and biochemically active natural compounds of biotechnological interest. Cyanobacteria are receiving more attention because of their non-toxic nature. They are recognized as an excellent source of vitamins and proteins. They are reported to be a source of fine chemicals, renewable fuel and bioactive compounds. Their role as antiviral, anti-tumor, antibacterial, anti-HIV and a food additive is well established. The toxins and other bioactive compounds produced by cyanobacteria include 40 per cent lipopeptides, 5.6 per cent amino acids, 4.2 per cent fatty acids, 4.2 per cent macrolides and 9 per cent amides (Singh *et al.*, 2005). Cyanobacterial lipopeptides include different compounds like cytotoxic (41 per cent), antitumor (13 per cent), antiviral (4 per cent), antibiotics (12 per cent) and the remaining 18 per cent activities include antimalarial, antimycotics, multi-drug resistance reversers, antifeedant, herbicides and immunosuppressive agents (Burja *et al.*, 2001). Iwata *et al.* (1990) reported the cholesterol-lowering effect of cyanobacteria due to presence of lipoprotein lipase, an enzyme for metabolism of triglyceride rich lipoproteins. *Aphanizomenon flos-aquae*, have been reported to show hypocholesterolemic effect due to its chlorophyll content, by stimulating the liver function and decreasing blood cholesterol level (Vlad *et al.*, 1995), inhibition of a maltase and sucrase in the digestive tract of rats (Kushak *et al.*, 1999) and accelerate recovery from mild traumatic brain injury (Valencia and Walker, 1999).

Additionally, cyanobacteria produce some high value bioactive natural compound active against different biosystems. Burja *et al.* (2001) reported the production of cyanovirin-N (CV-N) from *Nostoc ellipsosporum* that irreversibly inactivates diverse primary strains of HIV-1 and blocks the cell-to-cell transmission of HIV infection. Mori *et al.* (2002) proposed that CV-N, and its analogues might help in fighting dreadful disease of HIV in future. Another compound produced by cyanobacteria is borophycin. This compound was isolated from *Nostoc linckia* and *Nostoc spongiforme* var. *tenue* that show cytotoxicity against human epidermoid carcinoma and human colorectal adenocarcinoma cell lines (Burja *et al.*, 2001). Cryptophycins, a fungicide, isolated from *Nostoc* sp. ATCC 53789 showed toxicity against human tumor cell lines. Cryptophycins functions through tubulin-interaction thereby disrupting tubulin dynamics ultimately causing apoptosis of tumor cells (Panda *et al.*, 1998). Another major class of peptides produced by cyanobacteria is lipopeptides. Lipopeptides are amino acid derived fragment linked to fatty acids derived portion. Lipopeptides functions as anticancer, antibiotic, antiviral, enzyme inhibitor and antifungal activities (Burja *et al.*, 2001). Hapalosin, a cyclic desipeptide (lipopeptides) isolated from *Hapalosiphon welwitschii*, has a reversing activity against

MDR (multi drug resistance) derived from P-glycoprotein (Kashihara *et al.*, 2000). Cyanobacteria have been also reported to produce five classes of protease inhibitors such as micropeptins, aerugenosins, microginins, anabaenopeptins and microveridins. They are of significant value in a number of diseases like lung emphysema.

In addition, cyanobacteria produce a wide variety of neurotoxins, hepatotoxins and cytotoxins compounds. They produce cyclic heptapeptide hepatotoxins, microcystins (Carmichael, 1992; Luukkainen *et al.*, 1993), pentapeptide nodularins (Codd and Beattie, 1991) and three types of neurotoxins: (homo) anatoxin-a, anatoxin-a(s) and saxitoxins. Cyanobacterial neurotoxins block neurotransmission and hepatotoxins are inhibitors of serine/threonine specific protein phosphatases (PP1 and PP2A) (Carmichael, 1992; Skulberg *et al.*, 1992). Cytotoxins produced by cyanobacteria belong to classes Scytophycins that show toxicity against human epidermoid carcinoma (Moore *et al.*, 1986); cyanobacterin (anticyanobacterial activity) (Mason *et al.*, 1982), Hapalindole A (antialgal and antimycotic activity) (Moore *et al.*, 1984); acutiphycin (antitumour properties) (Barchi *et al.*, 1984) and lyngbyatoxin (Robinson *et al.*, 1991). In this way, cyanobacteria are working continuously for the benefits of human beings and can be genetically modified for maximum yield.

4. Future Prospects

Cyanobacteria are a promising but still unexplored natural resource offering a wealth of information for their utilization as biofertilizers, bioremediating agents and an array of bioactive compounds for new drugs. Reports suggest that deterioration in atmospheric physico-chemical properties leading to the threat of decline in soil and water productivity and their adverse effect on microbial diversity and on human health can be maintained healthy by utilizing cyanobacterial systems. The information gathered from literatures focus on the fact that traditional dependency on Actinomyctes and Hyphomycetes for drug discovery is now gradually decreasing whereas exploitation of photosynthetic cyanobacteria for peptides and other cytotoxic compounds are increasing. Despite the complexity of the algal genome, cyanobacteria are at the frontline of present day biotechnological programme including environmental biotechnology programme. Biochemical mystery behind ability of cyanobacteria to withstand diverse and stress habitats is being slowly unfolded and metabolic processes are going to be exploited for production of valuable products for human welfare. Scientists involved in exploitation of cyanobacterial potency are aiming towards understanding the metabolic and molecular strategies for development of genetically engineered rapid growing, higher substrate tolerance and high value products producing strains. Although, exploitation of cyanobacteria for biotechnological and ecological application is in stage of infancy, but the future is promising and may hold good in the near future.

References

Adams DG and Duggan PS (1999) Heterocyst and akinete differentiation in cyanobacteria. New Phytol 144: 3-33.

Allen MM (1984) Cyanobacterial cell inclusions. Ann Rev Microbiol 38: 1-25.

Barchi JJ, Moore RE, Furusawa E and Patterson GML (1984) Acutiphycin and 20,21-didehydroacutiphycin, new antineoplastic agents from the cyanophyte *Oscillatoria acutissima*. J Am Chem Soc 106: 8193-8197.

Bender J, Archibold ER, Ibeanusi V and Gould JP (1989) Lead removal from contaminated water by a mixed microbial ecosystem. Water Sci Technol 21: 1661-1665.

Bender J, Gould JP, Vatcharapijarn Y, Young JS and Phillips P (1994) Removal of zinc and manganese from contaminated water with cyanobacteria mats. Water Environ Res 66: 679-683.

Bergman B, Gallon JR, Rai AN and Stal LJ (1997) N_2 fixation by non-heterocystous cyanobacteria. FEMS Microbiol Rev 19: 139-185.

Bonilla I, Bolanos L and Mateo P (1995) Interaction of boron and calcium in the cyanobacteria *Anabaena* and *Synechococcus*; Physiol Plant 94: 31-36.

Borbely G, Suranyi GS and Kos P (1990) Stress responses of cyanobacteria and the pleiotropic effects of light deprivation. FEMS Microbiol Ecol 74: 141-152.

Bridle P and Timberlake CF (1997) Anthocyanins as natural food colour selected aspects. Food Chem 58: 103-109.

Burja AM, Banaigs EB, Abou-Mansour E, Burgess JG and Wright PC (2001) Marine cyanobacteria-a prolific source of natural products. Tetrahedron 57: 9347-9377.

Carmichael WW (1992). A Review: Cyanobacteria secondary metabolites–the cyanotoxin. J Appl Bact 72: 445-459.

Castenholz RW (1996) Endemism and biodiversity of thermophilic cyanobacteria. Nova Hedwigia Beih 112: 33-47.

Cerniglia CE (1984) Microbial metabolism of polycyclic aromatic hydrocarbons. Adv Appl Microbiol 30: 31-71.

Chattopadhyay P, Chatterjee S and Sen SK (2008) Biotechnological potential of natural food grade biocolorants. African J Biotech 7: 2972 -2985.

Choudhary K.K. (1999) *Ex-situ* conservation of cyanobacterial germplasm of north Bihar, India, Ph. D. thesis, BRA Bihar University, Muzaffarpur, Bihar, India.

Choudhary KK (2009) Ecological and biotechnological relevance of cyanobacteria. In: Soil Microflora. Eds Gupta RK, Kumar M and Vyas D, Daya Publishing House, New Delhi, pp 324-339.

Choudhary KK, Singh SS and Mishra AK (2007) Nitrogen fixing cyanobacteria and their potential applications. In: Advances in Applied Phycology. Eds Gupta RK and Pandey VD, Daya Publishing House, New Delhi, pp 142-154.

Clydesdale FM (1993) Color as a factor in food choice. Crit Rev Food Sci Nutr 33: 83-101.

Codd GA and Beattie KA (1991) Cyanobacteria (blue-green algae) and their toxins: awreness action in the United Kingdom. PHLS Microbiol Digest 8: 82-86. London.

De Philippis R, Paperi R, Sili C and Vincenzini M (2003) Assessment of the metal removal capability of two capsulated cyanobacteria, *Cyanospora capsulata* and *Nostoc* PCC 7936. J Appl Phycol 15: 155-161.

De PK (1939) The role of blue-green algae in nitrogen fixation in rice fields. Proc R Soc Lond B: 121-139.

Douglas SE (1994) Chloroplast origins and evolution. In: The Molecular Biology of Cyanobacteria. Ed Bryant DA, Kluwer Academic Publishers, Dordrecht, pp 91-118.

El-Enany AE and Issa AA (2000) Cyanobacteria as a biosorbent of heavy metals in sewage water. Environ Toxicol Pharmacol 8: 95-101.

El-Shehawy R, Lugomela C, Ernst A and Bergman B (2003) Diurnal expression of hetR and diazocyte development in the filamentous non-heterocystous cyanobacterium *Trichodesmium erythraeum*. Microbiol 149: 1139-1146.

Emodi A (1978) Carotenoid: Properties and Applications. Food Technol 32: 38-42.

Fay P and Van Baalen C (1987) The Cyanobacteria. Elsevier, Amsterdam, pp 534.

Fay P (1965) Heterotrophy and nitrogen fixation in *Chlorogloea fritschii*. J Gen Microbiol 39: 11-20.

FNB (1971) Food Colors. Food and Nutrition Board, National Academy of Sciences, Washington, D.C.

Gardea-Torresdey, JL, Becker-Hapak MK, Hosea JM and Darnell DW (1990) Effect of chemical modification of algal carboxyl groups on metal ion binding; Environ Sci Technol 19: 1372-1379.

Gardea-Torresdey, JL, Arenas JL, Webb R, Tiemann K and Gonzalez J (1996) Uptake of metal ions from solution by inactivated cells of cyanobacteria. Proceedings for 1996 HSRC-WERC Joint Conference of the Environment, pp. 48-58.

Gerasimenko LM, Dubinin AV and GA Zavarzin (1996) Alkaliphilic cyanobacteria from soda lakes of Tuva and their ecophysiology. Microbiology 65: 736-740.

Gupta AB and Lata KJ (1964) Effect of algal growth harmones on the germination of paddy seeds. Hydrobiologia 24: 430-434.

Hallberg GR (1987) Agricultural chemicals in groundwater: Extent and implications. Am J Alternative Agr 2: 3-15.

Hoffman L (1999) Marine cyanobacteria in tropical regions: diversity and ecology. Eur J Phycol 34: 371-379.

Iwata K, Inayama T and Kato T (1990) Effect of *Spirulina platensis* on plasma lipoprotein lipase activity in fructose-induced hyperlipidemic rats. J Nutr Sci Vitaminol 36: 165-171.

Kannaiyan S (1983) Studies on the effect of algal biofertilizer inoculation on rice crop. All India Applied Physiological Congress, Kanpur, 96.

Kashihara N, To-e S, Nakamura K, Umezawa K, Yamamura S and Nishiyama S (2000) Synthesis and biological activities of hapalosin derivatives with modification at the C12 position. Bioorg Med Chem Lett 10: 101-103.

Kaushik BD (1986) Response of cyanobacteria nitrogen fixation to exogenous nitrogenous source. In: Current status of biological nitrogen fixation research. Ed. Singh R, Nainawatee HS and Sawhney SK, HAU, Hisar, India pp 83.

Kaplan D, Christiaen D and Arad SM (1987) Chelating properties of extracellular polysaccharides from *Chlorella* spp. Appl Environ Microbiol 53: 2953-2956.

Kartikeyen N, Prasanna R, Sood A, Jaiswal P, Nayak S and Kaushik BD (2008) Physiological characterization of electron microscopic investigation of cyanobacteria associated with wheat rhizosphere. Folia Microbiol 54: 43-51.

Kolte SO and Goyal SK (1989). Natural variation in nitrogen potential of cyanobacteria. Phykos 25: 166-170.

Kushak R, VanCott E, Drapeau C, Winter H (1999) Effect of algae *Aphanizomenon flos-aquae* on digestive enzyme activity and polyunsaturated fatty acids level in blood plasma. Gastroenterol 116: A559.

Laamanen M (1996) Cyanoprokaryotes in the Baltic Sea ice and winter plankton. Arch Hydrobiol Suppl 117 Algol Stud 83: 423-433.

Lombi E, Zhao FJ, Dunham SJ and McGrath SP (2001) Phytoremediation of heavy metal-contaminated soils natural hyper accumulation versus chemically enhanced phytoextraction. J Environ Qual 30: 1919-1926.

Luukkainen R, Sivonen K, Namikoshi M, Fardig M, Rinehart KL and Niemela SI (1993) Isolation and identification of 8 microcystins from 13 *Oscillatoria agardhii* starins and structure of a new microcystin. Appl Environ Microbiol 59: 2204-2209.

Mason CP, Edwards KR, Carlson RE, Pignatello J, Gleason FK amd wood JM (1982) Isolation of chlorine-containing antibiotic from the freshwater cyanobacterium, *Scytonema hofmanni*. Science 215: 400-402.

Micheletti E, Colica G, Viti C, Tamagnini P and De Philippis R (2008) Selectivity in the heavy metal removal by exopolysaccharide-producing cyanobacteria. J Appl Microbiol 105: 88-94.

Mishra U, Choudhary KK, Pabbi S, Dhar DW and Singh PK (2005) Influence of Blue Green Algae and *Azolla* inoculation on specific soil enzymes under paddy cultivation. Asian Jr Microbiol Biotechnol Env Sc 7: 9-12.

Moore RE, Cheuk C, Yang XG and Patterson GML (1984) Hapalindoles: new alkaloids from the blue-green alga, *Hapalosiphon fontinalis*. J Am Chem Soc 106: 6456-6457.

Moore RE, Patterson GML, Entzeroth M, Morimoto H, Suganuma M, Hakii H, Fujiki H and Sugimura T (1986) Binding studies of [H-3] lyngbyatoxin A and [H-3] debromoaplysiatoxin to the phorbol ester receptor in a mouse epidermal particulate fraction. Carcinogenesis 7: 641-644.

Mori T, Barrientos LG, Han Z, Gronenborn AM, Turpin JA and Boyd MR (2002) Functional homologs of cyanovirin-N amenable to mass production in prokaryotic and eukaryotic hosts. Protein Expr Purif 26: 42-49.

Paerl HW, Steppe TF, Buchan KC and Potts M (2003) Hypersaline cyanobacterial mats as indicators of elevated tropical hurricane activity and associated climate change. AMBIO 32: 87-90.

Panda D, DeLuca K, Williams D, Jordan MA and Wilson L (1998) Antiproliferative mechanism of action of cryptophycin-52: Kinetic stabilization of microtubule dynamics by high-affinity binding to microtubule ends. PNAS 95: 9313-9318.

Pimentel D (1996) "Green Revolution Agriculture and Chemical Hazards." Scie Total Environ 188 Supplement: S86–S98.

Pimentel P, Hepperly P, Hanson J, Douds D and Seidel R (2005) Environmental, energetic, and economic comparisons of organic and conventional farming systems. Bioscience 55: 573-582.

Raungsomboon S, Chidthaisong A, Bunnag B, Inthorn D and Harvey NW (2008) Removal of lead (Pb^{2+}) by the Cyanobacterium *Gloeocapsa* sp. Bioresource Technol 99: 5650-5658.

Rai AN (1990). Cyanobacteria in symbiosis. In: Handbook of Symbiotic Cyanobacteria. Ed Rai AN, CRC Press, Boca Raton, FL, pp 1-7.

Reghukumar C, Vipparty V, David JJ and Chandramohan D (2001) Degradation of crude oil by marine cyanobacteria. Appl Microbiol Biotechnol 57: 433-436.

Robinson CP, Franz DR and Bondura ME (1991) Effects of Lyngbyatoxin A from the blue green algae *Lyngbya majuscula* on rabbit aorta contractions. Toxicon 29: 1009-1017.

Roger PA and Kulasooriya SA (1980) Blue-green algae and Rice, IRRI, Los Banos.

Sanchez-Baracaldo P, Haynes PK and Blank CE (2005) Morphological and habitat evolution in the Cyanobacteria using a compartmentalization approach. Geobiology 3: 145-165.

Saxena S, Kaushik BD (1992) Polysaccharide (biopolymer) from halotolerant cyanobacteria. Indian J Exptl Biol 30: 433-434.

Schopf JW (2000) The fossil record: tracing the roots of the cyanobacterial lineage. In: The ecology of cyanobacteria: their diversity in time and space. Eds Whitton BA and Potts M, Boston, Kluwer Academic, pp 13-35.

Singh PK (1973) Nitrogen-fixation by the unicellular blue-green alga *Aphanothece*. Arch Microbiol 103: 297-302.

Singh RN (1961). The role of blue-green algae in nitrogen economy of Indian Agriculture. ICAR Publication, New Delhi, pp 175.

Singh BV, Choudhary KK, Dhar DW and Singh PK 2001. Occurrence of some *Nostocales* from 24 Parganas of West Bengal. Phykos 40: 83-87.

Singh S, Kate BN and Banerjee UC (2005) Bioactive compounds from cyanobacteria and microalgae: an overview. Crit Rev Biotechnol 25: 73-95.

Skulberg OM, Carmichael WW, Anderson RA, Matsunaga, Moore RE and Skulberg R (1992) Investigation of a neurotoxic Oscillatorealean strain (Cyanophyceae) and its toxin. Isolation and characterization of homoanatoxin-a. Env Toxicol Chem 11: 321-329.

Slotton DG, Goldman CR and Frank A (1989) Commercially grown *Spirulina* found to contain low levels of mercury and lead. Nutr Rep Int 40: 1165-1172.

Stacey GH, Burris RH and Evans HJ (1992) Biological Nitrogen Fixation. New York. Chapman and Hall.

Taniuchi Y, Yoshikawa S, Maeda S-I, Omata T and Ohki K (2008) Diazotrophy under continuous light in a marine unicellular diazotrophic cyanobacterium, *Gloeothece* sp. 68DGA. Microbiology 154: 1859-1865.

Taton A· Grubisic S, Balthasart P, Hodgson DA, Laybourn-Parry J and Wilmotte A (2006) Biogeographical distribution and ecological ranges of benthic cyanobacteria in East Antarctic lakes. FEMS Microbiol Ecol 57: 272-289.

Thajuddin N and Subramanian G (1995) Aditional to the new reports of marine cyanobacteria from the east coast of India. Phykos 34: 33–37.

Thajuddin N and Subramanian G (2005) Cyanobacterial biodiversity and potential applications in biotechnology. Curr Sci 89: 47-57.

Trabelsi L, M'sakni NH, Ouada HB, Bacha H and Roudesli S (2009) Partial characterization of extracellular polysaccharides produced by cyanobacterium *Arthrospira platensis*. Biotechnol Bioprocess Eng 14: 27-31.

Turner JS and Robinson NJ (1995) Cyanobacterial Metallothioneins: Biochemistry and Molecular Genetics. J Ind Microbiol 14: 119-125.

Valencia A and Walker J (1999) A multi-axial treatment paradigm for mild traumatic brain injury to achieve reparative functional meta-plasticity. 3rd World Congress on Brain Injury, IBIA, Quebec City, June 1999.

Verma SK and Singh SP (1990) Factors regulating copper uptake in cyanobacterium. Curr Microbiol 21: 33–37.

Vlad M, Bordas E, Caseanu E, Uza G, Creteanu E and Pohnicenco C (1995) Effect of cuprofilin on experimental atherosclerosis. Biol. Trace Elem Res 48: 99- I 09.

Watanable A and Yamamoto Y (1971) Algal nitrogen fixation in the tropics. Plant Soil Spl vol: 403-413.

Whitton BA, Grainger SLJ, Hawley GRW and Simon JW (1991) Cell-bound and extracellular phosphatase activities of the cyanobacterial isolates. Microbial Ecol 21: 85-98.

Yoshida M, Yoshida T, Kashima A, Takashima Y, Hosoda N, Nagasaki K and Hiroishi S (2008) Ecological dynamics of the toxic bloom-forming cyanobacterium *Microcystis aeruginosa* and its cyanophages in freshwater. Appl Environ Microbiol 74: 3269-3273.

Zhang W and Majidi V (1994) Monitoring the cellular response of *Stichococcus bacillaris* to exposure of several different metals using *in vivo* 31P NMR and other spectroscopic techniques. Environ Sci Technol 28: 1577-1581.

Algal Biotechnology: New Vistas (2010) *Pages 277–287*
Editor: **Mihir Kumar Das**
Published by: **DAYA PUBLISHING HOUSE, NEW DELHI**

Chapter 19

Production of Nutraceuticals and Antioxidant Enzymes in a Tropical Food Alga *Nostochopsis lobatus*

*Usha Pandey**

Department of Botany, Faculty of Science and Technology, M.G. Kashi Vidyapith, Varanasi – 221 005, U.P., India

ABSTRACT

Recent global concern has shown increasing concern towards production of nutraceuticals, the functional foods, which provide additional physiological benefits in addition to the general energy need. The present study was an effort to investigate the production of nutraceuticals and antioxidant enzymes from a food alga *Nostochopsis lobatus*, a filamentous diazotrophic cyanobacterium (blue green algae), in a full factorial design in batch culture. Production of biomass, pigments, carbohydrates, proteins, amino acids and antioxidant capacity all were higher under immobilized cell cultures in comparison to free cell cultures. However, the activities of antioxidative enzymes showed an opposite trend. Under free cell culture, the alga showed significantly high activities of superoxide dismutase (2.00 unit μg protein^{-1}), peroxidase (12.05 unit mg protein^{-1}), catalase (1.95 unit mg protein^{-1}) and glutathione reductase (2.98 unit mg protein^{-1}). The study suggests that *N. lobatus* could be a promising bioresource for the production of nutritionally rich biomass, nutraceuticals, antioxidants as well as antioxidative

* E-mail: usha_pandey28@yahoo.co.in

enzymes. Use of immobilized cells in batch culture could be an effective approach for scaling–up production for commercial use. However, for production of antioxidant enzymes, free cell culture could be a more effective approach.

Keywords: Antioxidants, Cyanobacteria, Food algae, Nutraceuticals, Superoxide dismutase.

1. Introduction

Our body requires a number of compounds as added dietary supplements to meet the need for minimizing health risks from environment–induced human ailments. Nutraceuticals, often referred to as functional foods, are able to provide additional physiological benefits for human health in addition to the basic nutritional and energy needs. Often, functional foods are traditional foods enriched with ingredients able to provide specific beneficial actions for human health (Krishnaswamy, 1996). Antioxidants have gained importance in recent years due to their ability to neutralize reactive oxygen species which are implicated in several major human diseases including carcinogenesis–induced tumor promotion (Jose and Janardhanan, 2000; Devasagayam *et al.*, 2004; Mishra *et al.*, 2006). Generation of highly toxic reactive oxygen species (ROS) in tissues is a natural process, however, the imposition of stresses most often aggravates the production of ROS (Alscher *et al.*, 1997; Arora *et al.*, 2002). Naturally operating mechanisms coordinate the antioxidant responses in living tissues. However, over production of ROS, as happens under a variety of stress conditions, cause oxidative deterioration of cellular constituents that may lead ultimately to cell death. Antioxidants act as a major defense against ROS–induced oxidative damages.

A number of blue green algae (cyanobacteria) have been reported to be the rich source of proteins, carbohydrates, vitamins, minerals, amino acids and fatty acids (Thajuddin and Subramaniam, 2005). Species of *Nostoc, Aphanothece* and *Spirulina* are important edible blue greens. Of these, *Spirulina* received attention globally for its especially high protein content and nutritional values. Distributed in a number of saline and alkaline lakes of Africa, Mexico and India, this alga contains 60–70 per cent crude protein and 9–15 per cent lipids, the latter being made-up of unsaturated fatty acids. Edible blue greens contain certain bio-active compounds not found in higher plants and traditional therapeutic sources. Except for red algae, phycocyanins and phycoerythrins which are now known to have strong antioxidative potential are exclusive to blue green algae. Enriched in polyunsaturated fatty acids (Cohen and Vonshak, 1991; Mahajan and Kamat, 1995), β-carotene (Madhava *et al.*, 2000), phycobiliproteins (Pandey and Pandey, 2008a; 2008b), sulphated polysaccharides and sterols (Xue *et al.*, 2002), these algal organisms are receiving increasing attention as a bioresource for functional foods. Probably for these reasons this group of organisms has now received growing concern of nutraceutical industries.

Nostochopsis lobatus, a filamentous cyanobacterium (blue green algae), is used as nutritionally functional food supplement by tribes of India. This diazotroph grows luxuriantly attached to the rock surfaces in the form of mucilaginous balls in freshwater lakes and slow moving streams of Indian tropics and subtropics. The advantage of its natural growth in the form of mucilaginous balls of unialgal population is fully utilized by the local tribes who use it as a supplementary diet from long back (Tiwari, 1978). Recent researches conducted in our laboratory indicated *N. lobatus* to be a rich source of protein, carbohydrates and fatty acids (Pandey, 2003). In this study we reported that apart from high nutritional values, the aqueous extract of *N. lobatus* has potential antioxidant properties which can be exploited for commercial use.

2. Material and Methods

Nostochopsis lobatus (Wood) em. Gietler, a heterocystous cyanobacterium (blue green alga) was collected from a slow moving ftreshwater stream at Jharol, Udaipur, Rajasthan. An unialgal population was raised using axenic culture in nitrogen–free Allen and Arnon's medium (Allen and Arnon, 1955) at pH 7.5. Cultures were maintained at 25 ± 2°C with 1600/4000 Lux light intensity provided in 16/8 light/dark cycle. Growth was observed in free and immobilized cells in batch culture according to Yoon *et al.* (2000). Exponentially grown algal cells, after centrifugation and repeated washing, were suspended in 5 per cent solution of sodium alginate. The mixture was pumped drop–wise into $CaCl_2$ solution (0.2 mol L^{-1}) and the beads thus formed (average diameter, 5 mm) were allowed to harden for 2h. After repeated washing with sterile distilled water beads were resuspended in 1L growth medium. Similar growth cultures were also prepared for free cell.

Growth of the alga was measured in terms of dry weight per litre of culture medium. Ten day old harvested biomass was analyzed in triplicate for chlorophyll, carotenoids, phycocyanin, phycoerythrin, proteins and amino acids. Chlorophyll was extracted in N, N–dimethyl formamide and absorbance recorded at 664 nm following Inskeep and Bloom (1985). For carotenoids, acetone extraction procedure was followed. Phycoerythrin (PE) and phycocyanin (PC) were extracted in 0.05 M phosphate buffer (pH 6.8) and measured following Bennett and Bogorad (1973). Protein content was determined using Folin–ciocalteau reagent. For the measurement of free amino acids standard ninhydrin method was followed. Nutritive value was determined as : 4 X percentage of protein + 9 X percentage of fat + 4 X percentage of carbohydrate (Indrayan *et al.*, 2005).

Aqueous extracts of *N. lobatus* were used for determining antioxidant activity in free and immobilized cells. Standard assays included (*i*) free radical scavenging by oxygen radical absorption capacity (ORAC) and (*ii*) free radical formation by ferric reducing/antioxidant power (FRAP). Aqueous extract was prepared by stirring crushed algal balls in distilled water for 1h. For boiled water extract, the crushed algal material was boiled in distilled water for 30 min, so as to simulate conditions pertaining to their use for cooking purpose. The oxygen radical absorbance capacity (ORAC) was measured in terms of 2,2–azobis (2–amidinopropane) dihydrochloride (AAPH)–induced chemical damage to β-phycoerythrin. The decrease in fluorescence emission was recorded at every 5 min. Oxygen radical absorbance capacity was expressed in terms of μM trolox equivalent g^{-1} of fresh weight (CaO and Prior, 2002). For preparing FRAP regent, ferric chloride was mixed with 2,4,6–tripyridyl–s–triazine (TPTZ) to form Fe^{3+}–TPTZ complex. At low pH, in the presence of an antioxidant, this complex is reduced to Fe^{2+}–TETZ, which has an intense blue colour with peak absorption at 595 nm (Benzie and Strain, 1996). The values were expressed as mM ascorbic acid equivalent antioxidant capacity (AEAC).

For enzyme extraction, frozen algal samples (0.5 g) were ground to fine powder and then homogenized in 5 ml extraction buffer. After centrifugation, the supernatant was collected and used for determining enzyme activities. Superoxide dismutase (SOD) activity was measured following the method of Dhindsa *et al.* (1981). For catalase (CAT) and peroxidase (POD), the methods of Aebi (1984) and Kochbba *et al.* (1977) respectively, were followed. Glutathione reductase was estimated by the method of Foyer and Halliwell (1976).

3. Results and Discussion

Observations on growth of *Nostochopsis lobatus* in N–free basal medium indicated that the alga grew faster and accumulated high biomass under immobilized culture condition (Figure 19.1). Immobilized cells contained high amounts of various pigments and high nutritive value in comparison to the free cells (Table 19.1). Concentrations of chlorophyll, carotenoids, phycocyanin and phycoerythrin in free cells were 9.48, 2.06, 46.53 and 80.50 mg g^{-1} dry wt respectively. For immobilized cells, the respective values were 15.40, 3.38, 67.20 and 110.15 mg g^{-1} dry wt. Similarly, protein, free amino acids and carbohydrates were 23.5 per cent, 32.9 per cent and 41.9 per cent higher in immobilized cells over the free cells. On time scale, immobilized cells showed superiority with respect to the production of phycocyanin, phycoerythrin (Figure 19.2) and amino acids (Figure 19.3). On day ten, phycocyanin and phycoerythrin was found to be 67.20 and 110.15 mg dry mass L^{-1} respectively in immobilized cells and 46.53 and 80.50 mg dry mass L^{-1} in free cells. Immobilized cells possessed significantly high nutritive value (417) than the free cells (321) (Table 19.1).

Table 19.1: Pigment, Protein, Carbohydrate and Nutritive Values of *N. lobatus* under Free and Immobilized Cell Culture Condition. Values are mean (n = 3) ± 1SE. All values, except nutritive value, are in mg g^{-1} dry wt.

Variables	*Free Cells*	*Immobilized cells*
Chlorophyll	9.48 ± 0.98	15.40 ± 1.61
Carotenoids	2.06 ± 0.19	3.38 ± 0.40
Phcocyanin	46.53 ± 4.72	67.20 ± 6.95
Phycoerythrin	80.50 ± 8.05	110.15 ± 10.67
Protein	426.00 ± 41.32	528.0 ± 46.01
Free amino acids	4.61 ± 0.42	6.12 ± 0.58
Carbohydrate	99.50 ± 9.50	141.00 ± 15.30
Nutritive value	321.00 ± 27.80	417.00 ± 38.55

Table 19.2 presents data on antioxidant activity of *N. lobatus* expressed in terms of ascorbic acid equivalent and trolox equivalent. Both the measurements indicated higher antioxidant activity of aqueous extracts of immobilized cells than that of the

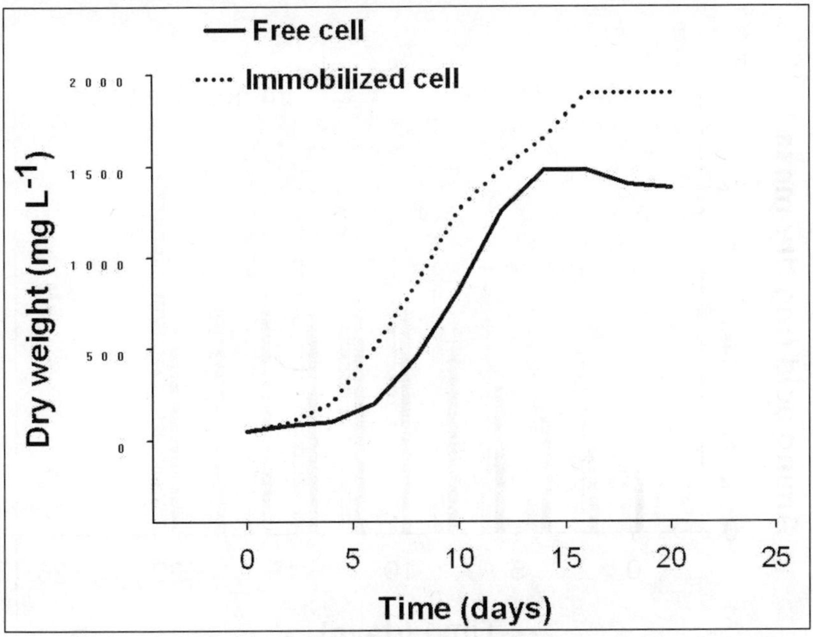

Figure 19.1: Growth of *N. lobatus* in Free and Immobilized Cell Cultures

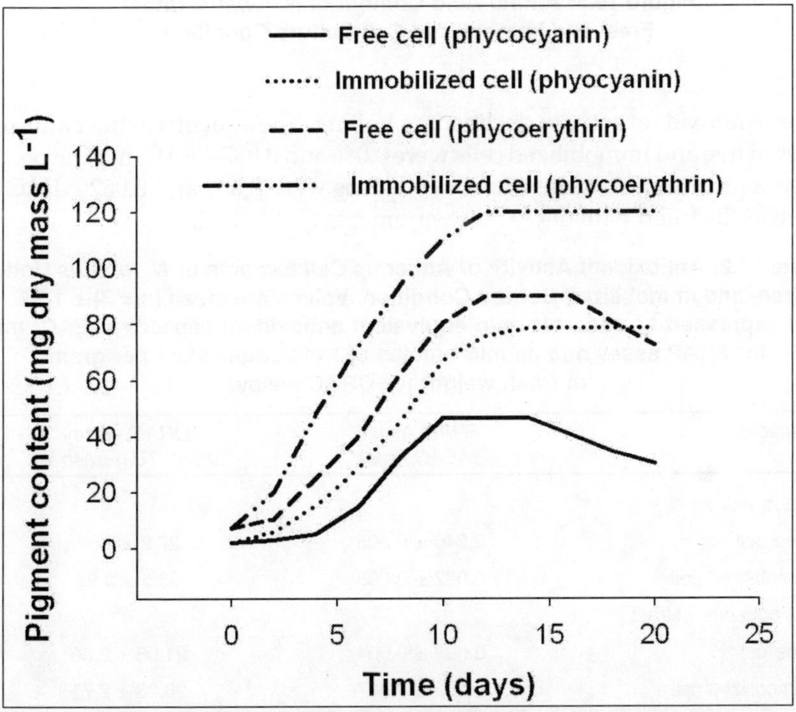

Figure 19.2: Phycocyanin and Phycoerythrin Contents of *N. lobatus* Under Free- and Immobilized Cell Culture Condition

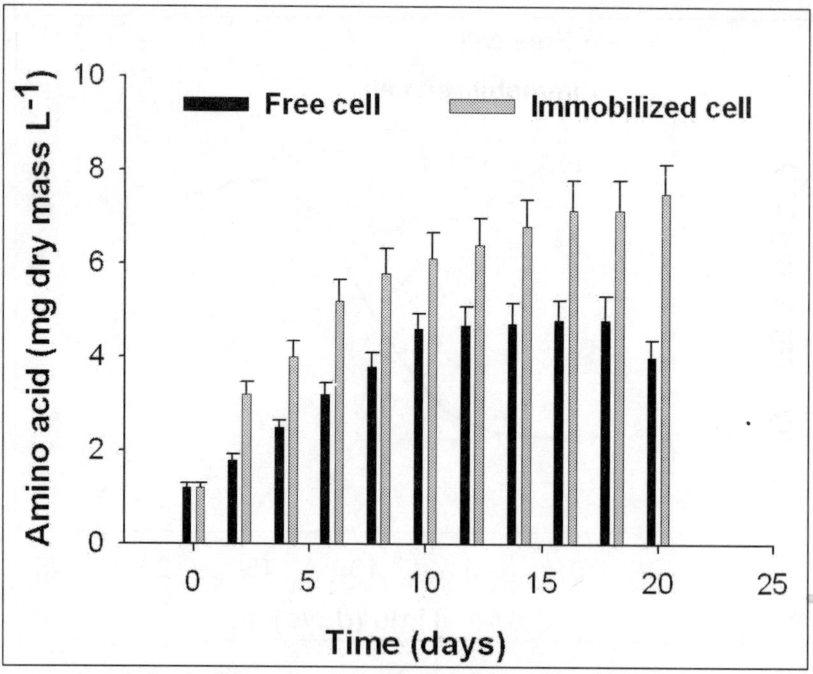

Figure 19.3: Amino Acid Content of *N. lobatus* Under Free- and Immobilized Cell Culture Condition

free cells. Antioxidant activity declined on boiling. The aqueous extract antioxidant activities in free and immobilized cells were 0.046 and 0.087 AEAC, mM respectively. In boiled aqueous extract, the respective values were 0.033 and 0.062 AEAC, mM. Similar was the trend with ORAC measurements (Table 19.2).

Table 19.2: Antioxidant Activity of Aqueous Cell Extracts of *N. lobatus* Under Free- and Immobilized Culture Condition. Values are mean (n = 3) ± 1 SE. Data expressed as ascorbic acid equivalent antioxidant capacity (AEAC, mM) for FRAP assay and as micromoles of trolox equivalent per gram of fresh weight for ORAC assay.

Variable	*FRAP Assay (AEAC, mM)*	*ORAC Assay (μ mol TE/g fresh wt)*
Aqueous extract		
Free cell	0.046 ± 0.005	27.2 ± 2.69
Immobilized cell	0.087 ± 0.008	38.5 ± 3.92
Boiled aqueous extract		
Free cell	0.033 ± 0.004	21.05 ± 2.30
Immobilized cell	0.062 ± 0.007	26.28 ± 2.71

Our observations indicated that, apart from its high protein (52.8 per cent), carbohydrate (14.1 per cent) and overall nutritive value (417), the aqueous extract of *N. lobatus* has significantly high antioxidant activity. Most of the antioxidant based pharmaceutical preparations including soft drinks and food additives rely on fruits and vegetables, although enzymatic antioxidants have been reported from a wide range of organisms including bacteria, algae, fungi and animals apart from higher plants (Bannister *et al.*, 1987). Antioxidative enzymes and a variety of metabolites such as ascorbic acid, glutathione, α-tocopherol, flavonoids and carotenoids, distributed widely in higher plants, help removing reactive oxygen species and thus, protecting tissues from oxidative stresses (Asada and Takahashi, 1987). Such plant sources with inherently high antioxidative properties are used in various therapeutic preperations. Algal organisms are probably the least explored biodiversity resource as far as their antioxidative properties concern.

Our results show that the antioxidant activity of aqueous extract (ORAC assay) of *N. lobatus* ranging from 27.2 to 38.5 μM of TE g^{-1} fresh wt was higher than those observed in Kokam (*Garcinia indica*; 29.3) (Mishra *et al.*, 2006), turmeric (*Curcuma longa*; 13.5) (Tilak *et al.*, 2004) and many fruits and vegetables (Cao *et al.*, 1996) which are widely used in cooking, cosmetics and several therapeutic preparations. Furthermore, high concentrations of carotenoids and phycobiliproteins as recorded in *N. lobatus* have high commercial value. These pigments of cyanobacteria are used as natural food colourants and food additives to enhance the color of the flesh of salmonid fish and to improve the health and fertility of cattle from long back (Emodi, 1978). In particular, concentration of phycoerythrin was potentially high in *N. lobatus*. Unlike other cyanobacteria where phycocyanin is the major phycobiliprotein, the content of phycoerythrin in *N. lobatus* amounts to about 11 per cent of dry weight, a value greater than that reported in *Anabaena* sp (8.3 per cent) (Rodriguez *et al.*, 1998). This adds further merit because this water soluble pigment is now increasingly being used in a variety of fluorescence immunoassays, diagnostics and biomedical research.

Production of high quality biomass as well as high phycoerythrin, protein, amino acid and carbohydrates and antioxidant capacity under immobilized condition indicate that a growth culture simulating the *in situ* growth condition of this alga would be beneficial for scaling–up its biomass production and antioxidant capacity on commercial scale. This alga grows in the form of mucilaginous balls attached to the rock surface in lakes and in slow moving streams. Probably due to this innate capacity, the biomass, pigments and antioxidant capacity of this alga appeared high under immobilized cell culture.

Antioxidant enzymes are the most important components in the scavenging system of ROS. Superoxide dismutase, which is a major scavenger of O_2^- was found high (2.00 unit μg protein^{-1}) in *N lobatus* (Table 19.3). This enzyme catalyzes the dismutation reaction of superoxide radical anions into O_2 and H_2O_2 (Scandalios, 1993). The H_2O_2 thus produced can be scavenged by catalase which also was found to be high (1.95 unit mg protein^{-1}) in this cyanobacterium. Peroxidase (POD) decomposes H_2O_2 by oxidation of co-substrates and glutathione reductase detoxifies hydroperoxides during oxidative stress. Both, peroxidase (12.05 unit mg protein^{-1}) and glutathione reductase (2.98 unit mg protein^{-1}) were found to be high in *N. lobatus* in free cell culture. Low antioxidatives enzyme activities in immobilized cells may be due to protective covering of calcium alginate and cell aggregation. This property of cyanbacterium make it unique for commercial exploitation. Activities of all the four enzymes increased at high light intensity. Interestingly, unlike pigment production, all the four antioxidant enzymes appeared high in free cell cultures (Table 19.3). This suggests that different culture condition can be used for producing algal metabolites depending upon requirement for commercial use. Furthermore, the synergy of biliproteins with antioxidant capacity of *N. lobatus* appeared interesting and has scientific as well as commercial significance. Both the biliproteins, phycocyanin and phycoerythrin were substantially high under immobilized culture condition. This suggest that under its natural growth condition, the antioxidant capacity of *N. lobatus* is due to its constitutive bilipigments. However, under modified growth condition, for instance under free cell growth, inducible antioxidative enzymes could play major role determining overall antioxidant capacity of *N. lobatus*.

Table 19.3: Antioxidant Enzymes of *Nostochopsis lobatus* at Different Light Intensities Under Free- and Immobilized Cell Cultures. Values are mean (n = 3) ± 1SE. Except for superoxide dismutase (unit µg protein^{-1}), all the measurements are in unit mg protein^{-1}.

Enzyme	*Free Cell Culture*		*Immobilized cell culture*	
	1600 Lux	*4000 Lux*	*1600 Lux*	*4000 Lux*
Superoxide dismutase	1.25 ± 0.10	2.00 ± 0.18	0.95 ± 0.05	1.35 ± 0.66
Peroxidase	5.06 ± 0.36	12.05 ± 0.10	0.65 ± 0.04	9.15 ± 0.68
Catalase	0.56 ± 0.48	1.95 ± 0.14	0.60 ± 0.04	1.24 ± 0.11
Glutathione reductase	1.26 ± 0.11	2.98 ± 0.15	0.98 ± 0.07	1.85 ± 0.14

Increasing consciousness about health risks associated with oxidative stress has brought a major shift in people's preference to opt functional foods which have additional physiological benefits. Since free radicals have been implicated in the etiology of several major human ailments including hepatotoxicity, cardiovascular disease, cancer and diabetes (Sie, 1996), naturally occurring free radical scavenger are now increasingly been used in modulating such human ailments (Jose and Janardhana, 2000). Our country which has been among pioneer in using natural dietary supplements for modulating chronic diseases, has witnessed increasing demand of such functional foods during recent years (Krishnaswami, 1996). Our observations indicated that *N. lobatus* has potential antioxidant capacity, higher than those reported in many fruits, vegetables and spices used in cooking, cosmetics and medicinal preparations. This alga can be grown on large scale for promoting its use for health associated benefits. Antioxidative property of *N. lobatus* adds an important positive attribute to its known nutritional value and hence its use in therapeutic preparation of soft drinks and food additives may be promoted through scaling–up its production for commercial use.

Acknowledgements

The author thanks Principal, G.N. Girls P.G. College and Prof. H. R. Tyagi, farmer convener, Department of Environmental Sciences, MLS University Udaipur, for laboratory facilities.

References

Aebi, M (1984) Catalase *in–vitro.* Methods Enzymol 105: 121–126.

Allen MB and Arnon DI (1955) Studies on nitrogen–fixing blue green algae I. Growth and nitrogen fixation by *Anabaena cylindrica.* Lemm Plant Physiol 30: 366–377.

Alscher RG, Donahue JL, Cramer CL (1997) Reactive oxygen species and antioxidants: Relationship in green cells. Physiol Plant 100: 224–233.

Arora A, Sairam RK, Srivastava GC (2002) Oxidative stress and antioxidative system in plants. Curr Sci 82: 1221–1235.

Asada K and Takahashi M (1987) Production and scavenging of active oxygen in photosynthesis. In: Photoinhibition. Eds. Kyle DJ, Osmond CB, Amtzen CJ, Elsevier Science Publisher, Amsterdam, pp. 227–287.

Banister JV, Bannister WH and Rotilio G (1987) Aspects of the structure, function and application of superoxide dismutase. Crit Rev BIochem 22: 111–180.

Bennett A and Bogorad L (1973) Complementary chromatic adaptation in a filamentous blue–green alga. J Cell Biol 58: 419–435.

Benzie IF, Strain JJ (1996) The ferric reduction ability of plasma (FRAP) as a measure of antioxidant power. The FRAP assay. Anal Biochem 239: 70–76.

Cao G and Prior RL (2002) Measurements of oxygen radical absorbance capacity in biological samples. Methods Enzymol 299: 50–62.

Cao G, Sofic E and Prior RL (1996) Antioxidant capacity of tea and common vegetables. *J Agric* Food Chem 44: 3426–3431.

Cohen Z and Vonshak A (1991) Fatty acid composition of *Spirulina* and *Spirulina* like cyanobacteria in relation to their chemotaxonomy. Phytochemistry 30: 205–206.

Devasagayam TPA, Tilak JC, Boloor KK, Sane KS, Ghaskadbi S and Lele RD (2004) Free radical and antioxidants in human health: Current status and future prospects. J Assoc Physicians India 7: 794–804.

Dhindsa RHR, Plumb–Dhindsa and Thorp JA (1981) Leaf senescence correlated with increased level of membrane permeability. Lipid peroxidation and decreased level of SOD and CAT. J Exp Bot 32: 93–101.

Emodi A (1978) Carotenoids, properties and applications. Food Technol 32: 38–42.

Foyer and Halliwell (1976) The presence of glutathione and glutathione reductase in chloroplasts. A proposed role in ascorbic acid metaboliam. Planta 133: 21–25.

Indrayan AK, Sharma S, Durgapal D, Kumar N and Kumar M (2005) Determination of nutritive value and analysis of mineral elements for some medicinally valued plants for Uttaranchal. Curr Sci 89: 1252–1255.

Inskeep WP, Bloom PR (1985) Extinction coefficient of chlorophyll a and b in N, N dimethylformamide and 80 per cent acetone. Plant Physiol 77: 483–485.

Jose N and Janardhanan K K (2000) Antioxidant and antitumor activity of *Pleurotus florida*. Curr Sci 79: 941–943.

Kochbba J, Lavee S, Roy SP (1977) Differences in peroxidase activity and isoenzymes in embryogenic and non–embriogenic 'Shamouti' orange ovular callus lines. Plant Cell Physiol 18: 463–467.

Krishnaswamy K (1996) Indian functional foods: Role in prevention of cancer. Nutr Rev 54: 5127–5131.

Madhava C, Bhat VB, Kiranmai G, Reddy MN, Reddanna P and Madyastha KM (2000) Selective–inhibition of cyclooxigenase–2 by C–phycocyanin, a biliprotein from *Spirulina platensis*. Biochem Biophys Research Comm. 277: 599–603.

Mahajan G and Kamat M (1995) γ-linoleic acid production from *Spirulina platensis*. Appl Microbiol Biotechnol 43: 466–469.

Mishra A, Bapat MM, Tilak JC and Devasagayam TPA (2006) Antioxidant activity of *Garcinia indica* (Kokam) and its syrup. Curr Sci 91: 90–93.

Pandey U (2003) Changes in biochemical composition of three cyanobacteria of commercial importance during one phase growth. Ind J Appl Pure Biol 18: 27–30.

Pandey U and Pandey J (2008a) Enhanced production of biomass, pigments and antioxidant capacity of a nutritionally important cyanobacterium *Nostochopsis lobatus*. *Biores Technol* 99: 4520–4523.

Pandey U and Pandey J (2008b) Enhanced production of high-quality biomass, Δ-aminolevulinic, billipigments, and antioxidant capacity of a food alga *Nostochopsis lobatus*. *Appl Biochem Biotechnol* 150: 221–231.

Rodriguez H, Rivas J, Guerrero MG and Losada M (1989) Nitrogen fixing cyanobacterium with a high phycoerythrin content. Appl Environ Microbial 55: 758–760.

Scandalios JG (1993) Oxygen stress and superoxide dismutase. Plant Physiol 101: 7–12.

Sies H (1996) Antioxdants in diseases, Mechanisms and Therapy. Academic Press, New York.

Thajuddin N and Subramaniam G (2005) Cyanobacterial biodiversity and potential applications in biotechnology. Curr Sci 29: 47–57.

Tilak JC, Banerjee M, Hari Mohan and Devasagayam TPA (2004) Antioxidant availability of turmeric in relation to its medicinal and culinary uses. Phytother Res 18: 798–804.

Tiwari DN (1978) The heterocysts of the blue–green alga *Nostochopsis lobatus*: Effects of culture conditions. New Phytol. 81: 653–656..

Xue C, Hu Y, Saito H, Zhang Z, Li Z and Cai Y (2002) Molecular species composition of glycolipids from *Spirulina platensis*. Food Chem 277: 9–13.

Yoon SH, Do JH, Lee SY and Chang HN (2000) Production of poly-γ glutamic acid fed–batch culture of *Bacillus licheniformis*. Biotechnol Let 22: 585–588.

Algal Biotechnology: New Vistas (2010) *Pages 287–296*
Editor: **Mihir Kumar Das**
Published by: **DAYA PUBLISHING HOUSE, NEW DELHI**

Chapter 20

Bioremediation of Heavy Metals by Microalgae

V.D. Pandey*

Department of Botany, Govt. Post-Graduate College
Rishikesh – 249201, Dehradun, Uttarakhand, India

ABSTRACT

Heavy metals constitute toxic, non-degradable and persistent environment pollutants which adversely affect the growth, development and survival of all life forms. Discharged from various mining and industrial sites, they find way into the food chain and drinking water. The detrimental effects of heavy metals result from their action on a number of cellular and biochemical processes, biomolecules and structures in living organisms, including human beings. They are known to cause various physiological disorders of hepatic, renal, respiratory and gastrointestinal system in human beings. The toxicity of heavy metals depends on their concentration, bioavailability, chemical forms and duration of exposure. The ever-increasing contamination of aquatic bodies and soil by heavy metals is an issue of serious concern and challenge world–wide. Bioremediation of heavy metal-contaminated water by employing various microorganisms, including microalgae have been recognized as a cheaper, more effective and an eco-friendly alternative to the conventional physico-chemical remediation methods. Many species of microalgae show heavy metal tolerance and possess efficient heavy metal removal capabilities from aqueous solutions.

Keywords: Bioremediation, Heavy metals, Microalgae.

* E-mail: vdpandey2@yahoo.co.in

1. Introduction

Heavy metals, such as cadmium (Cd), mercury (Hg), silver (Ag), lead (Pb), nickel (Ni), chromium (Cr), arsenic (As), copper (Cu), selenium (Se) zinc(Zn) and thallium (Tl) constitute toxic, non-degradable and persistent environment pollutants which adversely affect the growth, development and survival of all life forms- plants, microbes, animals and humans. Heavy metals are metallic elements with relative density greater than $5g/cm^3$, and are toxic or poisonous at low concentrations. Among 90 naturally occurring elements, 21 are non-metals, 16 are light metals and the remaining 53 are heavy metals (Weast, 1984). Most heavy metals are transition elements with incompletely filled orbital. They are natural components of the Earth's crust. In environment, they cannot be degraded or destroyed biologically, leading to their persistence. They enter our bodies via food, drinking water and air. As trace elements, some heavy metals, such as copper, selenium, zinc are essential to maintain the metabolism of the human body and many other living organisms as structural components of various biomolecules or as enzyme activators. However, at higher concentrations they can lead to poisoning. Heavy metals are dangerous because they tend to bioaccumulate in bodies of living organisms. Bioaccumulation refers to the gradual increase in the concentration of a chemical or element in a biological organism over time, compared to the chemical's concentration in the environment. They have ability to form complexes with wide range of compounds and biomolecules. There is increasing concern globally towards the contamination of water bodies and soil by heavy metals discharged from various industries, such as electroplating, smelting, tanning, canning, pesticides, film processing and mining. Heavy metals entering the ecosystem may undergo geoaccumulation, bioaccumulation and biomagnification. They cause various physiological disorders of hepatic, renal, respiratory, skeletal and gastrointestinal system in human beings.

Bioremediation of heavy metal-contaminated water by employing various microorganisms, including microalgae have been recognized as a cheaper, more effective and an eco-friendly alternative to the conventional physico-chemical remediation methods. Many microalgae and cyanobacteria show heavy metal tolerance and posses efficient heavy metal removal capabilities from aqueous solutions.

2. Heavy Metals: Pollution and Toxicity

Heavy metals occur in environment as hydrated ionic species or they may form a variety of complexes with inorganic and organic legands (organometals) or associated with colloids and suspended particulate matter. Mobility of heavy metals, and hence bioavailability and toxicity is dependent on pH of the water and soil. Acidification of aquatic bodies by acid rain amplifies the problem of heavy metal toxicity. Weathering and volcanoes are natural sources whereas anthropogenic sources include mining, smelting or metallurgical industries, pesticides, tanning, and automobiles. Heavy metals adversely affect ecosystem functions and biodiversity. Contamination of terrestrial and aquatic ecosystems with heavy metals disrupts the nutrient cycling. The toxicity; behaviour and environmental fate of heavy metals are closely related to their position in periodic table.

The pollution of various aquatic ecosystems by heavy metals is an important environmental problem because they constitute some of the most hazardous substances that can bioaccumulate (Rayms-Keller *et al.*, 1998). Metals that are deposited in the aquatic environment may accumulate in the food chain and cause ecological damage and posing a threat to human health (Grimanis *et al.*, 1978; Adams *et al.*, 1992). Anthropogenic activities including, industrial discharges, domestic sewage, non-point source runoff and atmospheric precipitation are the main sources of toxic heavy metals that enter aquatic systems. However, metals also occur in small amounts naturally and enter aquatic systems through ore-bearing rocks, wind-blown dust, forest fires and vegetation (Tao *et al.*, 1999; Fernandez-Leborans and Herrero, 2000). Sediments act as the most important reservoir or sink of metals and other pollutants in the aquatic environment (Van-Hattum *et al.*, 1993). Concentrations and bioavailability of metals depend on many different processes, which include (1) mobilisation of metals and their chemical speciation, (2) transformation (*e.g.* methylation) of metals including As, Hg, Pb and Sn (3) the control exerted by major sediment components (*e.g.* oxides of Fe and organics) to which metals are preferentially bound, (4) competition between sediment metals (*e.g.* Cu and Ag; Zn and Cd) for uptake sites in organisms, and (5) the influence of bioturbation, salinity, redox or pH on these processes (Bryan and Langston, 1992).

The heavy metal pollution of aquatic ecosystems is often most obvious in sediments, macrophytes and aquatic animals, than in elevated concentrations in water (Linnik and Zubenko, 2000). Therefore, aquatic ecosystems are typically monitored for pollution of heavy metals using biological assays (Yang and Rauckmann, 1987; Wong and Dixon, 1995). Many aquatic organisms have been used as bioindicators, including aquatic insects (Rayms-Keller *et al.*, 1998), fish (Burger *et al.*, 2002), protozoa (Fernandez-Leborans and Herrero, 2000) and plants (Mohan and Hosetti, 1999). Fish species are often the top consumers in aquatic ecosystems (Dallinger *et al.*, 1987) and thus metal concentrations in fish can act as an environmental indicator of the state of the environment (Jorgenson and Pedersen, 1994).

The toxicity of heavy metals depends on their concentration, bioavailability, chemical forms and duration of exposure. Some of the well-known clinical effects of heavy metals include liver damage, renal dysfunctions, pulmonary edema, bronchitis, osteomalacia, osteoporosis, arthritis and neurological damage. Few heavy metals like Cd have been reported to be carcinogenic (Zweig *et al.*, 1999). Food, air, water and dust/soil are the major potential exposure pathways for humans. At cellular level and biochemical level, heavy metals attack sulphur bonds in enzymes, thus disturbing their structure and function. Carboxylic and amino group of proteins are also chemically bound by heavy metals. Besides, many of them bind to cell membrane, hindering transport process. Some heavy metals induce the synthesis free radicals, such as superoxide, hydroxyl and hydrogen peroxide casing oxidative damage to cells.

3. Microbial Metal Resistance and Detoxification

The ability of microbes to sense and respond to alterations in the environment is crucial for their survival. Several microorganisms have been reported to possess effective mechanisms for metal tolerance and detoxification in response to elevated concentration of metals, including heavy metals in the environment. Resistance mechanisms may be general and specific. General mechanisms are independent of metal stress, and often serve several functions. Microbial slime layers or exopolymers provide an effective barrier against metal entry in to the cell. They contain a wide array of anionic groups, such as carboxylic, amino, thio, hydroxyl, phosphate and uronic acids that strongly bind metal cations. Siderophores, which are iron-complexing low molecular weight organic compounds, are produced by many microbes in response to low level of iron in the environment (Lankford, 1973). Siderophores facilitates iron transport in to the microbial cells. At the same time, siderophores may interact with other toxic metals, such as aluminium, gallium, chromium and copper. By binding such metals, siderophores can reduce metal availability resulting in reduced metal toxicity (Burd *et al.*, 2000). The best known specific or metal-dependent mechanism is the inducible synthesis of metallothioneins(MTs) in many microbes in response to the elevated concentration of heavy metals. Metallothioneins are low molecular weight, cysteine- rich proteins with a high affinity for Cd, Zn, Cu, Ag and Hg. They have been reported in algae, yeast, moulds, bacteria and cyanobacteria. The induction of MTs is primarily regulated at the level of transcription.

4. Heavy Metal Remediation Technologies

The growing concern towards the increasing contamination of soil and water bodies by heavy metals has stimulated an interest in developing systems and process that can effectively remove or detoxify toxic metals from them. The conventional physico-chemical processes used for the purpose includes precipitation, ion - exchange, membrane separation, chemical oxidation/reduction and reverse-osmosis. These processes are expensive and not eco-friendly. Bioremediation of heavy metal polluted soil and water, employing various microorganisms, is gaining importance as a cost-effective, efficient and eco-friendly alternative to the conventional physico-chemical remediation methods. Bioremediation is the process of removal, degradation or detoxification of toxic environmental pollutants using living organisms-typically plants and microbes. The actual and potential applications of bioremediation in controlling or mitigation of various environmental pollutants is well-recognized and widely accepted. The OECD (Organisation for Economic Cooperation and Development) in its report (1994) entitled 'Biotechnology for a Clean Environment' estimated that world market potential for all environmental biotechnologies would be $75 billion by 2000. Bioremediation can be broadly classified as *ex situ* and *in situ*. The *ex situ* technology refers to treatment involving the physical removal of the contaminated material to another area for treatment. In contrast, *in situ* technique involves treatment of the contaminated material in the original place itself. Many microorganisms are known to possess effective detoxification mechanism to overcome the detrimental effects of metals. Microbe-mediated bioremediation of toxic heavy metals is widely advocated and promoted owing to a number of additional potential advantages: (*i*) the rapid kinetics of metal removal, (*ii*) the capability of microbes to remove metal ions from the solution of even very low concentration, (*iii*) treatment of water containing multiple metals(De Phillips *et al.*, 2003). Both photosynthetic and non-photosynthetic microorganisms have been reported as effective agents for heavy metal bioremediation (De Phillips *et al.*, 2003; Blanco *et al.*, 1999; Rao *et al.*, 2005; Gaur and Adholeya, 2004; Brady and Duncan, 1994). Both viable and non-viable microbial biomass of bacteria, fungi and algae are utilized for removing toxic metal ions. There are two mechanisms involved in microbe-mediated bioremediation- biosorption and bioaccumulation.

Biosorption involves extracellular or cell surface accumulation of heavy metals by microbial cells. It is essentially a passive process which does not involve the expenditure of metabolic energy and is primarily controlled by physico-chemical factors. Since it is a passive process it occurs both in living and dead cells. Several mechanisms are known to be involved in cell surface sorption, which include cation exchange, co-ordination, chelation and microprecipitation. The cell wall of microorganisms consists of a wide variety of extracellular polysaccharides with negatively charged groups, such as carboxyl, hydroxyl, sulphyl and phosphyl. These groups have high affinity for metal cations and involved in metal biosorption. The process can effectively remove considerable amount of heavy metals from industrial effluents and wastewater. Different microbes show variation in their affinity for different heavy metals.. Microbial biosorbents, both living and dead microbial biomass of bacteria, moulds, yeasts and microalgae, are effective and efficient in decontamination of heavy-metal polluted effluents and water(Volesky, 1994). The metal sorption capacity of a biosorbent is evaluated by sorption or adsorption isotherms described by Langmuir and Freundlich (Volesky, 1994) The Freundlich adsorption isotherm is a curve relating the concentration of a solute on the surface of an adsorbent, to the concentration of the solute in the liquid with which it is in contact. The Freundlich Adsorption Isotherm is mathematically expressed as $q= KC_e^{1/n}$, where q is the solute(metal) adsorbed or sorbed per unit biomass of biosorbent (expressed in $mg.g^{-1}$), K is a constant, related to the adsorbent capacity, n is a constant, related to the energy of sorption, and C_e is the equilibrium(final) concentration of solute(metal) in the solution(expressed as $mg.L^{-1}$).The Langmuir adsorption isotherm describes quantitatively the build up of a layer of molecules or atoms on an adsorbent surface as a function of the concentration of the adsorbed material in the liquid in which it is in contact. In a modified form it can also described as bi-layer deposition. It often represents an initial surface adsorption followed by a condensation effect resulting from extremely strong solute-solute interaction. The Langmuir adsorption isotherm is mathematically expressed as $q=bC_e q_{max}(1+bC_e)^{-1}$, where q is the solute(metal) adsorbed or sorbed per unit biomass of biosorbent, q_{max} is the maximum amount of metal adsorbed(at saturation) per unit mass of biosorbent, b is a constant related to the energy of sorption and C_e is the equilibrium(final) concentration of solute(metal) in the solution.

Bioaccumulation in microorganisms refers to the uptake or transport and intracellular accumulation of heavy metals from ambient environment. In contrast to biosorption, it is an energy-dependent active process carried out only by living microbial cells. Due to the well developed heavy metal resistance, many microbes can accumulate and tolerate high concentration of heavy metals.(Nies, 1999). Most microbial cells use two types of uptake system for heavy metal ions. The first type is fast, unspecific and constitutively expressed. It is driven by the chemiosmotic gradient across the cytoplasmic membrane. The second type of uptake system is slower, has high substrate specificity, and uses ATP hydrolysis as the energy source, in addition to chemiosmotic gradient (Nies, 1999).The immobilization of microbial cells on solid supports is an important approach in effective bioremediation. Bioreactors using immobilized microbial cells have several advantages over conventional effluent treatment technologies.

5. Microalgae in Heavy Metal Bioremediation

Microalgae constitute a diverse group of prokaryotic and eukaryotic algae with great biotechnological applications (Apt and Behrens, 1999). Morphologically, they are unicellular, colonial and filamentous. Microalgae refer to all microscopic algae, planktonic or benthic, with size up to about 200μm. The size and shape of microalgae has important implications for a range of biological functions, such as surface exchange of materials, light absorption, ability for rapid growth and distribution in water column. Owing to their tremendous adaptability and metabolic versatility, they colonize diverse terrestrial and aquatic habitats, including extreme and polluted ones. For commercial and biotechnological applications, they can be grown in both open-culture systems and highly controlled closed culture systems. The prokaryotic microalgae include blue-green algae(cyanobacteria) and the eukaryotic microalgae is mainly represented by taxonomically diverse forms of green algae, diatoms, euglenoids, dinoflagellates and cryptomonads. Microalgae, because of their rapid growth rate, simple growth requirements, heavy metal tolerance and amenability to controlled laboratory culture, are the promising candidates for the removal of heavy metal contaminants from water. Many microalgae are components of phototrophic biofilms, which can be described as surface attached microbial communities driven by light energy. Microalgae such as diatoms, unicellular and filamentous cyanobacteria and green algae generate energy and reduce carbon dioxide, providing organic substrates and oxygen. The actual and potential applications of microalgal phototrophic biofilms in wastewater treatment and heavy metal bioremediation is gaining importance (Roeselers, 2008; Liehr *et al.*, 1994). Many microalgae have the capability to sorb or accumulate metals, and there is considerable potential in the detoxification of wastewaters polluted with heavy metals (Bender *et al.*, 1994; Mehta and Gaur JP, 2005; Wilde and Benemann, 1993). The amount of metal removed per unit microalgal biomass varies with the species and the type of metal.

Many cyanobacterial species are known to produce exocellular polysaccharides (EPS), which are anionic in nature due to the presence of uronic acids and/or other charged groups(De Phillips and Vincenzini, 1998). Owing to the anionic nature, these EPS possess a very high affinity for metallic ions. EPS-producing cyanobacteria are the organisms of intense research world-wide because of their actual and potential applications in diverse fields, including heavy metal bioremediation. Many cyanobacterial species/strains, such as *Cyanospira capsulata*, *Nostoc* PCC7936, *Oscillatoria anguistissima*, *Phormidium laminosum*, *Microcystis aeruginosa* and *Aphanothece halophytica* have been reported to possess efficient heavy metal removal capability (De Phillipis *et al.*, 2003; Ahuja *et al.*, 1997; Blanco *et al.*, 1999; Parker *et al.*, 2000). Many studies have documented the presence of metal-binding proteins, called metallothioneins, in several cyanobacteria(Robinson *et al.*, 2000). These proteins are rich in cysteine and bind the metal ions in metal-thiolate clusters. Metal sequestration by metallothioneins is one of the mechanisms in metal tolerance in many cyanobacterial species.

Among eukaryotic microalgae, the freshwater unicellular green alga *Chlorella vulgaris* is widely used in biosorption of heavy metals. It has been commercially developed as a potent algal biosorbent called AlgaSorb™ to treat wastewater or groundwater containing heavy metals (Gupta *et al.*, 2000). AlgaSorb™ contains algal cells immobilized in a polymer. It has a remarkable affinity for heavy metal ions and is capable of concentrating these ions by thousand fold.

6. Conclusion

Bioremediation of toxic heavy metals by use of cyanobacteria and other microalgae represents a cost-effective and eco-friendly remediation technology. It is emerging and fascinating area of research with tremendous application in environmental management. Over the last few years, a number of companies have developed and commercialize bioremediation technologies, employing various microorganisms, including microalgae. Now, bioremediation is a common buzzword in waste and pollution management. Removal and recovery of metals, including heavy metals from aqueous solutions by exploiting the biosorption potential of various microalgae and non-algal microbes is gaining importance world-wide. In view of the increasing level of heavy metal pollution, research efforts should be directed towards the enhancement of biosorption potential of microbes, including micro algae.

References

Adams WJ, Kimerle RA, and Barnett JW (1992) Sediment quality and aquatic life assessment. Environmental Science and Technology 26: 1865-1875.

Ahuja P, Gupta R and Saxena RK(1997) *Oscillatoria anguistissima*: a promising Cu^{2+} biosorbent. Current Microbiology 35: 151-154

Apt KE and Behrens PW(1999) Commercial developments in microalgal biotechnology. J Phycology 35: 215-226.

Bender J, Rodriguez- Eaton S, Ekanemesang UM and Phillips P (1994) Characterization of metal-binding bioflocculants produced by the cyanobacterial component of mixed microbial mats. Applied Environ Microbiol 60: 2311-2315.

Blanco A, Sanz B, Llama MJ and Serra JL(1999) Biosorption of heavy metals to immobilised *Phormidium laminosum* biomass. J Biotechnology 69: 227-240.

Brady D and Duncan JR(1994) Bioaccumulation of metal cations by *Saccharomyces cerevisiae*. Applied Microbiol Biotechnol 41: 149-154

Bryan GW and Langston WJ (1992) Bioavailability, accumulation and effects of heavy metals in sediments with special reference to United Kingdom estuaries: a review. Environ Pollut 76: 89-131.

Burd GI, Dixon DG and Glick BR (2000) Plant growth-promoting bacteria that decrease heavy metal toxicity in plants. Can J Microbiology 46: 237-245

Burger J, Gaines KF, Boring S, Stephens L, Snodgrass J, Dixon C, McMahon M, Shukla S, Shukla T and Gochfeld M. (2002) Metal levels in fish from the Savannah river: potential hazards to fish and other receptors. Environmental Research 89: 85-97.

Dallinger R, Prosi F, Segner H, and Black H (1987) Contaminated food and uptake of heavy metals by Rainbow Trout (Salmo gairdneri): a field study. Oecologia 73: 91-98.

De Phillipis R and Vincenzini M(1998) Exocellular polysaccharides from cyanobacteria and their possible applications. FEMS Microbiol Rev 22: 151-175.

De Phillips R, Paperi R, Sili C and Vincenzini M(2003) Assessment of metal removal capability of two capsulated cyanobacteria, *Cyanospira capsulata* and *Nostoc* PCC7936. J Applied Phycology 15: 155-161.

Fernandez-Leborans G. and Olalla-Herrero Y. (2000) Toxicity and bioaccumulation of lead and cadmium in marine protozoan communities. Ecotoxicology and Environmental Safety. 47: 266-276.

Gaur A and Adholeya A (2004) Prospects of arbuscular mycorrhizal fungi in phytoremediation of heavy metal contaminated soils. Current Science 86: 528-534.

Grimanis AP, Zafiropoulos D, Vassilaki R and Grimanis M. (1978) Trace elements in the flesh and liver of two fish species from polluted and unpolluted areas in the Aegean Sea. Environmental Science and Technology 12: 723-726.

Gupta R, Ahuja P, Khan S, Saxena RK and Mohapatra H (2000) Microbial biosorbents: meeting challenges of heavy metal pollution in aqueous solutions. Current Science 78: 967-973

Jorgenson LA and Pedersen B. (1994) Trace metals in fish used for time trend analysis and as environmental indicators. Marine Pollution Bulletin 28: 235-243.

Lankford CE(1973) Bacterial assimilation of iron. Crit Rev Microbiol 2: 273.

Liehr SK, Chen HJ and Lin SH(1994) Metal removal by algal biofilms. Water Sci Technol 30: 59-68.

Linnik PM and Zubenko IB (2000) Role of bottom sediments in the secondary pollution of aquatic environments by heavy-metal compounds. Lakes and Reservoirs: Research and Management 5: 11-21.

Mehta SK and Gaur GP (2005) Use of algae for removing heavy metal ions from wastewater: progress and prospects. Crit. Rev Biotechnol 25: 113-152

Mohan B S and Hosetti, BB (1999) Aquatic plants for toxicity assessment. Environmental Research 81: 259-274.

Nies, DH (1999) Microbial heavy-metal resistance. Appl Microbiol Biotechnol 51: 730-750.

Parker DL, Mihalick JE, Plude JL, Plude MJ, Clark TP, Egan L, Flom JJ, Rai LC and Kumar HD(2000) Sorption of metals by extracellular polymers from the cyanobacteria *Microcystis aeruginosa* f. *flos-aquae* strain C3-40. J. Applied Phycology 12: 219-224

Rao KR, Rashmi K, Latha JNL and Mohan PM(2005) Bioremediation of toxic metal ions using biomass of *Aspergillus fumigatus* from fermentative waste. Indian J Biotechnology 4: 139-143.

Rayms-Keller A, Olson KE, McGaw M, Oray C, Carlson JO and Beaty BJ (1998) Effect of heavy metals on *Aedes aegypti* (Diptera: Culicidea) larvae. Ecotoxicology and Environmental Safety 39: 41-47.

Robinson NJ, Rutherford JC, Pocock MR and Cavet JS (2000) Metal metabolism and toxicity: repetitive DNA. In: B.A. Whitton and M.Potts(eds.), The Ecology of Cyanobacteria: Their Diversity in Time and Space, Kluwer Academic Publishers, Dordrecht pp.443-463.

Roeselers G, van Loosdrecht MCM and Muyzer G (2008) Phototrophic biofilms and their potential applications. J Applied Phycology 20: 227-235

Tao S, Liang T, Cao J, Dawson RW and Changfa L (1999) Synergistic effect of copper and lead uptake by fish. Ecotoxicology and Environmental Safety 44: 190-195.

Van Hattum B, Korthals G, Van Staalen NM, Govers HAJ and Joosse ENG (1993) Accumulation patterns of trace metals in freshwater isopods in sediment bioassays: influence of substrate characteristics, temperature and pH. Water Research 27: 669-684.

Volesky B(1994) Advances in biosorption of metals: selection of biomass types. FEMS Microbiol Rev 14: 291-302

Weast RC (1984) CRC handbook of chemistry and physics, 64th edn., CRC, Boca Raton, Fla

Wilde EW and Benemann JR(1993) Bioremoval of heavy metals by the use of microalgae. Biotech Adv 11: 781-812

Wong PTS, and Dixon DG (1995) Bioassessment of water quality. Environmental Toxicology and Water Quality 10: 9-17.

Yang RSH, and Rauckmann EJ (1987) Toxicological studies of chemical mixtures of environmental concern at the National Toxicology Program: health effects of ground water contaminants. Toxicology 47: 15-34.

Zweig RD, Morton JD, Stewart MM (1999) Source water quality for aquaculture: a guide for assessment. The World Bank, Washington D.C.

Author Index

A

Ambashtha, V., 36

Adhikary, S.P., 111

B

Behera, G., 194

Behera, S.K., 194

Bhattacharya, T., 170

C

Chakraborty, S., 170

Choudhary, K.K., 259

D

Das, A., 133

Das, M.K., 62, 133, 240

Das, P., 62

Das, S., 222

E

Edwards, M., 1

G

Ganesan, K., 77

Gogoi, H.K., 176

Gupta, R.K., 159

K

Kalita, M.C., 176

Kaur, S., 176

Kumar, A., 93

Kumar, M., 252

Kumar, S., 93

M

Mandal, S., 25

Mishra, A.K., 62

P

Padhi, B.K., 25

Padhi, S.B., 194

Pandey, A., 205

Pandey, S.S., 36
Pandey, U., 276
Pandey, V.D., 287

R

Rath, J., 25

S

Singh, S., 222
Singh, Y.V., 145
Sinha, R.P., 93
Srivastava, R.B., 176
Subba Rao, P.V., 77
Suresh K.K., 77
Swain, P.K., 194

T

Tiwari, An., 222
Tiwari, Ar., 205
Tiwari, B.S., 36
Tripathy, P., 111
Tyagi, M.B., 93

Subject Index

A

Agriculture, 160, 229

Algal-

biodiesel, 176

cancer, 69

fatty acids, 194

industry, 01

lipids, 187

ponds, 10

toxins, 51

Antioxidants, 70, 77, 276

Antiviral, 69

Aquaculture, 195

Autophagy, 37

B

Bioactive compounds, 30, 85, 227, 259

Biocontrol, 159, 160

Biodeteriorating, 232

Biodiesel, 173, 176

Bioenergy, 159, 160, 166

Biofertilizer, 145, 159, 160, 230, 259, 262

Biofuels, 1, 170, 173

Biomimetic, 32

Biomineralization, 32

Bioprocess, 159, 160, 167

Bioremediation, 231, 259, 264, 287

Blooms, 206, 209

Blue-green algae, 149, 222

Brown seaweeds, 79

C

Colourants, 260, 266

Cosmetics, 227

Culture Room, 177

Cyanobacteria, 93, 111, 134, 222, 240, 253, 259

Cyanotoxins, 206, 253

D

Dermatotoxins, 206
Detoxification, 259
DNA damage, 94, 98
DNA repair, 94, 103

E

Ecological friends, 230
Energy security, 177
Essential fatty acids, 67

F

Fatty acids, 77, 170, 188, 195
Food, 67
Fuel, 227
Fungicides, 133

G

Global warming reducers, 230
Green seaweeds, 79
Growth response, 112

H

Health hazards, 253
Heavy metals, 287
Hepatotoxins, 206
Heterotrophic growth, 122

I

Immune system, 70
Insecticides, 240, 242
Instruments, 178

L

Light intensity, 112
Lipids, 170, 172, 177

M

MAAs, 94, 101
Macroalgae, 223
Marine algae, 171
Mass culturing, 183
Medicine, 62, 67
Metacaspases, 37
Microalgae, 177, 180, 189, 223, 287
Microbes, 159
Microcystins, 207
Minerals, 67
Mitochondria, 42
Mixotrophic, 122
Motility, 96
Mycorrhiza, 164

N

Nanobiotechnology, 29
Nanomaterials, 26
Nanotechnology, 26, 29
Neurotoxins, 206
Nitrogenase, 94
Nitrogen fixation, 145
Nitrogen metabolism, 97
Nutraceuticals, 19, 276
Nutrients, 62, 66, 80, 180

P

Pesticides, 135, 241
Pesticide hazard, 135
Pigmentation, 96
Photosynthesis, 97
Phycocyanin, 70
Polymerase Chain Reaction, 206, 212
Polyunsaturated fatty acids, 197
Programmed Cell Death, 36
Protein, 66, 77

R

Red seaweeds, 80

Renewable, 177
Rice field, 133, 240
ROS, 37

S

Salinity, 112, 117
Scytonemin, 94, 99
Seaweeds, 77
Spirulina, 62, 222
Superfood, 62

T

Therapeutic, 226
Tissue remodeling, 39
Toiletries, 227
Toxicity, 133, 240
Toxins, 231

U

UV-B, 94, 99

V

Vitamins, 67, 77, 180